The Living Universe

The Living Universe
NASA AND THE DEVELOPMENT OF ASTROBIOLOGY

STEVEN J. DICK AND JAMES E. STRICK

Rutgers University Press
New Brunswick, New Jersey, and London

First paperback printing, 2005

Library of Congress Cataloging-in-Publication Data

Dick, Steven J.
   The living universe : NASA and the development of astrobiology / Steven J. Dick and James E. Strick.
     p. cm.
Includes bibliographical references and index.
   ISBN 0-8135-3447-X (hardcover : alk. paper)   ISBN 978-0-8135-3733-7 (pbk : alk. paper)
   1. Exobiology—History. 2. Life on other planets—Research—History. 3. United States. National Aeronautics and Space Administration. I. Strick, James Edgar, 1956- II. Title.

QH325.D53 2004
576.8'39—dc22

2004004037

A British Cataloging-in-Publication record for this book is available from the British Library.

Copyright © 2004 by Steven J. Dick and James E. Strick
All rights reserved
No part of this book may be reproduced or utilized in any form or by any means, electronic or mechanical, or by any information storage and retrieval system, without written permission from the publisher. Please contact Rutgers University Press, 100 Joyce Kilmer Avenue, Piscataway, NJ 08854-8099. The only exception to this prohibition is "fair use" as defined by U.S. copyright law.

The publication program of Rutgers University Press is supported by the Board of Governors of Rutgers, The State University of New Jersey.

Manufactured in the United States of America

The NASA Vision
To improve life here,
To extend life to there,
To find life beyond
 — Announced by NASA Administrator,
Sean O'Keefe, April 12, 2002.

Astrobiology: The study of the living universe. This field provides a scientific foundation for a multidisciplinary study of (1) the origin and distribution of life in the universe, (2) an understanding of the role of gravity in living systems, and (3) the study of the Earth's atmospheres and ecosystems.
—NASA strategic plan, 1996
(First mention of *astrobiology* in a published NASA document, redefined from *exobiology*)

Contents

Acknowledgments *ix*
Abbreviations and Acronyms *xi*

Introduction   *1*

Part I *Before the Space Age*

1 *The Big Picture: Cosmic Evolution and the
   Biological Universe*   *9*

Part II *From Sputnik to Viking, 1957–1976*

2 *Organizing Exobiology: NASA Enters
   Life Science*   *23*

3 *Exobiology, Planetary Protection, and the
   Origins of Life*   *56*

4 *Vikings to Mars*   *80*

Part III *Broadened Horizons, 1976–2000*

5 *The Post-Viking Revolutions*   *105*

6 *The Search for Extraterrestrial Intelligence*   *131*

7 *The Search for Planetary Systems*   *155*

8	*The Mars Rock*	*179*
9	*Renaissance: From Exobiology to Astrobiology*	*202*
	Epilogue: Astrobiology Science: Into the Great Age of Discovery?	*221*

Appendix A Unpublished Sources *233*
Appendix B NASA Leadership in Exobiology *236*
Appendix C Topics at the First Astrobiology Science Conference *239*
Appendix D Objectives in the Astrobiology Roadmap (1999) *240*
Notes *243*
Selected Bibliography *287*
Index *295*

ACKNOWLEDGMENTS

Research, writing, and oral history interviews for this volume were supported by NASA grant NAG5-8594 from the exobiology program under Michael Meyer, by a grant from the NASA History Office under Roger Launius, and by a Visiting Scholar fellowship (for JS) from the Center for the History of Recent Science (CHRS), George Washington University, during 2000–2001. The SETI Institute also played an essential supporting role. Some of this work (for JS) was supported by a Dibner Postdoctoral Fellowship from 1996 to 1998 and by the Biology and Society Program at Arizona State University (ASU). Jim Collins and Jane Maienschein at ASU provided much advice and assistance. Maura Mackowski was a superb research assistant.

We wish to thank the numerous scientists listed in Appendix A, who gave freely of their time for oral history interviews. William Hagan freely shared transcripts of his interviews with Richard Young and Cyril Ponnamperuma. Susan Goldsmith not only ably transcribed interviews but was a fount of thoughtful criticism as well as broad and lively intelligence. We are also grateful to the individuals and institutions listed in Appendix A who provided access to archives, especially the late Sidney Fox, Imre Friedmann, the late Harold Klein, Joshua Lederberg, James Lovelock, Lynn Margulis, Harold Morowitz, Adolph Smith, Carl Woese, and the late Richard Young for allowing access to unpublished papers. Margulis also allowed access to papers by Elso Barghoorn in her possession. We acknowledge the National Library of Medicine for permission to quote from the Lederberg papers and the California Institute of Technology Archives for permission to quote from the Norman Horowitz papers. Dick first presented parts of chapter 1 at a session on "Evolution and Twentieth-Century Astronomy" at the History of Science Society Meeting, Denver, Colorado, 8 November 2001. Portions of chapter 5 are adapted from Steven J. Dick, "The Search for Extraterrestrial Intelligence and the NASA High Resolution Microwave Survey (HRMS): Historical Perspectives," *Space Science Reviews* 64 (1993): 93–139.

Strick first presented parts of chapter 2 at the History of Science Society meetings in Pittsburgh, 10 November 1999, and in Milwaukee, 8 November 2002, as well as at the National Air and Space Museum history lecture series, 18 January 2001. Portions of chapter 2 are adapted from James Strick, "Creating a Cosmic Discipline: The Crystallization and Consolidation of Exobiology, 1957–1973," *Journal of the History of Biology* 37 (2004). Earlier versions of some chapters also received substantial and helpful criticism from Nathaniel Comfort, Horace Judson, and the weekly seminar at the George Washington University's Center for the History of Recent Science and from Linda Caren, John Cronin, Michael Dietrich, Iris Fry, Keith Kvenvolden, Joshua Lederberg, Lynn Margulis, Stephen Pyne, J. William Schopf, Alan Schwartz, Grier Sellers, Matt Shindell, Adolph Smith, Mark Solovey, and Audra Wolfe.

Our thanks to the NASA History Office for unfailing support in providing resources and time, in particular from Stephen Garber, John Hargenrader, and Roger Launius. Finally, we wish to thank Audra Wolfe, our editor at Rutgers University Press, who has the unusual quality of knowing the subject thoroughly; we benefited greatly from her advice and support.

Personal support (for JS), through a long process and several changes of domicile, were as important as ever in completing a work of this size. JS thanks his wife, Wendy Sobey, and his children, Rachel and Alexander, for bearing up under the tensions involved in research and writing. Throughout the process they good-naturedly maintained a normal life, which helped him keep perspective and clear priorities. Friend and teacher David Brahinsky also helped JS push through blocks. SJD wishes to thank his wife, Terry, for her continued support through more books than she cares to count.

For both authors this has been a unique and rewarding collaboration between a historian of astronomy and a historian of biology. As with the science itself, astrobiology history has fostered interdisciplinary cooperation and has led to insights that would have been unachievable if pursued alone.

Abbreviations and Acronyms

AAMAT	Astrobiology Advanced Missions and Technology
ACME	Antarctic Cryptoendolithic Microbial Ecosystems
AEC	Atomic Energy Commission
AIBS	American Institute of Biological Sciences
ASEE	American Society of Engineering Education
ASTEP	Astrobiology Science and Technology for Exploring Planets
ASTID	Astrobiology Science and Technology Instrument Development
ATF	Astrometric Telescope Facility
AURA	Association of Universities for Research in Astronomy
AXAF	Advanced X-ray Astrophysics Facility
BIF	banded iron formation
CAN	Cooperative Agreement Notice
CASETI	Cultural Aspects of SETI
CCD	charge-coupled device
CETI	communication with extraterrestrial intelligence
CHRS	Center for the History of Recent Science
COMPLEX	Committee on Planetary and Lunar Exploration
CORAVEL	Correlation Radial Velocities
COSPAR	Committee for Space Research
DARPA	Defense Advanced Research Project Agency
DoD	Department of Defense
EASTEX	The East Coast branch of the National Academy of Sciences Space Sciences Board Panel on Extraterrestrial Life
ECD	electron capture detector
ECHO	Evolution of Complex and Higher Organisms
ExNPS	Exploration of Neighboring Planetary Systems
FAIR	Filled-Aperture Infrared
FEG	field emission gun
FY	fiscal year

GCMS	gas chromatograph–mass spectrometer
GEx	Viking gas exchange experiment
GHz	gigahertz
HRMS	High-Resolution Microwave Survey
HST	Hubble Space Telescope
ICBM	intercontinental ballistic missile
IDP	interplanetary dust particle
IOC	Initial Orbital Capability
IR	infrared
IRAS	Infrared Astronomical Satellite
ISSOL	International Society for the Study of the Origin of Life
JPL	Jet Propulsion Laboratory (Pasadena, California)
JSC	NASA Johnson Space Center (Houston, Texas)
LF	Life Finder
LPSC	Lunar and Planetary Science Conference
LR	Viking labeled release experiment
MAP	Multichannel Astrometric Photometer
MCSA	Multi-Channel Spectrum Analyzer
MOP	Microwave Observing Project
NAS SSB	Space Sciences Board of the National Academy of Sciences
NASA	National Aeronautics and Space Administration
NCAR	National Center for Atmospheric Research
NGI	Next Generation Internet
NGST	Next Generation Space Telescope
NIH	National Institutes of Health
NRA	NASA Research Announcement
NRAO	National Radio Astronomy Observatory
NRC	National Research Council
NSCORT	NASA Specialized Center of Research and Training
NSF	National Science Foundation
OAST	Office of Aeronautics and Space Technology
OLEB	*Origins of Life and Evolution of the Biosphere*
ONR	Office of Naval Research
OOL	origins of life
OSI	Orbiting Stellar Interferometer
OSSA	Office of Space Science and Applications
PAH	polycyclic aromatic hydrocarbon
PNAS	*Proceedings of the National Academy of Sciences* (USA)
POINTS	Precision Optical Interferometer in Space
PPLO	pleuropneumonia-like organisms
PPO	planetary protection officer
PPRG	Precambrian Paleobiology Research Group
PR	(Viking) pyrolytic release experiment
PSSWG	Planetary Systems Science Working Group

SEM	scanning electron microscopy
SETI	Search for Extraterrestrial Intelligence
SIM	Space Interferometry Mission
SIRTF	Space Infrared Telescope Facility
SISWG	Space Interferometry Science Working Group
SNC	Shergottite-Nakhlite-Chassignite (class of meteorites believed to be of Martian origin)
SOFIA	Stratospheric Observatory for Infrared Astronomy
SSEC	Space Science Exploration Committee
SSED	Space Science Exploration Division
SSWG	SETI Science Working Group
TEM	transmission electron microscopy
TOPS	Toward Other Planetary Systems
TOPSSWG	Toward Other Planetary Systems Science Working Group
TPF	Terrestrial Planet Finder
UFO	unidentified flying object
UV	ultraviolet
WBSA	Wide Band Spectrum Analyzer
WESTEX	The West Coast branch of the National Academy of Sciences Space Sciences Board Panel on Extraterrestrial Life

The Living Universe

Introduction

*I*n the opening weeks of 1998 a news article in the British journal *Nature* reported that NASA was about to enter biology in a big way. A "virtual" Astrobiology Institute was gearing up for business, and NASA administrator Dan Goldin told his external advisory council that he would like to see spending on the new institute eventually reach $100 million per year. "You just wait for the screaming from the physical scientists [when that happens]," Goldin was quoted as saying.[1] Nevertheless, by the time of the second Astrobiology Science Conference in 2002, attended by seven hundred scientists from many disciplines, NASA spending on astrobiology had reached nearly half that amount and was growing at a steady pace. Under NASA leadership numerous institutions around the world applied the latest scientific techniques in the service of astrobiology's ambitious goal: the study of what NASA's 1996 Strategic Plan termed the "living universe." This goal embraced nothing less than an understanding of the origin, history, and distribution of life in the universe, including Earth. Astrobiology, conceived as a broad interdisciplinary research program, held the prospect of being the science for the twenty-first century which would unlock the secrets to some of the great questions of humanity.

It is no surprise that these age-old questions should continue into the twenty-first century. But that the effort should be spearheaded by NASA was not at all obvious to those—inside and outside the agency—who thought NASA's mission was human spaceflight, rather than science, especially biological science. NASA had, in fact, been involved for four decades in "exobiology," a field that embraced many of the same questions but which had stagnated after the 1976 *Viking* missions to Mars. In this volume we tell the colorful story of the rise of the discipline of exobiology, how and why it morphed into astrobiology at the end of the twentieth century, and why NASA was the engine for both the discipline's founding and for its transformation.

Why did NASA plunge into "extraterrestrial biology" and origin of life research very soon after its formation in 1958? By this time American popular culture had for decades demonstrated a peculiar fascination with life beyond Earth, particularly on the red planet Mars. Remnants of the canals of Mars controversy—a theory promulgated by the renegade American astronomer Percival Lowell, holding that Martians had built canals on their parched and dying planet—still echoed from a half-century earlier. Orson Welles's 1938 radio dramatization of *The War of the Worlds,* which people found so believable that it induced panic in the streets, was only twenty years in the past. The modern UFO craze was only a decade old, and science fiction stories such as Ray Bradbury's *Martian Chronicles* were part of popular culture. All of these elements greatly stimulated American popular interest in the possibility of life on other worlds, including among some who became NASA scientists. In a more technical sense already in 1938 the Soviet biochemist Alexander Oparin, in his influential book *The Origin of Life,* suggested modern biochemical scenarios, testable in a laboratory, to account for the origin of life on a primitive lifeless earth. Scenarios from Oparin's book formed the basis for the origin of life scenes in Disney's *Fantasia* and thereby spread through popular culture. Oparin's book also triggered a generation of researchers who began devising laboratory experiments to simulate the initial steps in the origin of life. In 1953 University of Chicago graduate student Stanley Miller convinced his skeptical advisor, geochemist Harold Urey, that they should undertake an experiment simulating conditions of a primitive Earth atmosphere; to the astonishment of the experimenters, and scientists around the world, within a few days the experiment succeeded in producing amino acids—the first steps toward life.

All this was in the background when NASA was formed. NASA made real the search for what had heretofore been science fiction scenarios of life on other planets and brought with this reality a host of practical problems. Scientists interested in the search for life immediately pointed out that space probes must be sterilized, lest earthly life brought by the spacecraft themselves contaminate the Moon and planets or mix with traces of life detected on these worlds. The reverse problem of back-contamination of the Earth by extraterrestrial microbial pathogens also loomed as a possible frightening consequence of space exploration. Hard-nosed engineers at NASA were skeptical, but forward-looking biologists had a different point of view. Not only did they take seriously the contamination possibilities; some also saw that the possibility of finding life or its building blocks in space or on other planets offered an unprecedented new way to observe the experiment of prebiotic chemistry which had been run repeatedly under different chemical conditions. With the advent of the means to explore space, the prospect of developing a truly universal science of biology now seemed possible for the first time.

Although at first NASA had to be convinced of this point of view, once convinced, the agency acted quickly to bring personnel and their research problems together into a fledgling program of extraterrestrial biology. This program was centered around designing actual spacecraft and instruments as well as developing the basic science necessary to search for life on other planets. At the same time, NASA undertook to determine the necessary conditions for the origin of life anywhere in the universe. Planetary science, extraterrestrial life, and origin of life research quickly became melded, in less than a decade, into an unprecedented new scientific discipline: exobiology. Researchers who had previously had little or no contact were suddenly thrown together, sometimes uneasily, because of the technical breakthroughs of the Space Age.

Who were these researchers, this first generation of exobiologists? They included the likes of Carl Sagan, a young astronomer at Harvard and later Cornell; Stanley Miller, the chemist, fresh from his landmark experiment on the origin of life and already emphasizing its relevance to space research; and Joshua Lederberg, a young geneticist who received the Nobel Prize in the same year that NASA was formed. Three other biochemists were crucial to exobiology's early success: Melvin Calvin, soon-to-be Nobelist for his work on photosynthesis; Norman Horowitz, at CalTech, who brought a particular interest in Mars and a critical attitude toward Martian life; and Sidney Fox, whose laboratory was soon fueled by NASA funding for origin of life research. The goal of these scientists, among a growing number, was no less than a solution to the problem of the origin of life and where it might be found in the cosmos. In effect they began a process that would eventually produce a marriage between biology and astronomy, or at least certain parts of each discipline. As was the case for the manned lunar landing program, their vision of exobiology led to numerous spinoffs: technical breakthroughs, new insights in geology and astronomy, as well as some of the most important work in twentieth-century biology. Despite a deeply ambiguous role for biology within NASA, the exobiology program generated significant innovative ideas in biology, including Carl Woese's "three domain" classification for life, Lynn Margulis's heretical (but now widely accepted) endosymbiosis theory, and James Lovelock's Gaia hypothesis.

Despite its ambiguous role at NASA, the search for extraterrestrial life periodically became a driver for the American space program, exerting an influence that was disproportionate to its funding. From the beginning scientists and NASA administrators were fully aware of the enormous public relations potential of exobiology: they had grown up themselves enthralled by the promise of answering age-old questions about origins. Nothing short of putting men into space captivated public attention like searching for life on Mars. There was nothing more exotic, in all senses of the word, than the idea of extraterrestrial life or, most of all, extraterrestrial intelligence.[2]

Yet public relations is a double-edged sword. Almost immediately some biologists accused exobiology of being a science without a subject. How can one study extraterrestrial life when none is known to exist? they asked. (Never mind that those biologists had earthbound research programs and feared loss of funding if NASA poured large sums of money into exobiology programs, such as one billion dollars spent on the *Viking* missions to search for life on Mars.) Not that such opposition was completely surprising to the exobiology pioneers;

they realized from the beginning the double-edged nature of the public relations aspect of their subject. Since 1947, when the UFO fascination began to grip American culture, any discussion of extraterrestrial life or intelligent life straddled a very thin line between respectable science and a search for "little green men." Nowhere was this more evident than in the cancellation of congressional funding for the Search for Extraterrestrial Intelligence (SETI) program in 1993, when it was targeted as a fanciful waste of money.

Controversial or not, exobiology was not about to disappear. Exobiologists explicitly claimed as their territory some of the most fundamental questions of humanity. What *is* life? How could one claim to recognize life or its beginnings without a clear-cut definition? Yet in 1960 this was just as much a matter of contentious debate as it had been in 1660. Indeed, the exobiologists themselves produced some of the most sharply conflicting ideas, especially while debating what kind of life-detection devices to send to Mars on the *Viking* mission. Has almost a half-century of exobiological research led to any greater consensus in the centuries-old debate over what life is? This book will answer that question. It goes without saying that origin of life research has been fundamentally transformed by its incorporation into exobiology, not least because it never had a big funding patron before NASA in 1960.

Exobiology has also given major impetus to planetary science, in particular the study of Mars and, more recently, the Jovian moon Europa. The claims of fossilized life in the Martian meteorite ALH84001 played an important role in the rebirth of exobiology as astrobiology, a role that we shall examine in detail. Similarly, exobiology gave major impetus to the search for planets around other stars, a search that has intensified with new techniques in astronomy. Why? Because planets are needed for life, and, especially since the American astronomer Frank Drake first proposed the mathematical likelihood of intelligent life on other worlds in 1960, one of the variables needed to refine that calculation is the fraction of stars that have planetary systems. The discovery of new planetary systems in the mid–1990s has given a strong new push to efforts to search for life, including intelligent life, on other planets. Despite the congressional cancellation of the SETI program after less than a year of observations, SETI organizers quickly incorporated their work as a nonprofit group, the SETI Institute, and have continued largely with private donations. In their opinion the question was too important to be left to politicians.

Exobiology grew into a whole new scientific discipline by merging several previously quite disparate streams of research. Far from being a fluke or a short-lived creation that could only flourish under the relatively large infusion of money which NASA dispensed in the 1960s and 1970s for the *Viking* project, it has contributed significantly to viewing planetary scale processes such as global climate in a unified way. Exobiology actually favored interdisciplinary work that had great difficulty getting funded by the National Science Foundation (NSF) or the National Institutes of Health (NIH), the government agencies that fund most of the biological research in the United States. Since 1995 exobiology, under its new rubric of astrobiology, has expanded still further to embrace genomics, ecological research, and all science on the origin, history, and distribution of life in the universe. Today astrobiology remains a central driving force at NASA, a question of enduring popular interest, and one of the most important riddles of science. Given its fundamental questions, astrobiology is indeed here to stay.

Part I

Before the Space Age

CHAPTER 1

The Big Picture

COSMIC EVOLUTION AND THE BIOLOGICAL UNIVERSE

As we examine the details of NASA's central role in exobiology, we must not forget that our story takes place in the context of several grand themes. At one level it is, to be sure, a story of policy and politics, as government funding thrust an age-old idea into the arena of public policy. At another level it is a story of concepts, techniques, and the scientists who employ them at the outermost limits of the capabilities of science, impelled by high stakes that dwarf the controversy over Darwinian evolution on Earth. There is no doubt that the outcome of exobiology's studies will deeply affect humanity's sense of its place in the universe; as Darwinism placed humanity in its terrestrial context, so exobiology will place humanity in a cosmic context. That context—a universe full of microbial life, full of intelligent life, or devoid of life except for us—may to a large extent determine both humanity's present worldview and its far future.

None of these themes, however, is more central than the concept of cosmic evolution, which provides the grand context within which the enterprise of exobiology is undertaken. In setting the stage for the history of exobiology and NASA, it is important, then, that we understand how this concept arose and what it entails.

The idea of cosmic evolution implies a continuous evolution of the constituent parts of the cosmos from its origins to the present. Planetary evolution, stellar evolution, and the evolution of galaxies could in theory be seen as distinct subjects, in which one component evolves but not the other and in which the parts have no mutual relationships. Indeed, in the first half of the twentieth century scientists treated the evolution of planets, stars, and galaxies for the most part as distinct subjects, and historians of science still tend to do so.[1] But the amazing and stunning idea that overarches these separate histories is that the entire universe is evolving, that all of its parts are connected and interact, and that this evolution applies not only to inert matter but also to life, intelligence, and even culture. This overarching idea is what is called cosmic evolution, and the idea has itself evolved to the extent that some modern scientists even talk of a cosmic ecology, the "life of the cosmos," and the "natural selection" of universes.[2]

The concept of cosmic evolution gives rise to many questions. The scientist wants to know how far cosmic evolution proceeds: does it commonly end with planets, stars, and galaxies, or does it continue on to life, mind, and intelligence? We know of only one case of the latter—on planet Earth. The burning question is whether cosmic evolution commonly gives rise to life, resulting not only in an evolving physical universe but also in an evolving "biological universe." Scientists and historians have seen the idea of a universe full of life as a kind of worldview similar in status to the Copernican and Darwinian worldviews; some have even termed it "biocosmology."[3] These scientific questions immediately give rise to theological and philosophical questions: is life part of the "plan" of the universe, or, posed in a more secular way, is life the inherent outcome of a "biofriendly universe"? All of this is part of the history of the cosmic evolution debate, which makes the terrestrial evolution debate pale in significance, even though it involves us so directly. Cosmic evolution involves us directly, too, for, while terrestrial evolution addresses our place on Earth, cosmic evolution addresses our place in the universe. That is why the debate is so passionate and why philosophical and theological issues such as the nature of life, the probability of its origin, and the roles of chance and necessity are intertwined in the terrestrial and cosmic contexts.[4]

Such a broad scope dictates that any comprehensive history of cosmic evolution encompass everything from the Big Bang to intelligence and culture. One might say it would have to address not only the physical universe but also the biological universe and the cultural universe. Such a comprehensive history is, in fact, just what NASA embraced as part of its exobiology and Search for Extraterrestrial Intelligence (SETI) programs (fig. 1.1). It is important, therefore, to ask how the concept of cosmic evolution was first extended from the physical universe to the biological universe and how the idea of a biological universe evolved during the twentieth century to become a bona fide research program driven by NASA patronage.

The Birth of "Cosmic Evolution": Astronomers, Biologists, and Popularizers

Although the question of extraterrestrial life is very old, the concept of a full-blown cosmic evolution—the connected evolution of planets, stars, galaxies, *and* life on Earth and beyond—is much younger. As historian Michael Crowe has shown in his study of the plurality of worlds debate, in the nineteenth century a combination of ideas—the French mathematician Pierre Simon Laplace's "nebular hypothesis" for the origin of the solar system, the British naturalist Robert Chambers's application of evolution to other worlds, and Darwinian evolution on this world—gave rise to the first tentative expressions of parts of this

FIGURE 1.1. Cosmic evolution is depicted in this image from the exobiology program at NASA Ames Research Center, 1986. *Upper left:* the formation of stars, the production of heavy elements, and the formation of planetary systems, including our own. At left prebiotic molecules, RNA, and DNA are formed within the first billion years on the primitive Earth. At *center* the origin and evolution of life leads to increasing complexity, culminating with intelligence, technology, and astronomers, *upper right,* contemplating the universe. The image was created by David DesMarais, Thomas Scattergood, and Linda Jahnke at NASA Ames in 1986 and reissued in 1997.

worldview. The philosophy of Herbert Spencer extended it to the evolution of society, although not to extraterrestrial life or society. But some Spencerians, notably Harvard philosopher John Fiske in his *Outlines of a Cosmic Philosophy Based on the Doctrine of Evolution* (1875), did extend evolutionary principles to life on other planets.[5]

Neither astronomers nor biologists tended to embrace such a broad philosophical, and empirically unsupported, concept as full-blown cosmic evolution. Two astronomers, however, who are better known as popularizers of science, did propound the rudiments of the idea. In England and the United States Richard A. Proctor and in France Camille Flammarion were greatly influenced by Darwinian ideas. In Proctor's *Other Worlds than Ours* (1870), *Our Place among Infinities,* and *Science Byways,* the latter both published in 1875, the evolutionary view in which all planets would attain life in due time assumed a central role. By the 1872 edition of Flammarion's *La pluralité des mondes* the author shows the deep influence of Darwin. Life began by spontaneous generation, evolved via natural selection by adaptation to its environment, and was ruled by survival of the fittest, wherever it was found in the universe. In this scheme of cosmic evolution anthropocentrism was banished; the Earth was not unique, and humans were in no sense the highest form of life. Flammarion's *La pluralité* reached thirty-three editions by 1880 and was reprinted until 1921, while Proctor's *Other Worlds than Ours* reached twenty-nine printings by 1909, making him the most widely read astronomy writer in the English language. Historian Bernard Lightman makes the case that such popularizers used the concept of cosmic evolution to narrate an evolutionary epic long before it was accepted by scientists or incorporated into any research program. Thus were the general outlines of the idea of cosmic evolution spread to the populace.[6]

But a set of general ideas is a long way from a research program. In the first half-century of the post-Darwinian world cosmic evolution did *not* find fertile ground among astronomers, who were hard-pressed to find evidence for it. Spectroscopy, which displayed the distinct "fingerprints" of each of the chemical elements, revealed to astronomers that these elements were found in the terrestrial and celestial realms. This discovery confirmed the widely assumed idea of "uniformity of nature," that both nature's laws and its materials were everywhere the same. Astronomers recognized and advocated parts of cosmic evolution, as in the British astrophysicist Norman Lockyer's work on the evolution of the elements and the American astronomer George Ellery Hale's *Study of Stellar Evolution* in 1908; in this and his other published writings Hale stuck very much to the techniques for studying the evolution of the physical universe. Even Percival Lowell's *Evolution of Worlds* (1909) spoke of the evolution of the physical universe, not the biological universe, Martian canals notwithstanding. Although Lowell was a Spencerian, had been influenced by Fiske at Harvard and had addressed his graduating class on the "Nebular Hypothesis" two years after Fiske's *Cosmic Philosophy* (1874), he did not apply the idea of advanced civilizations to the universe at large. Even in the first half of the twentieth century astronomers had to be content with the uniformity of nature argument confirmed by spectroscopy. In an article in *Science* in 1920 the American astronomer W. W. Campbell (a great opponent of Lowell's canalled Mars) enunciated exactly this general idea of widespread life via the uniformity of nature argument: "If there is a unity of materials, unity of laws governing those materials throughout the universe, why may we not speculate somewhat confidently upon life universal?" he asked. He even spoke of "other stellar systems . . . with degrees of intelligence and civilization from which we could learn much, and with which we could sympathize." That was about all the astronomers of the time could say.[7]

For the most part biologists were also reluctant cosmic evolutionists. Two points of view at the turn of the century demonstrate this reluctance. The first was that of none other than the British naturalist Alfred Russel Wallace, cofounder with Darwin of the theory of natural selection, who wrote *Man's Place in the Universe: A Study of the Results of Scientific Research in Relation to the Unity or Plurality of Worlds* in 1903. Wallace concluded: "Our position in the material universe is special and probably unique, and . . . it is such as to lend support to the view, held by many great thinkers and writers today, that the supreme end and purpose of this vast universe was the production and development of the living soul in the perishable body of man." With regard to life on Earth, in stark contrast to Darwin, Wallace did not believe that the evolution of the human brain could be due to natural selection. And with respect to the biological universe, in an "additional argument dependent on the theory of evolution" added to the 1904 edition of Wallace's book, he argued that, because humanity is the result of a long chain of modifications in organic life, because these modifications occur only under special circumstances, and because the chances of the same conditions and modifications occurring elsewhere in the universe are very small, the chances of beings in human form existing on other planets is very small. Moreover, since no other animal on Earth approached the intelligent or moral nature of humanity, Wallace concluded that intelligence in any other form was also highly improbable. How improbable? He set the physical and cosmic improbabilities at a million to one, the evolutionary improbabilities at a hundred million to one, giving the total chances against the evolution of an equivalent moral or intellectual being to man, on any other planet, as a hundred million million to one. Clearly, for Wallace—for this pioneer in evolution by natural selection—there was no cosmic evolution in its fullest sense—that is to say, no biological universe.[8]

The second biologist especially relevant here is Lawrence J. Henderson, a professor of biological chemistry at Harvard and first president of the History of Science Society. In 1913, ten years after Wallace, he wrote a now classic book *The Fitness of the Environment,* subtitled "An Inquiry into the Biological Significance of the Properties of Matter." In it Henderson investigated how the environment on Earth became fit for life. He closed with a chapter on "Life and the Cosmos," which ended with these words: "There is . . . one scientific conclusion which I wish to put forward as a positive statement and, I trust, fruitful outcome of the present investigation. The properties of matter and the course of cosmic evolution are now seen to be intimately related to the structure of the living being and to its activities; they become, therefore, far more important in biology than has been previously suspected. For the whole evolutionary process, both cosmic and organic, is one, and the biologist may now rightly regard the universe in its very essence as biocentric." Clearly, Henderson grasped essential elements of cosmic evolution, used its terminology, and believed that his research into the fitness of the environment pointed in that direction. Yet, although he had a productive career at Harvard until his death in 1942, Henderson never enunciated a full-blown concept of cosmic evolution, nor did any of his astronomical colleagues.[9]

Henderson's idea of cosmic evolution in 1913 was largely stillborn, perhaps in part because just a few years later James Jeans's theory of the formation of planetary systems by close stellar encounters convinced the public, and most scientists, that planetary systems were extremely rare. The idea remained entrenched until the mid–1940s. Without planetary systems cosmic evolution was stymied at the level of the innumerable stars, well short of the biological universe. In the absence of evidence cosmic evolution was left to science fiction writers such as Olaf Stapledon, whose *Last and First Men* and *Star Maker* novels in the 1930s embraced it in colorful terms. But Henderson had caught the essence of a great idea—that life and the material universe were closely linked, a fundamental tenet of cosmic evolution which would lay dormant for almost a half-century.

Cosmic Evolution Becomes a Research Program

The humble and sporadic origins of the idea of cosmic evolution demonstrate that it did not have to become what is surely the leading overarching principle of twentieth-century astronomy, yet it did. Almost all astronomers today view cosmic evolution as a continuous story from the Big Bang to the evolution of intelligence, accepting as proven the evolution of the physical universe while leaving open the still unproven question of the biological universe, whose sole known exemplar remains the planet Earth. Today the central question remains how far cosmic evolution commonly proceeds. Does it end with the evolution of matter, the evolution of life, the evolution of intelligence, or the evolution of culture? But today, by contrast with 1950, cosmic evolution is the guiding conceptual scheme for a substantial research program.

When and how did astronomers and biologists come to believe in cosmic biological evolution as a guiding principle for their work, and how did it become a serious research program? The answer is that only in the 1950s and 1960s did the cognitive elements—planetary science, planetary systems science, origin of life studies, and SETI—combine to form a robust theory of cosmic evolution as well as provide an increasing amount of evidence for it. Only then, and increasingly thereafter, were there serious claims for disciplinary status for a field known alternatively as exobiology, astrobiology, and bioastronomy, the biological universe component of cosmic evolution. And only then did government funding become available, as the space program embraced the search for life as one of the primary goals of space science and cosmic evolution became public policy.

We have already hinted at why this coalescence had not happened earlier, Spencerian philosophy, and the ideas of Flammarion, Proctor, and Henderson notwithstanding. Although the idea of the physical evolution of planets and biological evolution of life on those planets in our solar system had been around for a while—and even some evidence in the form of seasonal changes and spectroscopic evidence of vegetation on Mars—not until the space program did the technology become available, resulting in large amounts of government funding being poured into planetary science so that these tentative conclusions could be further explored. Moreover, if evolution was truly to be conceived as a cosmic phenomenon, planetary systems outside our solar system were essential. Therein was the problem for much of the first half of the century. That innumerable planets might exist was an implication of Laplace's nebular hypothesis: if planets really formed as the normal by-product of a rotating cloud during stellar evolution, then they should be extremely common. The nebular hypothesis was eclipsed for the first four decades of the century, however, by a variety of hypotheses claiming that planets formed by the close encounter of stars—the so-called tidal theory, in which material was pulled out of the star to form planets. Because such close encounters would be extremely rare events, planetary systems would be extremely rare. Only in the 1940s, when the tidal theory was shown to be flawed and the nebular hypothesis came back into vogue, could an abundance of planetary systems once again be postulated. During a fifteen-year period from 1943 to 1958 the commonly accepted frequency of planetary systems in the galaxy went from one hundred to one billion, a difference of seven orders of magnitude. The turnaround involved many arguments, from the observations of a few possible planetary companions in 1943, to binary star statistics, the nebular hypothesis, and stellar rotation rates. Helping matters along was the dean of American astronomers, Henry Norris Russell, whose 1943 *Scientific American* article "Anthropocentrism's Demise" enthusiastically embraced numerous planetary systems based on just a few observations by Kaj Strand and others. By 1963 the American astronomer Peter van de Kamp announced his discovery of a planet around Barnard's star, and the planet chase was on, to be truly successful only at the end of the century.[10]

Thus was one more step in cosmic evolution made plausible by midcentury, even though it was a premature and optimistic idea, since only in 1995 were the first planets found around Sun-like stars, and those were gas giants such as Jupiter. But how about life? That further step awaited developments in biochemistry, in particular the Oparin-Haldane theory of chemical evolution for the origin of life. The first paper on the origins of life by the Russian biochemist Aleksandr Ivanovich Oparin was written in 1924, elaborated in the 1936 book *Origin of Life,* and reached the English world in a 1938 translation. By that time the British geneticist and biochemist J. B. S. Haldane had provided a brief independent account of the origin of life similar to Oparin's chemical theory. Both Oparin and Haldane were Marxists, and, as Loren Graham and others have pointed out, their worldview may have affected their science. By 1940, when the British Astronomer Royal, Sir Harold Spencer Jones, wrote *Life on Other Worlds,* he remarked, "It seems reasonable to suppose that whenever in the Universe the proper conditions arise, life must inevitably come in to existence."[11]

The contingency or necessity of life would be one of the great scientific and philosophical questions of cosmic evolution, but in any case the Oparin-Haldane chemical theory of origin of life provided a basis for experimentation, beginning with the famous experiment of Stanley Miller and Harold Urey in 1953 in which amino acids, the building blocks of proteins and life, were synthesized under possible primitive Earth conditions. By the mid–1950s another step of cosmic evolution was coming into focus: the possibility of primitive life. Again, optimism was premature, but the point is that it set off numerous experiments around the world to verify another step in cosmic evolution. Already in 1954 Harvard biochemist George Wald proclaimed the Oparin-Haldane process a natural and inevitable event, not just on our planet but on any planet similar to Earth in size and temperature. By 1956 Oparin had teamed with Russian astronomer V. Fesenkov to write *Life in the Universe,* which expressed the same view of the inevitability of life as Wald's.[12]

What remained was the possible evolution of intelligence in the universe. Although hampered by a lack of understanding of how this had happened on Earth, discussion of the evolution of intelligence in the universe was spurred on by the famous paper by the American physicists Giuseppe Cocconi and Philip Morrison in *Nature* in 1959. "Searching for Interstellar Communications" showed how the detection of radio transmissions was feasible with radio telescope technology already in hand. In the following year astronomer Frank Drake, a recent Harvard graduate, undertook just such a project (Ozma) at the National Radio Astronomy Observatory (NRAO), ushering in a series of attempts around the world to detect such transmissions. And in 1961 Drake, supported by NRAO director Otto Struve, convened the first conference on interstellar communication at Green Bank, West Virginia. Although it was a small conference attended by only eleven people including Struve, there were representatives from the astronomy community (Carl Sagan and Su Shu Huang, along with Drake), the biological community (Melvin Calvin, whose Nobel Prize for his work on photosynthetic mechanisms was announced while the meeting was in session), physicists (Cocconi and Morrison), an engineer (Barney Oliver, later of SETI fame), and even a medical doctor who had experimented with interspecies communications in the form of dolphins (John C. Lilly).[13] Thus, by 1961 the elements of the full-blown cosmic evolution debate were in place.

It was at the Green Bank meeting that the now famous Drake equation was first formulated. The equation $N = R_* \times f_p \times n_e \times f_l \times f_i \times f_c \times L$—purporting to estimate the number (N) of technological civilizations in the galaxy—eventually became the icon of cosmic evolution, showing in one compact equation not only the astronomical and biological aspects of cosmic evolution but also its cultural aspects. The first three terms represented the number of stars in the galaxy which had formed planets with environments suitable for life; the next two terms narrow the number to those on which life and intelligence actually develop; and the final two represent radio communicative civilizations. L, representing the lifetime of a technological civilization, embodied the success or failure of cultural evolution. Drake and most others in the field recognized that this equation is a way of organizing our ignorance. At the same time, progress has been made on at least one of its parameters; the fraction of stars with planets (f_p) is now known to be between 5 and 10 percent for gas giant planets around solar-type stars.

The adoption of cosmic evolution was by no means solely a Western phenomenon. On the occasion of the fifth anniversary of *Sputnik* Soviet radio astronomer Joseph Shklovskii wrote *Universe, Life, Mind* (1962). When elaborated and published in 1966 as *Intelligent Life in the Universe* by Carl Sagan, it became the bible for cosmic evolutionists interested in the search for life. Nor was Shklovskii's book an isolated instance of Russian interest. As early as 1964, the Russians convened their own meetings on extraterrestrial civilizations, funded their own observing programs, and published extensively on the subject.[14]

Thus, cosmic biological evolution first had the potential to become a research program in the early 1960s, when its cognitive elements—planetary science, planetary systems science, origin of life studies, and radio astronomy—had developed enough to become experimental and observational sciences and when the researchers in these disciplines first realized they held the key to a larger problem that could not be resolved by any one part but, rather, only by all of them working together. At first this was a very small number of researchers, but it has expanded greatly over the years, especially under NASA patronage. The idea was effectively spread beyond the scientific community by a variety of astronomers. As early as 1958, cosmic evolution was being popularized by Harvard astronomer Harlow Shapley in *Of Stars and Men*; it spread even more widely by the publication of Sagan's *Cosmos* (1980), Eric Chaisson's *Cosmic Dawn: The Origins of Matter and Life* (1981), and in France by Hubert Reeves's *Patience dans l'azur: L'évolution cosmique* (1981), among others.[15] By the end of the twentieth century cosmic evolution was viewed as playing out on an incomparably larger stage than what had been conceived by A. R. Wallace a century before.

The establishment of cosmic biological evolution as a research program can also be gauged by the claims of its practitioners, realizing, of course, that a certain amount of self-interest is at play in proclaiming one's subject a valid discipline if one is seeking federal funding. Even in the late 1950s one could argue that the study of cosmic evolution was not at all a connected research program in the sense that those interested in it had a common goal. Planetary science, planetary systems science, origin of life studies, and SETI remained largely separate research programs, undertaken by different groups of scientists. Aside from the shared general culture of astronomy, the planetary spectroscopy of Gerard Kuiper and William Sinton had little in common with Peter van de Kamp's astrometric studies of stellar motions or Frank Drake's radio astronomy in terms of technique, research programs, and even goals, while all three areas were removed from the biochemists and geochemists in their laboratories studying the origins of life. And, certainly, most members of all these groups disavowed the popular culture aspects of the debate, including UFOs—although many were interested in science fiction.

The catalyst for the unified research program of cosmic evolution, and for the birth of a new scientific discipline, was the space age. No one would claim that a field of extraterrestrial life studies, or cosmic evolution, existed in the first half of the twentieth century. Even by 1955, when Otto Struve pondered the use of the word *astrobiology* to describe the broad study of life beyond the Earth, he explicitly decided against establishing a new discipline: "The time is probably not yet ripe to recognize such a completely new discipline within the framework of astronomy. The basic facts of the origin of life on Earth are still vague and uncertain; and our knowledge of the physical conditions on Venus and Mars is insufficient to give us a reliable background for answering the question" of life on other worlds. But the imminent birth of "exobiology" was palpable in 1960, when Joshua Lederberg coined the term and set forth an ambitious but practical agenda based on space exploration in his article in *Science,* "Exobiology: Experimental Approaches to Life beyond the Earth." Over the next twenty years numerous such proclamations of a new discipline were made. By 1979 NASA's SETI chief, John Billingham, wrote that "over the past twenty years, there has emerged a new direction in science, that of the study of life outside the Earth, or exobiology. Stimulated by the advent of space programs, this fledgling science has now evolved to a stage of reasonable maturity and respectability."[16]

The extent to which NASA had served as the chief patron of cosmic biological evolution is evident in its sponsorship of many of the major conferences on extraterrestrial life, although the Academies of Science of the United States and the USSR were also prominent supporters. It was NASA that adopted exobiology as one of the prime goals of space science, and it was from NASA that funding would come, despite an early but abortive interest at the National Science Foundation.[17] As we shall see, pushed by prominent biologists such as Joshua Lederberg, beginning already in the late 1950s, soon after its origin, NASA poured a small but steady stream of money into exobiology and the life sciences in general. In the early 1960s Lederberg, Sidney Fox, Melvin Calvin, and Wolfgang Vishniac were only the most prominent among a rapidly expanding number of researchers receiving grants of hundreds of thousands of dollars, prompting evolutionist George Gaylord Simpson to complain about "exbiologists" siphoning off funding for more realistic research. In the same paper he opined that exobiology was a "'science' that has yet to demonstrate that its subject matter exists!"[18] By 1976 $100 million had been spent on the *Viking* biology experiments designed to search for life on Mars from two spacecraft landers. Even as exobiology saw a slump in the 1980s in the aftermath of the *Viking* failure to detect life on Mars unambiguously, NASA kept exobiology alive with a grant program at the level of $10 million per year and with the largest exobiology laboratory in the world at its Ames Research Center. Cosmic evolution's potential by the early 1960s to become a research program was converted to reality by NASA funding.

This is true not only of NASA's exobiology laboratory and grants program but also of its SETI program. Born at Ames in the late 1960s quite separately from the exobiology program, NASA SETI expended some $55 million prior to its termination by Congress in 1993. It was the NASA SETI program that was the flag bearer of cosmic evolution. As it attempted to determine how

FIGURE 1.2. Cosmic evolution, as it appeared in the Roadmap for NASA's Office of Space Science Origins theme, 1997. The origins theme there is described as following the fifteen-billion-year-long chain of events from the birth of the universe at the Big Bang, through the formation of chemical elements, galaxies, stars, and planets, through the mixing of chemicals and energy that cradles life on Earth, to the earliest self-replicating organisms—and the profusion of life. (Courtesy NASA.)

many planets might have evolved intelligent life, all of the parameters of cosmic evolution, as encapsulated in the Drake equation, came into play.

With the demise of a publicly funded NASA SETI program in 1993, the research program of cosmic evolution did not end. The remnants of the NASA SETI program were kept alive with private funding, and similar, if smaller, SETI endeavors are still carried out around the world. Within NASA a truncated program of cosmic evolution continued, with its images subtly changed. In 1995 NASA announced its Origins program, which two years later it described in its Origins Roadmap as "following the 15 billion year long chain of events from the birth of the universe at the Big Bang, through the formation of chemical elements, galaxies, stars, and planets, through the mixing of chemicals and energy that cradles life on Earth, to the earliest self-replicating organisms—and the profusion of life." Any depiction of "intelligence" is conspicuously absent from the new imagery (fig. 1.2), for, thanks to congressional action, programmatically it could no longer be supported with public funding. With this proclamation of a new Origins program, cosmic evolution became the organizing principle for most of NASA's space science effort.

In 1996 the *Astrobiology program* was added to NASA's lexicon. The NASA Astrobiology Institute, centered at NASA's Ames Research Center, funds some fifteen other centers for research in astrobiology at the level of several tens of millions of dollars. Its paradigm is also cosmic evolution, even if it also carefully avoids mention of extraterrestrial intelligence. No such restriction is evident at the SETI Institute in Mountain View, California, headed by Frank Drake. The institute has under its purview tens of millions of dollars in grants, all geared to answering various parameters of the Drake equation, the embodiment of cosmic evolution, including the search for intelligence.

As we enter the twenty-first century there is no doubt about the existence of a robust cosmic evolution research program. NASA is its primary patron, and even many scientists without government funding now see their work in the context of this research program. Other agencies, including the European Space Agency, are also funding research essentially in line with the Origins and Astrobiology programs. Beginning in the 1960s, all the elements of a new discipline gradually came into place: the cognitive elements, the funding resources, and the community and communications structures common to new disciplines. In 1979 a new Commission on Bioastronomy was formed in the prestigious International Astronomical Union; the International Society for the Study of the Origin of Life routinely incorporates exobiology in its meetings; and a variety of other societies also embrace exobiology. Already in 1968 the journal *Origins of Life* (now *Origins of Life and Evolution of the Biosphere*) began publication, and in the new century two new journals devoted to the more general field of astrobiology have begun publication. Numerous universities offer courses on life in the universe, and there is at least one university (the University of Washington in Seattle) now offering a graduate program in astrobiology. In the early years of the twenty-first century cosmic evolution is a thriving enterprise, providing the framework for an expansive research program, drawing in young talent sure to perpetuate a new field of science which a half-century ago was nonexistent.

Part II

From Sputnik *to* Viking, *1957–1976*

CHAPTER 2

Organizing Exobiology

NASA ENTERS LIFE SCIENCE

*E*xobiology did not exist, either in name or substance, before the dawn of the Space Age. Nonetheless, in less than two decades it had become a fully fledged scientific discipline. How could such a transformation come about so rapidly, and who were the major players involved in creating this new discipline? In an era in which "big science" had become the acknowledged standard, large-scale patronage was crucial. For exobiology the new American space agency, the National Aeronautics and Space Administration, played a key role, though not nearly at the same level as in its manned space program or even its other space science projects. The story of how individual scientists tailored their careers to encompass research in exobiology as they attempted to negotiate the increasingly complex landscape of large federal science agencies is as colorful as the varied personalities involved. This chapter introduces some of these personalities while describing the evolving landscape of federal science grants, the science it supported, and some of the larger questions surrounding the creation of exobiology.

Beginnings

In early November 1957 the microbiologist Joshua Lederberg visited the famous geneticist J. B. S. Haldane at Haldane's new home in India. Lederberg, only thirty-two, would win the Nobel Physiology / Medicine Prize in less than a year for his pioneering work on bacterial genetics, and he held a long-standing interest in the origin of life. Haldane, much the senior of the two scientists, was one of the British scientific socialist circle of the 1930s and 1940s, and he had written a seminal paper on the chemical origin of life in 1929. Both men were awed by the rapid advent of rocketry and the recent launch of the first two Soviet *Sputniks*. As Lederberg tells the story, over dinner on the evening of 6 November, waiting to see a lunar eclipse that night, they speculated on whether the Soviets might detonate a nuclear explosion on the darkened part of the moon, "put a red star on the moon," to mark the fortieth anniversary of the Bolshevik Revolution. Although their fear did not materialize that night, the potential for reckless use of the new technology continued to disturb both men.[1]

A month later Lederberg was back in the United States, circulating two memos to a hundred or more prominent scientists and to the National Academy of Sciences (NAS), speculating on the possibilities of "cosmic microbiology" and "lunar biology." Lederberg was concerned that a totally unique opportunity, the scientific search for life, including microorganisms, on the moon and other planets was in real danger of being thrown away because of a politically motivated stunt. Crashing a spacecraft on the moon as quickly as possible to prove technological prowess would hopelessly contaminate the moon with earthly organisms and/or their chemical building blocks, Lederberg argued.[2] If the spacecraft, like *Sputnik 2,* contained a live dog, the problem would be a million times worse. Although doing so would slow down the attempt to be "first to reach the moon," it was vital to develop procedures to sterilize lunar and interplanetary satellites, he argued forcefully, lest a priceless scientific opportunity be irretrievably lost.

In the wake of *Sputnik* and the opening of the Space Age, biologists around the world began to speculate about what this new technology would mean for the life sciences. Those interested in extraterrestrial life, of course, saw immediately that for the first time their subject could be studied in more than just a theoretical way. But origin of life research got as much or even more of an electrifying stimulus from the launching of space vehicles. And within a decade, with NASA as matchmaker, the two fields had been wed, merged together to create a new discipline, exobiology. So exhilarating was the wedding that by the early 1970s hardly anyone could imagine that working on the origin of life problem had not always been part and parcel of the search for life on other planets.

On 29 July 1958 President Dwight D. Eisenhower signed the National Aeronautics and Space Act, creating NASA as the U.S. space agency. By that time Lederberg had interested Hugh Dryden, the first NASA deputy administrator, in the problem of preventing extraterrestrial contamination and searching for native life forms on the moon and planets. Dryden immediately asked the National Academy of Sciences to set up a Space Sciences Board (SSB) to advise NASA. And Lederberg was made head of the SSB's subpanel on extraterrestrial life. Lederberg had not been idle; he kept up his campaign to alert scientists to the contamination threat in an article in *Science* called "Moondust."[3] And he recruited like-minded scientists to staff the NAS SSB, looking especially for young talents who were coming up during the space age, such as the astronomer Carl Sagan.

Lederberg was frustrated with the stodgy, conservative, nationalistic attitudes of many of the older scientists. He was constantly having run-ins with curmudgeonly physicist Phil Abelson, for many years editor of *Science,* because of Abelson's skepticism that there was any life out there in the cosmos.[4] In April 1961 Abelson declared to the National Academy of Sciences: "In looking for life on Mars we could establish for ourselves the reputation of being the greatest Simple Simons of all time."[5]

Sagan was another matter. He met Lederberg when living in Madison, Wisconsin, in 1958, while still a twenty-four-year-old doctoral student at nearby Yerkes Observatory. Ever since his science fiction reading days as a child, Sagan had been an enthusiast of extraterrestrial life. He sat in on Harold Urey's lectures as an undergraduate at the University of Chicago, just at the time when Urey's graduate student Stanley Miller was making international headlines for his experiment producing amino acids under primitive Earth conditions. And ever since Sagan had viewed his own mission in science as nothing less than "extending Miller's results to astronomy."[6] Sagan had also shown a talent from his undergraduate years as an explainer and popularizer of science as well as a scientist.[7] And, surely, given the extraordinary level of public interest in NASA and the international "space race" and the extraordinary level of funds at NASA's disposal, Lederberg saw that advancing his scientific agenda would benefit most if both the public and NASA accepted the importance of understanding the origin of life and the search for life on other worlds. Lederberg introduced Sagan to NASA people and got him involved on the ground floor of developing exobiology in 1959.[8]

With his new Nobel Prize in hand, in the fall of 1958 Lederberg had moved to the Stanford Medical School to set up a new genetics department. From there he argued that the NAS SSB subpanel on extraterrestrial life would work most effectively if it met as an East Coast group (EASTEX) and a West Coast group (WESTEX), and he urged the groups to get to work as quickly as possible.[9] EASTEX first met 19–20 December 1958 at MIT.[10] WESTEX convened shortly thereafter, on 21 February 1959 at the Stanford Biophysics Department, with Lederberg as the prime moving force. The WESTEX group also included, among others, Harold Urey, Carl Sagan, molecular biologists Gunther Stent and Matt Meselson, geneticist Norman Horowitz, biochemist Melvin Calvin, and microbiologist C. B. Van Niel. Several of them had written important papers on origin of life, Calvin, Horowitz and Urey having given papers at the first International Conference on the subject, in Moscow in 1957. Van Niel was among those who had first emphasized the importance of the distinction between prokaryotic and eukaryotic cells; from 1930 to 1962 dozens and dozens of students who would later become the most influential biologists of two generations took his summer course on General Microbiology at Stanford University's Marine Station.[11]

The August 1957 Moscow conference, just before *Sputnik*, shows that origin of life research had been growing, if ever so slowly, before the Space Age. In the spring of 1953, just three weeks after Watson and Crick's famous paper on DNA structure was published, Stanley Miller and Harold Urey's equally famous paper on creating the chemical building blocks of life in the laboratory appeared.[12] Miller had simulated the presumed atmosphere of the early Earth in a closed flask, added heat and a spark discharge, and found that after only a few days amino acids and other complex organic molecules had formed in the flask (fig. 2.1). At about the same time Sidney Fox was working on the reactions that amino acids undergo, once formed, under conditions relevant to the early Earth.[13] And Alexander Oparin in the Soviet Union had been working since the 1930s on experiments with chemical systems called coacervates, trying to model early stages of the origin of complex membrane-bounded structures from simple precursor molecules (fig. 2.2).[14] All three men were in on the beginning of a new upsurge of interest in exploring the origin of life question experimentally. Indeed, Oparin's book *The Origin of Life,* first appearing in English in 1938, had been a major stimulus in the early thinking of Fox, Horowitz, Lederberg, and a handful of others who revived this research in the years after World War II.[15] But the field had been sparsely funded, to put it mildly. Stanley Miller has written that his entire experiment was carried out largely by "bootlegging" funds from other grants that his advisor Urey had received; the equipment and supplies did not exceed a thousand dollars. In addition, Miller himself had a teaching assistantship from the University of Chicago his first year and an NSF graduate student fellowship of about fifteen hundred dollars for his second and third years.[16]

It was Oparin who organized the 1957 Moscow conference, bringing together for the first time the scattered workers around the globe who saw their research as relevant to the origin of life question.[17] The conference convened in August of that year, amid the tensions of the Cold War. Oparin had explicitly stated that dialectical materialism was important to his research agenda and had been a supporter of the Soviet biologist Trofim Denisovich Lysenko, as Loren Graham has shown.[18] It was only two years since the first Soviet megaton-scale hydrogen bomb explosion, only three months since the first British thermonuclear bomb test, and the conference had barely ended when the TASS News Agency announced that the Soviet Union had just successfully tested the first intercontinental ballistic missile (ICBM), launching it over four thousand miles. Barely six weeks later those tensions heightened to a fever pitch with the launching of *Sputnik 1.* Yet, even before the Moscow meeting took place, the scientists attending could not fail to see it in a Cold War context. For the Americans who had been invited, the most palpable evidence of this involved visits from U.S. government intelligence officers inquiring about their intentions and requesting that they bring back any information about Soviet science which might be useful to their country. Erwin Chargaff, the distinguished biochemist, described being approached by these figures with his usual sarcastic wit. He found their request insulting and their low level of comprehension of science appalling.[19] One can only guess that Linus Pauling's reaction may have been similar, since he had been denied a visa to go to a conference just a few years before because of his activities publicizing fallout dangers from nuclear weapons testing. Stanley Miller on the other hand, only twenty-seven years old at the time, agreed to keep

FIGURE 2.1. Stanley L. Miller with one of his flasks enclosing a simulated primitive Earth atmosphere, February 1970. (Courtesy S. L. Miller.)

FIGURE 2.2. Aleksandr Oparin (*left*), the preeminent Russian origin of life researcher, and Cyril Ponnamperuma (*right*), head of the chemical evolution branch of exobiology at NASA Ames, c. 1964 (NASA photo, courtesy Linda Caren.)

his eyes and ears open and report whether he learned anything interesting. In the event there was little to learn except the names and personalities of the Soviet scientists at the conference, according to Miller.[20]

But, whatever the skepticism of the scientists about such notions, the fear of CIA agents that the Soviets, led by the world-famous Oparin, might possess some important lead in origin of life research, might even be close to creating life in the laboratory, was in the air. Thus, when NASA was formed in 1958, the epitome of Cold War science institutions, with the goal of catching up to the Russians in science, it is perhaps not quite so surprising that Lederberg and others so quickly convinced the new space agency that origin of life was an important area to investigate.[21]

It was Lederberg who first coined the term *exobiology* to include research into the origins of life on Earth and the development of instruments and methods to search for signs of life in the cosmos. He reasoned that one needed to know what conditions were necessary for life to begin on Earth in order to know how and where to search for life on other worlds. The term neatly encompassed the areas Lederberg found interesting in a package he felt sure would be funded from NASA's abundant coffers. He first used the term in private letters as early as June 1959, in a public talk in January 1960, and in print (in *Science*) in August of that year.[22] Lederberg contrasted *exobiology* with *eobiology* (Earth's own), but, whereas the former term caught on very quickly, the latter never did. The very popularity of the term *exobiology* shows what keen instincts Lederberg had for recognizing that the time was right to combine two previously unrelated, and relatively offbeat, areas of research and to do so under the aegis of NASA in a way that gave to both high prestige, copious funding, and a cutting-edge profile. Exobiology had its critics, some from the very outset, but it made newspaper headlines immediately, and it has remained prominent in the public imagination ever since.

Thus, when NASA first officially created a Life Sciences office on 1 March 1960, the field as Lederberg defined it was assumed from the beginning, and under the name exobiology, to be firmly within its purview. This included making research on sterilizing space vehicles to avoid contaminating other worlds a priority. And, as soon as missions to return from the moon began to be planned, the same expertise was directed toward protecting against "back contamination," or the inadvertent return of possible cosmic microbes to Earth that could perhaps allow *Andromeda Strain* scenarios to develop.[23] Few scientists, surely, have ever seen their objectives, both scientific and policy-oriented ones, converted into reality so completely and so quickly by a government agency as happened with Lederberg and exobiology. The question still remained, however: could an entire scientific discipline, just because it was dreamed up by one man (even if a very smart man) flourish for long? How would workers in many different disciplines, from astronomy to geochemistry to microbiology, come together to establish journals, professional societies, and the other trappings usually thought necessary for a scientific discipline to become established?[24]

From the start many academic biologists criticized the putative discipline, saying that, because there is no known life on other worlds, its creation amounted to establishing a field of science that has no subject matter.[25] Chief among these critics was George Gaylord Simpson, who called advocates "ex-biologists turned exobiologists." He noted, not incidentally, that such a chase after pure imaginings would divert resources away from Earth-bound biology research. This debate took place at the time when E. O. Wilson has described the evolutionary biology he and Simpson practiced as already in danger of extinction because of competitive pressure from the newly burgeoning field of molecular biology.[26] Then, too, because the UFO craze had been sweeping the country since 1947, from its inception exobiology walked a fine line between being perceived as being at the cutting edge of futuristic science and seeming to be, in the public eye, a "search for little green men."[27]

Lederberg had worries that the relationship with NASA and the publicity that went with it could cut both ways. As his diary records in mid-1959: "I wanted to avoid as far as possible contact with and support of the Man in Space program. . . . I don't want to see exobiology tag along after the military."[28] The very size and political nature of much of NASA's Cold War mission made some of its programs unwieldy behemoths more subject to the capriciously changing winds of Congress. And the man-in-space effort of all NASA projects was perhaps most obviously political rather than scientific, as Audra Wolfe has pointed out.[29] By emphasizing that exobiology was a pure science program, Lederberg hoped to keep its science from being manipulated in the interest of national prestige, as Project Mercury was from start to finish.[30] In this respect Norm Horowitz heartily concurred with him, helping bring to bear pressure from CalTech big shots, through science advisors George Kistiakowsky and Lee DuBridge. By May 1960 he wrote to Lederberg, quite concerned about any good science (such as serious exobiology) disappearing from view in the public's wild ideas of NASA and its programs. "I think this is a good time to put pressure on [NASA administrator Keith] Glennan from all sides," Horowitz concluded.[31]

The first exobiology grant money from NASA was awarded in March 1959, before the Life Sciences office even got organized. Microbiologist Wolf Vishniac of Yale Medical School, a member of the EASTEX committee, was awarded forty-five hundred dollars to begin developing a device that could detect microorganisms living in the soil of another planet.[32] Vishniac developed the device in response to a challenge from the astronomer Thomas Gold at the very first EASTEX meeting; he called it the "Wolf Trap." Like everyone else, Vishniac imagined the first place the device might actually detect extraterrestrial life was on Mars. And, indeed, Vishniac's design was one of four selected a decade later to fly on the *Viking* Mars lander mission.

Like many young scientists, Vishniac may have had some qualms about becoming involved in NASA work because it did not fit very neatly within the established disciplines that usually evaluate one's work for tenure, promotion, and grants from more traditional agencies such as the National Science Foundation (NSF) and the National Institutes of Health (NIH). This was a tension that persisted for at least twenty years. But the early generosity of NASA to academic scientists willing to join in the exobiology venture was more than enough motivation for many of the best and brightest in their fields to take the plunge. Among them were several Nobel Prize winners, including Lederberg, Calvin, Urey, H. J. Muller, Fritz Lipmann, George Wald, M. Keffer Hartline, and Manfred Eigen. Many of them had read and been deeply impressed by Oparin's book *The Origin of Life* after it first appeared in English in 1938, as had biochemical geneticist Norman Horowitz and protein chemist Sidney Fox. The prominent CalTech geochemist Harrison Brown, who first got Harold Urey interested in the study of meteorites, was among the very first grantees. There were many differences of opinion among them about approaches to the questions posed by exobiology, but there was no shortage of talent.

When Simpson attacked, Wolf Vishniac immediately responded in a letter to *Science,* as did Sidney Fox soon afterward.[33] A few months later, in August 1964, microbiologist and sanitary engineer Gilbert Levin, an early exobiology grantee, writing a "significance and status report" on exobiology, said: "The *significance* of the term exobiology is in dispute and there are those who declare that the subject has no *status*. . . . The subject is too important to permit such 'sea-lawyer' rationalization to impede its investigation. . . . The true significance of exobiology is best revealed by the questions it can help answer."[34]

Levin went on to discuss the search for life on Mars, with NASA's first Mars probe scheduled for launch in November, Project Ozma (a search for an artificial extraterrestrial radio signal), and the search for life in the cosmos generally. He argued that the science of exobiology was still in its infancy, yet the data from Project Ozma, from U.S. lunar probe *Ranger 7,* and from recent chemical studies of the Orgueil meteorite served as examples putting the lie to the claim that the field had "no data."[35] On 23 May 1965 the well-known science and science fiction writer Isaac Asimov published a piece about exobiology in the *New York Times Magazine* called "A Science in Search of a Subject." Although the title definitely played off of the publicity Simpson had drawn, the article was highly sympathetic to exobiology, citing "big guns" Urey and Lederberg but also the vocal young Carl Sagan as authorities who saw exobiology as the most exciting scientific challenge of the generation.[36]

First Projects: Academia and the Ames Research Center

How much money, exactly, was Simpson talking about? Before a Life Sciences office existed, NASA had already funded at least two scientists whose work was more or less directly relevant to exobiology. Microbiologist Wolf Vishniac had received a grant to begin developing his Wolf Trap, as already mentioned. Gilbert Levin, the other respondent who rose to defend exobiology against Simpson's challenge, had also received a small grant to begin developing a life detection device he called "Gulliver," based on bacterial respiration of detectable radioactive CO_2 from a radioactive $_{14}C$-labeled substrate in the nutrient broth. Levin was a sanitary engineer who first developed the technique in the mid–1950s as a means of detecting even minute amounts of sewage contamination in water. In a conversation with NASA chief Keith Glennan over drinks at a Christmas party in 1958, he was urged to apply for a NASA grant to develop a version of the test which could be sent to Mars. Levin followed up and got support by late 1959.[37] A year later, from the new Office of Life Sciences, he had been granted $141,173 for full-scale development of the Gulliver device for a Mars mission before the end of the decade; in 1963 he received another $221,000 and in 1964 an additional $156,500.[38]

Other big recipients during the first granting period after the creation of the Office of Life Sciences included CalTech geophysicist Harrison Brown, Richard Ehrlich of the Armour Research Foundation, Wilmot Castle Company (for "research on sterilization of space probe components"), Lederberg's group at Stanford (for "cytochemical studies of planetary microorganisms," i.e., developing the Multivator life detection laboratory), and Sidney Fox at Florida State (for study of "chemical matrices of life") (see table 2.1).[39] By the second semiannual period of grants under the Life Sciences office, Fox's group had become the biggest exobiology grantees, receiving a hefty $784,000 for their work on proteinoid microspheres, and Wolf Vishniac's grant was also renewed.[40]

But the stable of talent was expanding as word got out that NASA was a new pool of money for this kind of work. Other new grantees included Harold Morowitz at Yale (see chap. 3), James Lovelock (to begin developing gas chromatographs that could be sent on lunar and Mars landing probes), and M. Scott Blois of the Stanford Biophysics Lab.[41] Charles R. Phillips of the army's Fort Detrick chemical and biological warfare labs received a grant as well, for research on sterilizing space probes and to "determine contaminants of spacecraft components and materials."[42] In 1962 University of Houston biochemist John Oró and Berkeley biochemist and 1961 Nobel laureate Melvin Calvin both received substantial grants.[43]

During this period Gerald Soffen, a young biologist trained at Princeton under Harold Blum and now with NASA at the Jet Propulsion Laboratory (JPL), also persuaded Norman Horowitz of CalTech to be a consultant on Levin's Gulliver project.[44] The fact that Levin was a sanitary engineer without a doctorate, and not an academic scientist, was worrisome to NASA officials; they feared the Gulliver experiment would not be taken seriously in the scientific community without a Ph.D.-level scientist as part of the team.[45] As Horowitz put it: "I have agreed to serve as co-experimenter on Gil Levin's 'Gulliver' life-detection project. It seems that the Gulliver has had no official standing up to now, i.e., it was not even on the tentative list of experiments being considered for Mariner B [renamed *Voyager* in 1963]. NASA wanted to have a professional biologist attached to the experiment, in order to give it status with the scientific community and with themselves. I have agreed to take this responsibility, since I think the Gulliver is a well-designed device that deserves to be considered for a Mars mission. I am sure that you agree with this, even though you may personally prefer the Multivator."[46]

The Multivator was a portable biochemical laboratory, capable of performing a battery of biochemical tests on a Martian soil sample. Along with his Stanford associate Elliott Levinthal, Lederberg was developing it to fly on the same mission as Levin's Gulliver and Vishniac's Wolf Trap.[47] (The Mars life

TABLE 2.1 Selected Early NASA Exobiology Grants, 1959–1964

Date	Investigator(s)	Amount	Subject of Research
March 1959	Wolf Vishniac	$4,485	Development of "Wolf Trap" life detector
October 1960–June 1961	Harrison Brown	$86,850	Problems of lunar and planetary exploration
	Joshua Lederberg	$380,640	Development of Multivator biochemical lab.
	Samuel Silver	$173,800	Biochemistry of terrestrial microbes in simulated planetary environments
	Gilbert Levin	$141,173	Development of Gulliver life detector
	Wilmot Castle Co.	$106,879	Sterilization of space probe components
	Sidney Fox	$103,804	Study of proteinoid microspheres
	Richard Ehrlich	$27,766	Life in extraterrestrial environments
July–December 1961	Sidney Fox	$784,000	Study of proteinoid microspheres
	Harold Morowitz	$38,196	Study of *Mycoplasma* as minimal cell
	James Lovelock	$30,100	Develop gas chromatograph for *Surveyor*
	M. Scott Blois	$86,800	Molecular evolution in proto-biological systems
	Charles Phillips	$30,000	Sterilization of spacecraft components
	Wolf Vishniac	$15,155	Development of Wolf Trap
January–June 1962	Juan Oró	$71,250	Organic cosmochemistry
	Norman Horowitz	?	Added as consultant on Gulliver, 18 May 1982
July–December 1962	University of California–Berkeley (Samuel Silver)	$1,990,000	Construct space sciences research building
	Stanford University (J. Lederberg)	$535,000	Construct biomedical instrumentation facilities
	Melvin Calvin	$252,500	Reflection spectra as basis for studying ET life
	Wilmot Castle Co. (C. W. Bruch)	$105,297	Sterilization of space probe components
	Gilbert Levin	$87,556	Development of Gulliver
	Gustaf Arrhenius	$83,018	Composition and structure of meteorites
	Harold Urey	$73,054	Meteorite inert gases and isotopic abundances
January–June 1963	Sidney Fox	$550,000	Study of proteinoid microspheres, hosting Wakulla Springs, Fla., conference

(continued)

TABLE 2.1 *(continued)*

Date	Investigator(s)	Amount	Subject of Research
	Gilbert Levin	$221,000	Design and build prototype of Gulliver device
	James Lovelock	$55,000	Develop Surveyor lunar gas chromatograph
	Richard Ehrlich	$49,139	Survival of algae in simulated Martian conditions
	Carleton Moore	$28,978	Study and curation of meteorite specimens
July–December 1963	H. Jones	$403,548	The chemistry of living systems
	Joshua Lederberg	$132,000	Multivator ("cytochemical studies of planetary microorganisms")
	Harold Urey	$78,974	Meteorite organic and inorganic compounds
	John Lilly	$36,475	Feasibility of communication between man and other species [dolphins]
January–June 1964	Joshua Lederberg, Elliot Levinthal, Carl Djerassi	$485,000	Cytochemical studies of planetary microorganisms
	Peter Bulkeley (on related grant)	$62,984	Cytochemical studies of planetary microorganisms
	Wolf Vishniac	$215,950	Microbiol. and chemical studies of planetary soils
	Ernest Pollard	$193,625	Physics of cell synthesis, growth, division
	H. H. Hess (NAS)	$172,675	Study of exobiology
	Gilbert Levin	$156,496	Continue development of Gulliver device
	Sidney Fox	$100,000	Study of proteinoid microspheres
	Colin Pittendrigh	$66,318	Circadian rhythms on a biosatellite and on Earth
	Charles Phillips	$30,000	Studies on sterilization
	Ralph Slepecky	$19,458	Study of spore-forming bacteria
July–December 1964	Sidney Fox	$197,600	Study of proteinoid microspheres
	Wolf Vishniac	$138,441	Microbiological studies of planetary soils
	Harold Urey	$94,000	Study of meteorite organic compounds
	Klaus Biemann	$73,117	GCMS for detection of life-related organics
	J. R. Vallentyne	$46,880	Paleobiochemistry of amino acids and polypeptides

Note: Information, including project titles, taken from NASA Semiannual Reports to Congress; dollar amounts (in 1962–1964 dollars) are given from each six-month grant period.

detection projects will be discussed further in chap. 3.) In addition to his first grant Lederberg soon received much more funding for Multivator and other, related work on exobiology projects, especially related to life detection on Mars. In 1962 Lederberg received $535,000 to construct a new research facility for biomedical instrumentation, in addition to research grants.[48]

By fiscal year 1963 NASA's Life Sciences total expenditures had reached $17.5 million, with an additional $3.5 million for medical science, fully half of what the NSF spent on those areas during the same period. NASA had become a significant player, along with the NSF, NIH, and the Atomic Energy Commission (AEC), among others, in funding life sciences research.[49] Grants included capital expenditures for new research buildings; for example, a two million–dollar building at the University of California–Berkeley. The new building at Stanford expressly dedicated to exobiology was reported to be 35 percent completed by January 1965.[50] The big players drawing from this new pot of money were Lederberg, Calvin, and Fox. Fox's group received large amounts in 1963 and 1964; on the strength of his accumulated grants, Fox was able to set up an entire freestanding research institute at the University of Miami in 1964, with the university supplying only the buildings, teaching salaries, and administrative infrastructure.[51] Vishniac was also continuously funded.[52] In addition, Urey, Sagan, Harrison Brown, and others in the inner circle were regular grantees, and Princeton biologist Colin Pittendrigh joined this group. Pittendrigh became involved through Lederberg in early planning efforts for life detection on Mars.[53] His research on the effect of being in orbit on circadian rhythms was funded by NASA in this period. Many smaller grants went out to the academic research community during these years as well. Microbiologist Ralph Slepecky at Syracuse University, for example, got support for studies on the survival of bacterial spores.[54] Biologist Richard Young did related work, at NASA Ames Research Center, on the survival of bacterial spores under simulated Martian conditions.[55] Carleton Moore at Arizona State University received a grant to study meteorites.[56] One report said that "NASA grantees have made noteworthy progress in understanding how life can grow and exist in hostile and extreme environments."[57] Even John Lilly, the researcher studying communication with and among dolphins, got a grant for "a study of the feasibility and methodology for establishing communication between man and other species."[58] He had first come in contact with NASA at an October–November 1961 meeting on the search for extraterrestrial intelligence (SETI), organized by the National Academy of Sciences.

Grants to the academic community, however, were only about half of what NASA spent on exobiology research. At the beginning, the Office of Life Sciences intended for about half of the general research work and facilities construction to be funded (much more than half, if one included the budgets for actual development, launch, and operation of exobiology hardware on space missions such as *Viking*) to be in-house. By the early 1970s and throughout the subsequent history of the program, the split was closer to one-third for in-house work and two-thirds to the university community.[59] Two NASA-affiliated facilities quickly developed large exobiology research groups: the Jet Propulsion Laboratory in Pasadena, California, and the Ames Research Center in Moffett Field, California. (The JPL Exobiology program will be discussed further in chap. 4). Richard S. "Dick" Young was a young biologist who had worked at the rocketry center in Huntsville, Alabama, in the late 1950s while completing his Ph.D. degree at Florida State University in Tallahassee, so that he could put experiments into nose cones and get them flown. In 1960 he came to work at the new NASA Life Sciences office, and by late 1961 (after exobiology had been moved from Life Sciences to Space Sciences under new administrator James Webb's reorganization)[60] he was sent to the Ames Research Center to begin building up a life sciences lab and research group there, particularly specializing in exobiology.[61]

By September 1962 Young had hired a biologist, Vance Oyama, and recruited two young postdocs to come as the first nucleus of the research group, Cyril Ponnamperuma and George Akoyunoglou, who had just completed doctoral degrees on chemical evolution studies under Melvin Calvin (see fig. 2.2).[62] The National Research Council collaborated with NASA to create several postdocs per year in exobiology and other topics from 1962 onward; the postdoctoral students worked at NASA Ames under one of the staff scientists there (this became a major recruiting mechanism, to attract young scientists into the field of exobiology).[63] Ponnamperuma later recalled:

> When I got there . . . there were only two people in the Life Sciences: Dick Young and Vance Oyama. My intention was to stay for one year and then hopefully get back to Berkeley. But on the second day, Dick Young said, "Why don't you stay and set up a lab for the study of the origin of life?" And that's what we did immediately.
>
> So my personal involvement there I would say was primarily because of Dick Young. And then our first laboratories were in rented quarters, and then they put up this new building [1965]. As a matter of fact, the name at the time was "Life Synthesis" Branch; I was the one who changed it to chemical evolution. I was a bit horrified to find, when I first got there, a secretary answering the telephone with "Life Synthesis." Well, our goals were high at that time, you see.[64]

Why should NASA see chemical evolution as an obvious part of its brief? According to Ponnamperuma:

> In the early days, there is no question about it, NASA felt that if it wanted to search for life beyond the earth—you see, the search for extraterrestrial life had been given to NASA as the prime goal of exobiology. That is more or less a direct quote from the National Academy

[of Sciences] document [of January 1963]. Part of that is the study of the origin of life: if you are going to look for life somewhere else, you want to establish the processes, the fact that life appears to be an inevitable result of evolution in the universe. You can't go and look for life elsewhere unless you know it will originate somewhere. . . . The other thing is that if you want to do something on the surface of Mars, you need to know what kinds of things to look for.

So to NASA, it was always subservient to the search for life beyond the earth. It was tied to the planetary missions. This is the trouble we are having right now [1982]. Dick Young mentioned, I think, today that tying exobiology to a NASA objective has become difficult. It was hung on the Viking program: as long as we were looking for life on Mars, exobiology was very safe. Now, they need to know where to stick it in.[65]

The reconceptualizing of exobiology and NASA's relations with the field after the 1976 *Viking* missions to Mars will be discussed at much greater length in chapter 5.

In January 1964 NASA hired Harold P. "Chuck" Klein, a well-established microbiologist and chair of the Brandeis University Biology Department, to come to Ames and become its first formal head of the Exobiology Division there. By year's end Klein had shown sufficient talent as an administrator (and had survived the transition from academia's freedoms to the account-for-every-paperclip mind-set of government bureaucracy) that he became the chief of all Life Sciences operations at Ames, replacing the distinguished neurologist Webb Haymaker.[66] Richard Young was then promoted to replace Klein as head of the Ames Exobiology Division; Young remained in that post until 1967, when he was promoted to Washington, D.C., to replace Freeman Quimby as NASA headquarters head of Exobiology, overseeing funding to the Ames group as well as to the nationwide university exobiology community (fig. 2.3). At that time L. P. "Pete" Zill replaced Young as head of Exobiology at Ames.

Klein's tenure at Ames encompassed the "boom days" of NASA, when the *Apollo* program was in full swing and planetary missions began to multiply, including *Mariners* to Mars and Venus, *Pioneers* and *Voyagers* to the outer planets, and *Vikings* to Mars. He oversaw the construction of a new laboratory building (completed in December 1965) and the training of many NRC postdoc scientists; in addition, Klein presided over the division at a time when a great many staff scientists were hired as civil servants. In the Exobiology (soon to be called Planetary Biology) Division of Life Sciences alone, there were three bureaucratic branches: Chemical Evolution, Biological Adaptation, and Life Detection Systems. Hires included microbiologist Ruth Mariner Mack, chemist Fritz Woeller, chemist Katherine Pering, and, in 1966, geochemist Keith Kvenvolden. (By July 1970, under Zill's supervision, the scientific staff of the Exobiology Division had reached sixty [table 2.2]). Kvenvolden was hired by Ponnamperuma,

FIGURE 2.3. Four successive chiefs of the NASA Exobiology Program. *Left to right:* Richard S. Young, Donald DeVincenzi, John Rummel, and Michael Meyer. Photo taken at the 1993 ISSOL meeting in Barcelona and captioned "The Dynasty." (Courtesy D. DeVincenzi.)

head of the Chemical Evolution branch, to set up a lab specifically for the purpose of doing high-purity, extremely clean analysis of lunar samples, which it was anticipated would be arriving within three years or so from *Apollo* missions. As it turned out, this was also an ideal lab for analyzing the native organics from new meteorite infalls, since its high cleanliness standards made possible for the first time reliable blanks, analyses with the absolute minimum possible contamination from Earthly organic compounds. The timely fall of the Murchison meteorite in Australia in September 1969 gave the Ames clean lab the chance to compare such an extraterrestrial sample with the moon rocks they were analyzing.[67]

One of Klein's NRC postdocs, a biochemist named Don DeVincenzi, was hired on as a staff scientist / civil servant in October 1969, when his postdoc was coming to an end. By 1971 DeVincenzi had been hired into an administrative position at Ames. He spent a year, 1973–1974, at NASA headquarters in Washington, D.C., as assistant to Richard Young. Keith Kvenvolden, meanwhile, became chief of the Chemical Evolution branch of Ames Exobiology (upon the departure of Ponnamperuma in 1971). He was appointed to replace Pete Zill as head of the entire Exobiology (now called Planetary Biology) Division at Ames in August 1974; whereupon DeVincenzi, just back from Washington, became Kvenvolden's deputy.[68]

TABLE 2.2 *Personnel of NASA Ames Exobiology Division, July 1970*

L. P. Zill, Chief
E. B. Cushman, Secretary
R. Johnson (*Viking* Project)
Walter O. Peterson

Chemical Evolution Branch	*Biological Adaptation Branch*	*Life Detection Systems Branch*
C. Ponnamperuma, Chief	M. Heinrich, Chief	V. Oyama, Chief
D. Avery, Secretary	D. Rittenberg, Secretary	R. Woodworth, Secretary
K. Pering, Chemist	N. Willetts, Chemist	B. Tyson, Chemist
F. Woeller, Chemist	C. Volkmann, Microbiologist	G. Carle, Chemist
J. Flores, Chemist	R. Rasmussen, Microbiologist	B. Berdahl, Chemist
J. Lawless, Chemist	L. Jahnke, Microbiologist	O. Whitfield, Technologist
J. Williams, Biology Lab Technician	L. Hochstein, Microbiologist	C. Johnson, Chemist
J. Mazzurco, Bio Lab Tech	B. Dalton, Bacteriologist	M. Lehwalt, Microbiologist
M. Romiez, Chemist	H. Mack, Microbiologist	M. Silverman, Microbiologist
K. Kvenvolden, Geochemist	M. Stevenson, Microbiologist	G. Pollock, Chemist
E. Peterson, Chemist	H. Ginoza, Chemist	A. Miyamoto, Chemist
S. Chang, Chemist	P. Deal, Plant Physiologist	E. Merek, Plant Physiologist
	D. DeVincenzi, Chemist	J. Coleman, Bio Lab Tech
	K. Souza, Microbiologist	E. Munoz, Biologist
M. Chada, Chemist (assoc.)	L. Kostiw, Microbiologist	P. Kirk, Chemist
W. Saxinger, Microbiologist (assoc.)	J. Lanyi, Microbiologist	
P. Banda, Biophysicist (assoc.)	E. Bugna, Chemist	
V. Schramm, Biochemist (assoc.)	R. Mack, Zoologist	C. Boylan, Bacteriologist (assoc.)
S. Morimoto, Chemist (assoc.)	C. Turnbill, Electron Technician	
L. Replogle, Chemist (assoc.)	R. MacElroy, Biochemist	
	N. Bell, Microbiologist	
	A. Mandel, Microbiologist	
	S. Kraeger, Microbiologist (assoc.)	
	Y. Asato, Genetics (assoc.)	
	R. Ballard, Microbiologist (assoc.)	
	M. Lieberman, Microbiologist (assoc.)	

Source: Information provided here courtesy of Harold P. Klein.

Early Tensions: Fox and Proteinoids versus the "Nucleic Acid Monopoly"

One of the early big beneficiaries of NASA exobiology patronage was protein chemist Sidney Fox. He had been working on amino acid chemistry relevant to the origin of life since 1953 or 1954. Fox ran a lab at Florida State University from 1955 to 1964 with perhaps four or five graduate students at any given time, several of whom might be working on origin of life–related problems.[69] He was asked by NASA administrator Freeman Quimby to organize a second international conference on the origin of life in 1963, and Fox eagerly assented. The conference, in Wakulla Springs, Florida, succeeded in attracting many of the biggest names in the field, including Oparin, J. B. S. Haldane, N. W. Pirie, J. D. Bernal, and others. Fox showcased his own work, and even one of his senior doctoral students presented a paper; Richard Young presented some work done in conjunction with Fox's lab.[70] Fox quickly applied for more money for his lab and was favored by NASA. Fox received enough money to establish an Institute for Space Biosciences at Florida State. A year later he persuaded the University of Miami to hire him and help him build an entire freestanding research institute there, with as many as a dozen graduate students and visiting postdocs as well. Fox's Institute of Molecular Evolution thrived until his retirement in 1988, largely on NASA funds, though by the 1970s Fox had also begun to fill in with money from some private donors. He ceaselessly publicized his lab's efforts and solicited donations. During those two decades a great many origin of life researchers were trained in Fox's lab, many of whom have since become leaders in the field, such as Kaoru Harada, James Lacey, and Alan Schwartz.

Considering his success as an institution builder, we would do well to step back and look at Fox's research program for a moment. Fox was trained as a protein chemist. He often liked to emphasize that famed geneticist T. H. Morgan was on his dissertation committee and frequently told him, "Fox, all the important problems of life are problems of proteins."[71] Fox set out in the 1940s to develop amino acid–sequencing techniques and made important contributions but in the end was "scooped" by Fred Sanger's sequencing of insulin. By the early 1950s Fox was experimenting with what products mixtures of amino acids would react to form under hot, dry conditions. He found that they polymerized to form a substance he called "proteinoid," which was not a straight chain polypeptide like protein but did seem to form in a nonrandom way, given known conditions and starting mixtures of amino acids. Proteinoids were also shown to exhibit a range of enzymatic activities (though to a degree much less than that of true protein enzymes), and Fox emphasized that the structures having this property were created by spontaneous but nonrandom chemistry. After the 1953 Miller-Urey experiment, Fox described this process as a likely next step on the road to complex biological molecules, in a lifeless chemical world where amino acids had already formed.

By late 1958 Fox's group found that, when hot water was added to proteinoid, it spontaneously produced tiny spheres of 1–5 μm in diameter, about the size of small bacteria. In a paper in *Science* in May 1959 Fox and his group described proteinoid microspheres and suggested that they gave the first clear-cut experimental answer to how one could get, by spontaneous chemical and physical processes, from simple amino acids, formed Miller-Urey style, to membrane-bounded structures with some critical "lifelike" properties, saying they had developed "a comprehensive theory of the spontaneous origin of life at moderately elevated temperatures."[72] They showed that the microspheres absorbed biological stains and showed differential permeability to some compounds so that their inner content was soon different from that of the surrounding medium. As time went on, Fox and his students further characterized the microspheres, observing that their membrane, while not lipid, did have a bilayer structure. The microspheres spontaneously budded and sometimes divided, reproducing themselves and increasing in number. All these were lifelike properties, and Fox claimed more and more forcefully that development through proteinoids represented the most likely model by which life had developed. He spoke of the microspheres more and more as lifelike or even as alive in a rudimentary way, declaring that his group had solved the origin of life problem, at least in principle.[73] By the 1970s Fox emphasized that a differential electrical charge was maintained across the membrane of the microsphere. He referred to this characteristic as the most rudimentary beginning of the electrical charge difference across the membrane of neurons and said that in that sense the microspheres also had "rudimentary consciousness."[74]

From the beginning Urey and Miller were skeptical of the relevance of proteinoids to the origin of life. They pointed out that amino acids formed in their experiment only in aqueous solutions, whereas proteinoids required almost total removal of water from the system in order to form. Then for microspheres to form required adding water back into the system. Given the time frame in which organic compounds might remain stable at high temperatures, Miller and Urey considered such a sequence of hydration, dehydration, and rehydration a geologically unlikely event. They published this criticism in *Science* in July 1959.[75] Fox responded in a letter to *Science,* saying that in a tidal area at the sea edge with underground volcanism such repeated wetting and drying could indeed be a common set of conditions.[76]

Urey and Miller expanded their criticisms in a reply to Fox's letter, accusing Fox of linguistic sleight of hand in using terms such as *proteinoid* and *lifelike* to try to smooth over big gaps and difficulties. They insisted that all living things today are made out of proteins, which, they stressed, were very different chemically from Fox's proteinoids. They considered the somewhat nonrandom composition with which proteinoids formed completely meaningless compared to the precision of the genetic code in determining amino acid sequences in proteins.[77] Fox's conception also violated their epistemological commitment to a random chemistry model of origins.[78] Their tone was one of barely concealed derision for what they considered slipshod, sloppy scientific thinking. Miller's anger grew over the next few years as NASA Life Sciences administrators found Fox's work not only interesting but also worthy of large-scale funding. He and Norman Horowitz became more convinced than ever that explaining the steps to the origin of DNA were crucial and that Fox was thus avoiding perhaps the central issue in the question of how life originated. Both had such disrespect for Fox that they boycotted the 1963 international conference he organized, even though the likes of Oparin and Haldane were present.[79]

Miller challenged the relevance of Fox's "thermal peptides" (he refused any longer to even call them proteinoids) for the origin of life more forcefully than ever in his 1973 text, cowritten with Leslie Orgel. Horowitz congratulated Miller for taking a "firm stand on Fox . . . I think Leslie sometimes tends to be more tolerant of him than is necessary." Horowitz particularly liked Miller's statement that, "except for holes or cracks in the cooling lava which might get hot enough, a volcano is not a suitable place to conduct a thermal synthesis of polypeptides."[80]

Here we see that NASA patronage may have had a significant effect in giving Fox and his "proteinoid theory" a considerably longer lease on life than they might have enjoyed in its absence. Fox's lab was responsible for some further discoveries, though its claim that moon rocks returned by *Apollo* astronauts contained amino acids turned out to be the result of earthly contamination.[81] Many of Fox's peers grew steadily more skeptical, however, about whether proteinoids were indeed a separate phenomenon with little relevance to living systems.

Fox responded with confident pronouncements that his group had solved the origins problem and that most resistance to accepting that fact came from deep intellectual prejudice, such as the belief that nucleic acids had primacy over protein as master molecules. Fox called this a dogma and labeled the growing body of researchers who believed it the "nucleic acid monopoly." He argued that membrane-enclosed "protocells" probably came first and that the development of complex heredity molecules such as nucleic acids came only much later. Fox invoked historian of science Thomas Kuhn's conception of paradigm shift to explain the intellectual change needed to accept that so simple a solution as the proteinoid model could be correct.[82] And Fox never ceased predicting, up until his death in August 1998, that the shift would soon come.[83]

The deep conviction that had guided a research program, founded an institute, and trained a generation of workers was seen as unyielding bias and egotism when it continued in the face of any and all criticism. A postdoc from Fox's own lab wrote a devastating critique of the proteinoid theory in 1979, which was republished and widely read.[84] Fox's insistence that microspheres had consciousness and his increasingly loose and playful use of metaphoric language about their "mating," for example, were too much for even the most broadminded of his peers, and by the mid–1980s Fox had become highly marginalized, considered to have made his worthwhile contributions long ago.[85] Whether this could have occurred substantially earlier had Fox not benefited from early, large-scale NASA patronage seems a question worth asking, since that assumption is explicitly believed by many in the field.

Does this mean that in the zeal with which NASA threw money at its Cold War mission in the early years the result was a lot of bad science? Certainly not. Some might wish to interpret the story of Fox's proteinoid theory of life that way (though Albert Lehninger, in his widely respected biochemistry text, still cited Fox's "protein-first" view as an alternative to "nucleic acids first" in

1970 and again in 1975).[86] One charge raised by his opponents, Fox's nucleic acid monopoly, is that NASA administrators who were attracted to Fox's work in the early 1960s were bureaucrats; had they been cutting-edge research scientists, his opponents claim, Fox, despite being "an excellent self-promoter," would never have received such large grants, and his funding would have been cut off more quickly.[87] Quimby and, later, Richard Young and Donald DeVincenzi certainly continued to believe that Fox's work might be important longer than many in the research community (and far longer than Miller or Horowitz).[88] The tenability of this claim will be discussed later.

Exobiology Arrives

Early in 1967 Richard S. Young moved from NASA Ames, where he had been head of the Ames Exobiology Division, to NASA headquarters in Washington, D.C., to replace Freeman Quimby as director of the Exobiology Program at the national level. At this time NASA began to instigate a whole series of meetings on origin of life and exobiology broadly (table 2.3). A series of five meetings was planned in conjunction with the Smithsonian Institution and the New York Academy of Sciences, beginning with meetings in May 1967 and May 1968 at Princeton. Only four of these meetings took place, but they had an impact more for bringing together a wide range of scientists, along with NASA funding, than for any other outcome. Cyril Ponnamperuma, among others, claimed that these NASA-sponsored meetings were some of the most essential glue holding together the nascent field of exobiology, until more formal structures such as a journal and professional organization (the International Society for the Study of the Origin of Life [ISSOL]) came along: "Scientists need a framework in which to work. So they [NASA] have helped that. The rails have been pretty well greased all along. More than the initial catalytic effect, more than giving the objective, the constant stimulus has been from [NASA]."[89]

NASA sponsored meetings on more specialized topics as well. Between 1968 and 1971 radio astronomers discovered two dozen organic molecules, such as formaldehyde, in giant molecular clouds in interstellar space, where they had previously been unknown. In addition, in December 1970 extraterrestrial amino acids and hydrocarbons were found by a NASA Ames team under Ponnamperuma, for the first time unequivocally, on a recent, uncontaminated sample of a meteorite. By February 1971 a meeting was convened at Ames to assess the implications of interstellar organic molecules for the origin of life.[90]

In January 1970 and January 1971 NASA convened scientific meetings in Houston to report and discuss findings coming in on the lunar samples brought back by the *Apollo 11* and *12* missions. In October 1971 the scientists who specialized in carbon chemistry—that is, extraterrestrial organics—convened a meeting of their own with NASA sponsorship, at the University of Maryland in College Park. Cyril Ponnamperuma had just moved from Ames that fall to set up a Laboratory of Chemical Evolution in the chemistry department, and he was

TABLE 2.3 *Selected Origin of Life / Exobiology Meetings through 2002*

Meeting	Published Proceedings or Papers
1953, Society for Experimental Biology, Cambridge	1954, *New Biology* special issue (April)
1955, Brooklyn Polytech	1956, papers in *American Scientist*
December 1956, New York Academy of Sciences	1957, *Annals of New York Academy of Sciences* special issue (August)
August 1957, First International Conference, Moscow	1959, Clark and Synge, eds. Proceedings
January 1960, First COSPAR meeting, Nice, Fr.	1960, Bijl, ed. *Space Research*
1961, Second COSPAR mtg., Florence	
1961, Woodring Conference	
April 1962, New York Academy of Sciences, on organics in meteorites	
May 1962, Third COSPAR meeting, Washington, D.C.	1963, *Life Sciences and Space Research*, vol. 1
June 1963, Fourth COSPAR meeting, Warsaw	1964, *Life Sciences and Space Research*, vol. 2, Florkin and Dollfus, eds.
October 1963, Second International Conference, Wakulla Springs, Fla.	1965, Fox, ed. *Origins of Prebiological Systems*
Spring 1964, Woodring follow-up meeting, Carnegie Institute, Geophysics. Lab, Washington, D.C.	
May 1964, Fifth COSPAR meeting, Florence	1965, Florkin, ed., *Life Sciences and Space Research*, vol. 3
May 1965, Sixth COSPAR meeting, Mar del Plata, Arg.	1966, *Life Sciences and Space Research*, vol. 4, A. Brown and Florkin, eds.
Summer 1965, Mars meetings	1966, Pittendrigh, ed., *Biology and the Exploration of Mars*
May 1966, Seventh COSPAR meeting	1967, *Life Sciences and Space Research*, vol. 5
May 1967, Princeton Conference I	1970, Margulis, ed., *Origins of Life*
1967, Eighth COSPAR meeting	1968, *Life Sciences and Space Research*, vol. 6
November 1967, Royal Society of London, Aspects of Biochemistry of Possible Significance for Origin of Life	1968, *Proceedings of the Royal Society of London*, vol. 171B, no. 1, Pirie, ed.
May 1968, Princeton Conference II	1971, Margulis, ed., *Origins of Life II*
April 1970, Third International Conference, Pont-á-Moussan, Fr.	1971, Buvet and Ponnamperuma, eds., *Molecular Evolution I*
February 1971, NASA Ames, Interstellar Organic Molecules and the Origin of Life	
May 1970, Santa Ynez, Calif., Conference III, planetary astronomy	1972, Margulis, ed., *Origins of Life III*
May 1971, Elkridge, Md., Conference IV, chemical evolution / radio astronomy	1973, Margulis, ed., *Origins of Life IV*

TABLE 2.3 *(continued)*

Meeting	Published Proceedings or Papers
October 1971, College Park, Md., organics in lunar samples	1972, *Space Life Sciences,* vol. 3, special issue
August 1972, Symposium on Cosmochemistry, Smithsonian Astrophysical Observatory, Cambridge, Mass.	1973, A. G. W. Cameron, ed., *Cosmochemistry*
2–3 April 1973, Roussel UCLAF conference on "The Origin of Life," Paris	
June 1973, Fourth International Conference/ 1st ISSOL meeting, Barcelona	1974, Oró, Miller, Ponnamperuma, and Young, eds., *Cosmochemical Evolution and the Origin of Life*
May 1974, Royal Society of London, Discussion on the Recognition of Alien Life	1975, *Proceedings of the Royal Society,* ser. B 189, no. 2, Pirie, ed.
August 1974, Conference at Bakh Institute, Moscow	January and April 1976, *Origins of Life,* special issues
October 1974, College Park Colloquium 1	1976, Ponnamperuma, ed., *Giant Planets*
October 1975, College Park Colloquium 2	1976, Ponnamperuma, ed., *Precambrian Early Life*
October 1976, College Park Colloquium 3	1978, Ponnamperuma, ed., *Comparative Planetology*
1977, Fifth International Conference / Second ISSOL meeting, Kyoto	
1978, Amino Acid Biogeochemistry, Airlie House, Va.	
October 1978, College Park Colloquium 4	1980, Ponnamperuma and Margulis, eds., *Limits of Life*
June 1979, NASA Ames 1	1981, Billingham, ed., *Life in the Universe*
May 1979–August 1980, UCLA PPRG	1983, Schopf, ed., *Earth's Earliest Biosphere*
1980, Twenty-first COSPAR meeting	1981, *Life Sciences and Space Research,* vol. 19
June 1980, Sixth International Conference / Third ISSOL meeting, Jerusalem	1981, Y. Wolman, ed., *Origin of Life*
October 1980, College Park Colloquium 5	1981, Ponnamperuma, ed., *Comets and the Origin of Life*
June 1981, NATO Advanced Study Institute, Maratea, It.	1983, Ponnamperuma, ed., *Cosmochemistry and the Origins of Life*
July 1981, January and May 1982, ECHO Workshops	1985, Milne, Raup, and Billingham, eds., *Evolution of Complex and Higher Organisms*
October 1981, College Park Colloquium 6	1982 papers in *OLEB*
1982, First GRC OOL	
July 1983, Seventh International Conference / Fourth ISSOL meeting, Mainz, Ger.	1984, Dose, Schwartz, and Thiemann, eds., Proceedings

(continued)

TABLE 2.3 *(continued)*

Meeting	Published Proceedings or Papers
July 1983, Clay Minerals and Origin of Life, Glasgow	1985, Cairns-Smith and Hartman, eds., *Clay Minerals and the Origin of Life*
1985, Second GRC OOL	
July 1986, Fifth ISSOL meeting / Eighth International Conference, Berkeley, Calif.	1987 papers in *OLEB*
June 1987, IAU Colloquium on Bioastronomy, Balaton, Hungary	1988, G. Marx, ed., *Bioastronomy: The Next Steps*
1987, Third GRC OOL	
1988, Prebiotic Syntheses, Okazaki Conference, Japan	
July 1989, Sixth ISSOL meeting, Prague, Czech.	1990 papers in *OLEB*
July 1990, NASA Ames 4	
August 1990, Fourth GRC OOL, Plymouth, N.H.	
October 1991, NATO ASI, Edice, Sicily	1993 Greenberg et al., eds., *Chemistry of Life's Origins*
October 1992, First Trieste Conference on Chemical Evolution	
1993, Fifth GRC OOL	
July 1993, Seventh ISSOL meeting, Barcelona	1994 papers in *OLEB*
April 1994, NASA Ames 5	
August 1994, Sixth GRC OOL, Newport, R.I.	
July 1996, Eighth ISSOL meeting, Orléans, Fr.	1997 papers in *OLEB*
1996, Fifth International Conference on Bioastronomy, Capri	1997 Cosmovici, Bowyer, and Wertheimer, eds., Proceedings
1997, Seventh GRC OOL,	
September 1997, Fifth Trieste Conference on Chemical Evolution	1998, Chela-Flores and Raulin, eds., *Exobiology: Matter, Energy, and Information . . . Universe*
1998, Amino Acid and Protein Geochemistry, Washington, D.C.	
February 1999, Eighth GRC OOL, Ventura, Calif., Schopf and Lazcano, cochairs	
July 1999, Ninth ISSOL meeting, San Diego, Calif.	2000, papers in *OLEB*
April 2000, First Biennial Astrobiology Science Conference, Ames	
July 2000, Tenth GRC OOL, Plymouth, N.H.	
January 2002, Eleventh GRC OOL, Ventura, Calif., Kenneth Nealson, NASA, chair	
April 2002, Second Biennial Astrobiology Science Conference, Ames	2003, papers in *International Journal of Astrobiology,* vol. 2
June–July 2002, Tenth ISSOL meeting, Oaxaca, Mex.	2003 papers in *OLEB*

instrumental in setting up the meeting. It was the first of what came to be a whole series of what Ponnamperuma called "College Park Colloquia on Chemical Evolution." A third International Conference on the Origin of Life was convened in April 1970 in Pont-à-Mousson, France, and a fourth in June 1973 in Barcelona, Spain, partly with NASA funds. Independent of NASA, a "roundtable" conference on "origin of life" was held 2–3 April 1973 at Maison de la Chimie in Paris, by Roussel UCLAF. In addition, the Bakh Institute of Biochemistry in Moscow sponsored an origin of life / exobiology meeting in August 1974 to mark the fiftieth anniversary of Oparin's original 1924 pamphlet.

In his first few years as program chief, Richard Young was everywhere. He turned up at almost every meeting and was always recruiting. Young approached scientists whose work he thought promising (or they approached him) and suggested that they apply for Exobiology Program funding at a modest level; "seed money" was what he had in mind. Thus, in 1971 Ponnamperuma recommended Lynn Margulis's work on serial endosymbiosis theory to Young after the NSF had turned her down, and he encouraged her to apply for a program grant. At the April 1973 Paris meeting Young approached Carl Woese and suggested that he apply. Young funded both of them immediately, though modestly, and NASA has been a critical means of support for both (almost the sole means for Margulis) ever since.

In particular, Young was looking for ideas so interdisciplinary in their breadth that they were having difficulty getting funding from the NIH or NSF. (The first person with origin of life or exobiology as a major research focus to be elected to the National Academy of Sciences, not until 1973, was Stanley Miller. So the field was still perceived as an odd "borderland" area, not fitting comfortably into biochemistry, geochemistry, microbiology, cell biology, or any other existing disciplinary niche.) The large federal science-funding agencies were organized to review proposals pretty much along disciplinary lines. Thus, something far from central to cell biology, such as Margulis's 1970 proposal for work related to endosymbiosis, was likely to be rejected by NSF's Cell Biology Division, often out of hand. As Jan Sapp has shown, by 1970 the study of cytoplasmic inheritance (such as Margulis's study of DNA in mitochondria, chloroplasts, kinetosomes, and other organelles) had been marginalized by the rising power of nuclear (chromosomal) inheritance work, especially after Watson and Crick's research on DNA structure and the consolidation of molecular biology.[91] Margulis recalls:

> I applied for a three-year grant for $36,000, I remember distinctly, to continue this work—we had been productive, we did publish a paper or two on that [seeking DNA in kinetosomes] at that point. And that was exactly when *Origin of Eukaryotic Cells* first edition, Yale University Press, came out. . . . My grant officer calls me up and he says "I'm sorry to tell you we've turned down your proposal," it was a three year proposal. And he went on to say, "you didn't suggest the following

controls," he was telling me what was wrong, "you didn't have the following experiment." I said, "look on page seven, that's exactly the experiment we have there so I don't understand." He said, "well, frankly I haven't read the proposal but let me tell you that there are some very important molecular biologists who think your work is shit." He said that on the phone . . . he said, "your work appeals to the small minds in biology." And I said, "well who are the small minds in biology?" And he said "well, natural historians." And I said "that's quite a compliment." Anyway, he said "don't ever apply to [NSF] Cell Biology again."[92]

Margulis was stymied and quite eager when Young encouraged her to apply to NASA Exobiology. She recalls that, even with the American Institute of Biological Sciences (AIBS) review panels, the Exobiology grant application process had a "small town" feel. Young had a fair amount of latitude, if he wanted to encourage a particular investigator, at least with some modest initial funding. In 1971 Margulis received a grant of fifteen thousand dollars.[93] Both she and Woese attest that this early seed money was critical to sustaining their research programs, and it gradually increased year by year, as their research proved more fruitful and fulfilled Young's hopes.[94]

The search for and nurturing of interdisciplinary "diamonds in the rough" which had been passed over by NSF and NIH soon became Dick Young's trademark. And the tradition was very much handed down by apprenticeship to his successors, DeVincenzi, Rummel, and Meyer (see fig. 2.3). People who first met origin of life workers or first got connected with NASA through these meetings, in addition to Margulis and Woese, include Jeff Bada (1967, 1971), Elso Barghoorn (1967, 1971), David Buhl (1971), H. D. "Dick" Holland (1968), Sol Kramer (1967, 1968), James Lovelock (1968), Leslie Orgel (1967, 1968, 1970, 1971), Carl Sagan (1963, 1967, 1968, 1971), J. W. Schopf (1967, 1970, 1971), Alan Schwartz (1963), and many others. Barghoorn was a well-established geologist, but Schwartz and Bada were still graduate students when they first attended these meetings, and Schopf had only just finished his Ph.D. work. Many others were still quite young scientists (e.g., Margulis, Ponnamperuma, and Sagan) or were unknown to the few who had dedicated their research primarily to origin of life / exobiology. (Stanley Miller himself was still only thirty-seven when he attended the NASA-sponsored meeting in Princeton in 1967.)

Barghoorn and his student Schopf specialized in identifying Precambrian fossils of microorganisms in ancient rock samples (beginning with the two billion–year–old Gunflint chert from the northern shore of Lake Superior).[95] Their involvement brought to the attention of origin of life researchers a reverse line of work: the examination of steadily older and older fossil bacteria could work backward toward the origin of the first life on Earth. That way the gap could steadily be narrowed between what was known of later, complex life forms and others much more similar to the original, most primitive living things. Furthermore, once one could narrow the time window in the Earth's geologic past, during which life must first have appeared, one could also know much more about the specific chemical and geological conditions under which the initial formative steps must have occurred. Precambrian paleontology, an uncommon specialty before the 1960s, proliferated and flourished under NASA support (more on this in chap. 5).

Buhl was one of the astronomers who first detected organic molecules in interstellar space, and NASA has continued to support the search for further details about how much and what kind of potential precursor molecules of life are to be found in comets, meteorites, and other planets as well as in interstellar space. Woese's work on the origins of the genetic code led directly to the discovery of the Archaea, a third "domain" of life as different (in their ribosomal nucleotide sequence) from bacteria and eukaryotes as those two are from each other. Woese's work also led to highly sophisticated molecular methods for constructing lineages ("family trees") of all known living organisms, which give highly suggestive hints about the nature of the last common ancestor of all forms living today.

James Lovelock, initially hired by JPL as a consultant on life detection strategies for the moon and planets, met Margulis, Holland, and Lars Gunnar Sillén at the 1968 origin of life meeting.[96] Thinking comparatively about the atmospheres of Mars, Earth, and Venus, he went on over the next few years, and after 1970 in collaboration with Margulis, to develop the controversial Gaia hypothesis. First published in a developed form in 1974, this amounted to the claim that all living things on Earth, along with the lithosphere, oceans, and atmosphere, act as a unified, synergistic system (which Lovelock named "Gaia," after the ancient Greek Earth goddess) analogous to the body of a single organism, which homeostatically controls environmental conditions in the oceans, the atmosphere, and so on, so that they remain within the range needed to support life.

This sampling gives an idea of the broad range of interdisciplinary research programs spawned, supported by, and/or spun off from NASA Exobiology funding. As John Rummel, one of Young's successors put it, from the beginning exobiology had no choice but to seek and encourage interdisciplinarity:

> All the interesting questions [in exobiology] are interdisciplinary. Certainly all the leaders appreciated that and . . . it was always important to people who were in program management in exobiology that they not be replaced by somebody who was narrowly focused. Because that person would never be successful. And any attempt to narrowly influence the field in a particular discipline would have serious repercussions in terms of the scientific quality of the results. So if I brought anything to the program it was a desire to have good inconsistencies in the people who were funded so that they could have a much better time arguing with each other.[97]

As Rummel observed, mixing bright, talented people from such diverse fields of inquiry was not without intellectual fireworks and personality clashes.

On a higher level Richard Young's patronage of interdisciplinary work had the potential to backfire in academia. The criticisms of Sidney Fox's research mentioned earlier resonate with a documented history of tension between the academic life science research community and NASA. Cutting-edge researchers in the academy criticized the work of NASA Life Sciences programs from the inception through the entire first decade of their existence, and Young inherited this legacy when he came to head the Exobiology Program in 1967. The chief criticisms were that NASA management priorities always put life sciences research (unlike physical sciences and engineering) at the bottom, far below engin-eering and technical support to launch missions and catch up in the space race. NASA bureaucrats even split up life sciences research under several different offices in November 1961, less than two years after the Life Sciences division had been created. Most of exobiology research was put under the Office of Space Sciences at that time, where it has remained for most of the years since. Furthermore, academic scientists repeatedly criticized the design of experiments funded by NASA, saying that improper or insufficient controls rendered the results ambiguous. These charges were repeated in multiple reviews, up through the early 1970s.[98] On the other hand, every time NASA sought a more qualified person from the life sciences research community to fill a managerial position, no highly qualified, cutting-edge academic showed any interest in giving up the freedom of his or her lab for the managerial headaches of a bureaucratic position (recall Klein's experiences in moving from Brandeis University to NASA Ames). Thus, the situation seemed unlikely to improve, even when a new NAS report in August 1970 offered some more tactfully worded versions of the long-standing criticisms.[99]

This negative stance cannot be taken, however, to validate fully the claims of Fox's opponents. Of the top Life Sciences officials involved in the early 1960s, the three most involved in exobiology were all men who came from the research community and were lauded for their competence, notwithstanding the fact that none of them had been involved in origin of life or other exobiology-related fields prior to coming to NASA. (This is not much of a substantial criticism at a time when a small handful of scientists were just inventing "exobiology," and by definition at first very few could claim any competence in that field.) Consider their backgrounds: Richard Young, a Ph.D. embryologist who had flown sea urchin eggs in missile nose cones to study the effect on development before starting the first NASA Life Sciences laboratory at the Ames Research Center in late 1961; Freeman Quimby, a Ph.D. physiologist who had been at the San Francisco Office of Naval Research before coming to head the Washington, D.C., headquarters NASA Life Sciences office in February 1960; and Orr Reynolds, also a physiologist, who had been head of research at the Office of Defense Research and Engineering before taking charge of the biology division of the new Office of Space Sciences (as Quimby's superior) in early 1962. All were established researchers first, though Quimby and Reynolds had shown managerial ability. Richard Young, as the first head of the Exobiology Program, came to be more widely lauded on all sides of the exobiology research community for having a good sense of sound science than almost any other figure in the history of the field. Thus, a simplistic story that paints the managers of the early years of Life Sciences as bureaucrats who did not understand the science cannot explain away the appeal of NASA-funded research programs such as Fox's nor prove that they were scientifically weak.

Nor can the small town atmosphere of the early days, with a portrait of exobiology managers almost single-handedly picking and choosing what to fund, serve as a simple scapegoat for any research retrospectively judged less valuable. Even under Freeman Quimby, by 1965 at the latest, a system of review panels for exobiology grant applications had been put in place, administered through the American Institute for Biological Sciences.[100] Carleton Moore, a meteorite geologist who was a member of AIBS review panels from the beginning, recalled making site visits to labs such as Fox's to evaluate the quality of work being done.[101] It seems true that Quimby and Young exercised a fair amount of discretion, as the Exobiology director had the right to take the review panel's findings into account and then himself make the final decision about any given proposal. Donald DeVincenzi, working as deputy under Young in NASA headquarters for a year, from 1973 to 1974, recalled: "After the review by a 15-member panel from the American Institute of Biological Sciences, he would look at them and add his own comments, and then funded them. It was that aspect of it that I found interesting; that is that he did not have to blindly follow the peer review results, strictly on the peer review scores. He was able to put his own emphasis on it, he could for example, fund a proposal that had slightly lower scores if *he* thought that that proposal was promising and worthwhile."[102]

Young had to provide written justification for overriding peer review scores, but, as with DeVincenzi when he took over upon Young's departure in August 1979, these cases were the exception rather than the rule, so that "it was based on sound peer review panels, supplemented by [the director's] own evaluations." Furthermore, during his tenure, says DeVincenzi: "I knew from the feedback of panel members when there was any problem. Whether a project was their idea or my idea, we followed up on how it went. That was the way we got people willing to be reviewers; it was a sort of hallmark of the [Space Sciences and, within it, Exobiology] program that people talked with one another freely."[103]

Above and beyond evaluations of NASA and its methods for selecting work to be funded, historian and philosopher of science Iris Fry has come to similar conclusions to those described here about Fox's work itself. She also notes that Fox's research program made some important philosophical contributions as well as technical ones:

> though major parts of Fox's theory were later challenged by many researchers, his influence at the time was instrumental in turning the problem of the origin of life into a scientific subject. Though the relevance

of his microspheres to the process of emergence is dismissed by many, this is not the case as far as the proteinoids are concerned. . . . Various scenarios, metabolic as well as genetic, rely on the possibility of the prebiotic formation of proteinlike polymers possessing enzymatic activity as a crucial step in the origin of life [Stuart Kauffman's scenario in his 1993 *The Origins of Order,* e.g.]. Fox's philosophical contribution to the subject is no less important than his empirical contribution. Against the chance approach, Fox helped formulate the philosophical anti-chance conception, pointing to the role of strong constraints channeling the emergence of life and its evolution.[104]

Beyond intellectual matters at least some of the hostility from the academic science community was due to what we might call "NASA envy." This is illustrated clearly in the case of the microbiologist Wolf Vishniac, the first scientist to receive a NASA grant for exobiology research. Vishniac designed one of the four experiments originally selected in 1969 to fly in the *Viking* biology package. Rising costs caused his experiment to be cut from *Viking* in March 1972, rather suddenly depriving Vishniac's lab of its major source of external funding. He was asked to remain part of the Viking Biology Planning Team, but he began to write rather exasperated apology notes for missing some meetings. He was scrambling to re-tailor his research program on microbial life in extreme environments, so that it would be mainstream enough to be funded by the NSF and/or NIH. But Vishniac, like others in his position, found that he was being punished by those agencies for accepting "space dollars." The NIH had turned down a grant application; according to Vishniac, "I was told unofficially that it received a low priority because I was 'NASAing' around."[105] The NSF had also decided not to renew a grant of his, "partly because of his association with NASA. The exobiologist told [*Viking* team leader Gerald] Soffen that 'it is essential that I recapture some sort of standing in the academic world and I must therefore limit my participation in *Viking* to essentials only.'"[106] Clearly, to some extent NASA officials internalized this attitude about their exobiology science, at least in the early years: witness Soffen's felt need to include Horowitz, a Ph.D. scientist and a "real biologist" on the Gulliver experiment, "in order to give it status with the scientific community *and with themselves* [NASA]."[107]

Regarding the perception that exobiology was tossed around like a bureaucratic football under the new NASA administrator James Webb, exobiology scientists say this seriously misunderstands the actual situation within NASA. DeVincenzi, Rummel, and most of the exobiology scientists are convinced that exobiology is actually much more appropriately housed with Space Sciences than lumped together arbitrarily with astronaut physiology and space medicine, just because "those things are also biology." Exobiology work requires the closest interdisciplinary interaction, they point out, with planetary astronomy and geology, climatology and atmospheric physics and chemistry, oceanography, and so forth, and thus belongs in a nonarbitrary, rational way, administratively, with those sciences. Furthermore, much of the stigma of very poorly done NASA science, with poor or no controls, they agree, *did* belong to the loose field of astronaut medicine, which they were glad to part company with.[108]

Despite this hostile climate, exobiology had a sufficiently broad group of scientists, the continued impetus of liberal NASA funding, and a secure enough place in the public imagination that the field continued to grow. Soon, in addition to increasingly regular meetings with a stable (if expanding) core group, some of the most clear-cut features materialized which mark a consolidating scientific discipline: namely, a disciplinary journal and a professional society. The journal that began publication in 1968 was called *Space Life Sciences*. Its subject matter constituted all of what the NASA Life Sciences office had lumped together at its creation: all topics exobiological—plus the effects of such things as space flight and zero gravity—on living organisms and metabolic processes. This was much to the distaste of Lederberg, Cyril Ponnamperuma, and other "pure" exobiologists, but, as in the Office of Life Sciences, it was an artifact of the serendipitous events that had led to the journal's founding.

A highly enterprising and wealthy Armenian immigrant to the United States, Gregg Mamikunian, became a naturalized citizen and was involved in the chemical evolution programs at JPL in the early 1960s. He was interested, for instance, in the analysis of meteorites for traces of life or its precursor molecules.[109] According to Ponnamperuma:

> One day he got the idea that space life sciences needed a journal. So he telephoned Reidel; Pergamon was producing [a journal] in some other discipline. so he called Reidel up, and Reidel said they would be delighted. And that's how the journal began. It went through a bad history at the beginning. Mamikunian held up the manuscripts and people started complaining. Then a man named Lovelace . . . who was an M.D., took it over and it was still primarily space life sciences. He asked me at the time whether I would be an associate editor, and I agreed to do that, just to look over the origins of life / chemical evolution articles.[110]

After only a year or two Lovelace wanted to give up the journal, being too busy with other pursuits, so he suggested to Reidel that Ponnamperuma become full-time editor. Ponnamperuma had little interest in zero-gravity work; in late 1972 he agreed to it, but only on the condition that the journal be devoted solely to chemical evolution and exobiology. Reidel agreed, so, beginning with volume 5 in 1974, the journal's name was changed to *Origins of Life: An International Journal Devoted to the Scientific Study of the Origin of Life*.[111] Ponnamperuma overhauled the editorial board accordingly, staffing it with exobiology regulars such as Barghoorn, Klein, Lederberg, Oró, Sagan, and Young. In 1983 the editorship passed to chemist James Ferris at Rennselaer Polytech. A new publisher, Kluwer, took over soon after, and the name was changed to *Origins of Life and Evolution of the Biosphere* to indicate the extent to which studies of the early history of life on Earth, early ecosystems, and so forth, were now included under the exobiology umbrella. This trend continued with the creation of the Astrobiology Institute in 1997. Alan Schwartz at Nijmegen University in the Netherlands assumed editorship of the journal.

No doubt part of what gave Ponnamperuma the confidence to insist that the journal be devoted exclusively to exobiology was the sense that the field had grown and matured sufficiently that it needed (and could more than fill the pages of) a journal entirely its own. In 1971 another journal had begun publication, the *Journal of Molecular Evolution,* which included origin of life research as one of its major areas of coverage. But two signal events in 1972 contributed to this sense as well. First, early in the year Oparin, Fox, Oró, Young, Marcel Florkin, and others had founded the International Society for the Study of the Origin of Life and began planning its first meeting, which was to be the Fourth International Conference on Origin of Life, in Barcelona in 1973.[112] Subsequently, ISSOL meetings were planned with considerable regularity in every third year (see table 2.3). The society and its regular meetings on an international scale showed that the field had achieved stability. Norman Horowitz cited the new journals and the society as evidence that the field had become a consolidated research area in a prominent 1974 review article. He added that, even considering only the literature since 1970 or so, "a large number of review articles, critical and theoretical discussions, books, and conference proceedings dealing with the origin of life have appeared in recent years."[113]

Shortly before, in the summer of 1972, Horowitz formed a committee to nominate Stanley Miller for membership in the National Academy of Sciences, the most prestigious scientific body in the United States. Horowitz realized that the stringent nominating process, historically centered mostly on existing, well-established disciplines such as the Biochemistry Section of NAS, was a barrier to a scientist in a new borderland area such as exobiology. Thinking Miller highly deserving, he felt that nominating him for membership would simultaneously serve as a "good test case" for other top-notch workers in the new field (though he had strong ideas about who they were and, even more clearly, who they were not).[114] Miller, it should be noted, had received most of his funding from NSF and other non-NASA sources up to this time, making him immune to the kind of NASA envy which was so destructive for Wolf Vishniac at just this time.[115] One of those who signed the nominating petition, the biochemist John Edsall of Harvard, agreed, saying Miller's work "is certainly outstanding and he makes an excellent candidate for a nomination of this sort [requiring a Voluntary Nominating Group], since his field of research does not fit neatly into any of the regular categories." In a letter trying to assuage possible opposition by the Biochemistry Section, Horowitz added: "As you know, Stanley inhabits a sparsely populated interdisciplinary area between biochemistry and geochemistry and has contributed to both."[116] Miller was successfully voted into the National Academy in early 1973. If it was a test case, then exobiology had passed the test and gained a de facto foothold among the highest ranks of the nation's scientists. George Gaylord Simpson, now retired near Tucson, Arizona, might still persist in his opinion.[117] But "this view of life" had been rendered moot by the passage of events; exobiology had arrived.

CHAPTER 3

Exobiology, Planetary Protection, and the Origins of Life

*I*n the first fifteen years of the NASA exobiology program the largest expenditures by far were mission oriented: developing experiments to travel on space probes, especially to Mars, and constructing "clean lab" facilities to analyze meteorites or returned samples from the Moon for organics that might be relevant to the origin of life. NASA funding pushed origin of life research in new directions, including the study of life in extreme environments and the development of the field of theoretical biology. At the same time NASA expanded work under existing approaches. Cyril Ponnamperuma and the chemical evolution team he assembled at NASA Ames carried out many new variations on Miller-Urey synthesis experiments, as did other labs.[1] A great deal of energy and brainpower also went into debating the best policies and procedures to protect against microbial contamination from one world to another, which could vitiate all attempts to measure native organic compounds, let alone determine the possible existence of any biota native to the Moon or planets. Both forward contamination (Earth organisms carried to another world on an insufficiently sterilized spacecraft) and back contamination (return of alien life to Earth with returning astronauts and/or samples) were considered. While most researchers considered back contamination from the Moon an extremely unlikely possibility, it was still thought that the consequences could be so severe that a quarantine effort was justified, both on samples and astronauts. More challenging was the development of analytic labs so free of any earthly organics that results from extraterrestrial samples could be reliably attributed to the sample itself. From many different directions, through an astonishing variety of often seemingly unrelated activities, NASA was gradually building the new discipline of exobiology.

The Mars Program, through June 1965

Although early talk about life on other planets had focused on Venus as well as Mars, by 1962 space probes and ground-based astronomers had shown the surface of Venus to be as astronomer Carl Sagan had predicted: a runaway greenhouse at a temperature of hundreds of degrees, far too hot for any life to survive. Thus, while concern about forward contamination still applied to all other moons and planets, the attention of those eagerly seeking life on other worlds focused almost exclusively on Mars. There is one sense in which exobiologists were thereby vindicating G. G. Simpson's critique of their zealous crusade. In theory exobiology could benefit as much or more from the comparative study of other planets where life did *not* appear; the comparison would highlight the factors necessary for the origin of life most strikingly by their absence. (Lovelock's comparison of Venus, Earth, and Mars was precisely this kind of broad-based approach [see chap. 4].) A truly systematic exobiology would therefore have focused equal amounts of resources on as many different solar system bodies as could be practicably reached by the available technology. Nevertheless, resources shifted quickly and overwhelmingly toward Mars exploration. This was a big risk: if the search for life on Mars turned out to be a bust, the scientific reputation of exobiology would suffer, and Congress's willingness to continue pouring in millions of dollars would be the first victim.[2]

During the early 1960s, however, the free flow of money from Congress to NASA and from NASA to the research community made such worries seem excessively fussy: there would be enough money to do everything in the end, it seemed. A report in the 24 August 1962 issue of *Science* on a "Soviet Space Feat" of the previous week very much captured this attitude: the feat would not result in more funds for NASA, the author opined, because the tap was already open full bore. "Thus, the Soviet feat is not likely to result in more funds for NASA, since under Kennedy NASA has been told to think big and has received everything it has requested."[3]

As described in chapter 2, among the very first exobiology grantees were Wolf Vishniac, Gilbert Levin, and Joshua Lederberg, who were developing life detection devices to be sent to Mars. Vishniac's Wolf Trap was based on using the light-scattering property of multiplying microbial cells in a nutrient solution. It would mechanically introduce soil from another world into a nutrient broth, incubate the mixture, and look over time for the typical light-scattering reaction as the broth became cloudy with growth. Levin's Gulliver (see fig. 4.2) incubated soil in a nutrient broth that included carbon sources (formate, lactate, and glutamate) radioactively labeled with $_{14}C$ then measured the gas over the solution over time with a Geiger counter, seeking to detect $_{14}C$-labeled CO_2 given off by any microbes as they oxidized the carbon sources.[4]

Lederberg's Multivator was a more ambitious device, with a rotating chamber containing fifteen separate chemical test chambers, so that many different biochemical analyses could be carried out on a soil sample, all directed from an Earth-based lab. Dust-bearing air was drawn into the device and "combined with appropriate reagents or biological materials. The resulting reactions are then detected with a photomultiplier . . . for detection of biologically important macromolecules by fluorimetry, turbidimetry, nephelometry, absorption spectroscopy, or absorption spectral shifting in a test substrate."[5] The primary biochemical assay with which the device was first being tested was for the enzyme phosphatase. (Such a large automated lab made sense in the context of the large Mars lander mission called *Mariner B* and later *Voyager*, as it was envisioned between 1960 and late 1965. As costs escalated for such a large spacecraft, the mission was scaled back considerably, so the experiments had to be sent designed to operate in a largely preprogrammed sequence, with very little of the flexibility designed into a device such as Multivator. It was essentially discontinued at that time.)

In June 1964 the Space Sciences Board of the National Academy of Sciences (NAS SSB) sponsored a series of meetings, through the summer of 1965, to plan Mars exploration strategy, especially with biology in mind. A Mars launch window was coming up in November 1964; both the United States and the Soviets launched Mars probes at that time for July 1965 encounters with Mars. As it turned out, only the U.S. *Mariner 4* was still operational when it flew by Mars. But the Cold War competition atmosphere still very much surrounded deliberations. Lederberg, Vishniac, Princeton biologist Colin Pittendrigh, and NAS administrator J. P. T. Pearman (who had been a supporter of the 1961 Green Bank SETI meeting) were prominent forces at the meetings. The proceedings were published in early 1966 as the volume *Biology and the Exploration of Mars*.[6] According to a journalist's account (brushing quickly past the qualifiers), life on Mars was judged by these panels to be "so likely, in fact, that a group of eminent astronomers, physicists, biologists and chemists . . . urged [NASA] to underwrite an elaborate Martian research program that will find out for sure."[7] Norman Horowitz tended to be a devil's advocate in these discussions; it is not surprising, however, that a reporter would pick up on the underlying enthusiasm of the Lederbergs and Sagans and minimize the reservations of the "stodgy."

Horowitz felt he was only maintaining the skeptical attitude proper to a scientist; he was extremely wary of the emotional factor in science, having been burned by it early on, when he was one of the first advocates of the controversial one-gene, one-enzyme hypothesis in the early 1940s.[8] In an interesting exchange that sheds light on both men, Lederberg wrote to Horowitz in January 1963: "I don't know whether I've had any chance to say this out loud. . . . In recent years I have had a chance to reflect back on the noise I used to make about the one-gene, one-enzyme theory, and I now see that I was not only factually wrong in opposing it, even as an intellectual exercise, but showed rather poor judgment in failing to defend it. Perhaps I was reacting to the idea (that no one else *ever* had) that it was the *ultimate* Truth; what in science ever is!"[9] Horowitz responded: "I am happy to have your note in re: one-gene, one-enzyme. You did use to give me a hard time in those discussions. I used to go home from those meetings wondering whether I was the victim of some monstrous self-delusion—the case seemed so clear to me and yet so murky to others whose opinions I respected. I sensed, of course, that an emotional factor was involved also, but I could never quite make out the basis for it. I am glad to have your comment on that, too."[10] Horowitz, himself a victim of prejudice, was thus sensitized early in his career to the "emotional element" behind science. But, ironically, he was to become the "power that be" with his own philosophical investment in no life on Mars. One cannot fault his basic skeptical attitude, only proper in science. But the way it manifested in specific cases was such that Fox or Sagan must have felt very much like the young Horowitz when faced, during the early- to mid-1970s, with the mature Horowitz.

A quite similar series of developments occurred in the SSB's deliberations about interplanetary contamination. From the early meetings of the WESTEX subcommittee in 1959–1960, Lederberg and Sagan argued for high priority for anti-contamination efforts for outgoing U.S. planetary probes. They argued almost as forcefully for efforts to prevent back contamination from sample return missions when those began, presumably first with lunar samples returned by *Apollo* and/or by the Russians. They wanted the NAS SSB's official position represented as such to the international Committee on Space Research (COSPAR), which began in 1958 and quickly became a forum for exobiology discussions. When COSPAR formed an anti-contamination panel at its 1963 Warsaw meeting, it was at their urging, and the Americans who became involved were Allan H. Brown, Wolf Vishniac, Colin Pittendrigh, Lawrence Hall, and Carl Sagan.[11] NASA Exobiology began its own Planetary Quarantine Program in the second half of 1963.[12] Allan Brown was also on the NASA Biosciences subcommittee and was a strong advocate of taking back contamination seriously. He still argued thus at the 1964–1965 Mars meetings, claiming that, even if the risk was very small, the scale of harm could be very great, so all prudent precautions had to be taken.[13]

As early as February 1960, however, Horowitz found himself again the dissenting voice, especially on back contamination. He thought some concern for sterilization might be warranted, though as time went by during the planning of the *Viking* mission he came to believe it was superfluous for Mars, as he thought conditions there so harsh that no imported Earth microbes would survive. But from the beginning he considered worry about back contamination to be losing all sense of perspective on space exploration, getting priorities out of order. In a memo to Lederberg dated 6 February 1960, Horowitz argued that:

> Against the slight risk of pandemic disease and the perhaps greater one of economic nuisance, one must weigh the potential benefits to mankind of unhampered traffic with the planets. The present situation may be likened to that which obtained in Europe in the decades before Columbus set forth on his voyage of discovery. If men had known then that Columbus would bring back with him a disease—syphilis—that was to plague Europe for centuries, they might well have prevented him from ever leaving Spain. Suppose, however, that they had known also of the tremendous benefits that were to flow from the discovery of the New World. Can there be any doubt what their decision would have been then?

> In view of the small risk involved in the premature return of planetary probes, it would be inadvisable to adopt a position—e.g., an embargo on returning spacecraft—which might prejudice the development of the necessary technology for return flights. Also to be considered is the probably deleterious effect on public opinion of an excessively cautious policy. (By this I mean that the public may be frightened out of any interest in space exploration.) . . . The procurement of . . . samples should therefore be the *primary goal* of exobiological research. It should be understood that the biological exploration of the planets by instrumented robot payloads is not a substitute for this primary objective, but is only a step toward it. This and all other aspects of the exobiological research program should be subordinate to the attainment of the primary goal.[14]

Horowitz asked Lederberg to present his views at the upcoming WESTEX meeting of 29 February, which he would not be able to attend. Lederberg said he would certainly do his best, though he could not argue for such views as eloquently as Horowitz himself could; he urged Horowitz to reconsider attending to present them in person. Further:

> I think I do agree that the acquisition of planetary samples is, and should be stated to be, a primary goal of planetary exploration. . . . On the other hand, I also feel that we should go just as far as we can with instrumental analysis partly to see what insights this will give on the kinds of hazards discussed. I think that when the preliminary experiments . . . have been done, we will then be in a much better position to decide which, if any, precautionary measures are still justified.
>
> I think your remarks about Columbian exploration and the return of syphilis to the Old World are quite apropos. But I think we are in a better position than Columbus was to have our cake and eat it too. I think it is unfair to suggest that the choice is between syphilis and America when a little caution and patience could give us the best of both worlds.
>
> I don't believe it would be possible, without a well financed public relations campaign, to frighten the public out of space exploration. Judging by the way things have been going, a rash blunder motivated by no policy at all is a more likely danger.[15]

Lederberg circulated Horowitz's memo to the rest of the WESTEX committee, suggesting it be a topic for discussion at the upcoming meeting, with or without Horowitz present. If Lederberg's reply seems like polite disagreement, not all WESTEX members reacted so cordially. Aaron Novick of the University of Oregon was angry: Horowitz's memo and attitude "demand comment," he wrote in a memo of his own to the committee.

> In the case of the problem of contaminating other planets with Earth life, most people apparently believe that this is largely a scientific prob-

lem. Contaminating a planet would be a scientific catastrophe and would otherwise not affect mankind. Back contamination as we agree poses a threat to everyone. Admittedly the probability of back contamination is very small indeed, but quite possibly the product of this small probability times the measure of all possible catastrophes is finite. . . .

The analogy to Columbus, like most analogies, only creates confusion. Perhaps I enlarge upon this confusion, but it is not inconceivable—witness the myxoma virus in the rabbit population in Australia—that syphilis might have erased pretty much all of the population of Europe. Had this occurred, it would be agreed that restraint of Columbus would have been a good idea. . . . Alternatively, it might have been worthwhile to wait until Fleming's discovery of penicillin.[16]

Evidently, the subject remained a disputed one at WESTEX. Although Horowitz seems to have withdrawn and placed his energies into other areas, he does not seem to have changed his opinion much.[17]

One of the lasting outcomes of the contamination debate was the creation of a U.S. government administrative position called the "planetary protection officer" (PPO), charged with oversight of planning to avoid any contribution from the U.S. space program to such problems. This development occurred during Dick Young's tenure as NASA headquarters Exobiology chief; Young became the first planetary protection officer. This dual set of duties continued to be combined in the same position with Young's successors, Don DeVincenzi and John Rummel. But Michael Meyer was planetary protection officer for only the first few months after he took over. By early 1993 the two jobs were separated, and the PPO job was advertised. John Rummel was rehired as PPO on 1 November 1997, as a non–civil service contractor; he continues in that role as of this writing (December 2003).[18]

Morowitz, the Minimal Cell Approach, and Theoretical Biology

In the spring of 1960 Ernest Pollard, head of the eclectic Biophysics Program at Yale, was showing Melvin Calvin around the department, looking in on the labs and the research currently going on there. One of the labs was run by Harold Morowitz, who, like Carl Woese, had earned his Ph.D. degree in biophysics at Yale; Morowitz had returned in 1955 as an assistant professor in the program. He was working on *Mycoplasma,* the simplest prokaryotic cells, then known as pleuropneumonia-like organisms (PPLOs). Morowitz had an interest in understanding what was the minimal complement of things needed for a fully functional living cell and was studying *Mycoplasma* as the case closest to that minimal border.[19] When Morowitz showed Calvin his work, the Berkeley biochemist's reaction was: "You know, NASA would be interested in that. You should apply to Freeman Quimby for exobiology funds." Morowitz did, and within a year he had received his first NASA grant, for $38,196. He was steadily supported by NASA Exobiology money from that time until 1992.[20]

In light of his broad humanities interests and how involved with NASA Morowitz was, his absence from the 1967 and 1968 Princeton origin of life meetings (where the focus was exceptionally broad) seems odd. Morowitz explains: "I haven't been part of the origin of life Establishment. I'm not a joiner." During the process of looking, with his wife, for unusually stimulating schools for their five children, Morowitz was asked to review a National Science Foundation (NSF) grant application by Clair Folsome to set up a mycoplasma research program at the University of Hawaii–Honolulu.[21] He then applied to spend his sabbatical year (1967) with Folsome's group; working on the minimal cell line of reasoning, they addressed the properties a minimal cell membrane must have for life.[22] In addition, Morowitz wrote most of *Energy Flow in Biology* at that time, a book that quickly became a classic in the origin of life community for its thorough thermodynamic treatment of the problem.[23] Having fallen in love with Hawaii, the family spent two more sabbatical years in Maui, where Morowitz wrote *Life on the Planet Earth* (1975) and another book.[24]

Morowitz's research on the minimal cell approach has been remarkably productive for four decades. As Carl Woese wrote to him in 1977: "You epitomize that rigorous Yale biophysics approach; I was influenced by it, but have never mastered it. When I see how most biologists are trained today, I appreciate even more how important our training was; and you are perpetuating it."[25] By November 1976 Joshua Lederberg was taking cues from the mycoplasma approach. Writing to Dick Young for NASA funding, he proposed a new initiative to "look for eobionts," that is, to characterize the earliest living forms. Lederberg suggested that the most fruitful approach would be to start from mycoplasmas and work backward.[26]

Although Morowitz was also interested in halobacteria and in bacterial photosynthesis,[27] it is his work in mycoplasma studies which has paid off the most. Indeed, Morowitz's work has made *Mycoplasma* such a well-known benchmark for studies of the minimal cell that the *Mycoplasma* genome was among the early ones to be sequenced fully. Now, in addition to the catalog of basic metabolic processes and building blocks Morowitz cataloged, it is known that a suite of about 470 genes are needed for this simplest prokaryotic cell. A remarkably precise "recipe" can be spelled out at this point for a cell close to the hypothetical "minimal cell." Although this work does not directly address what steps must have come before to assemble this recipe, it nonetheless represents a clear benchmark of progress in the overall state of the origin of life problem.

By the late 1970s, however, Morowitz suspected mycoplasmas were probably not the first cells. His colleague Clair Folsome, of the Exobiology Laboratory of the University of Hawaii, described the "Onsager-Morowitz" definition of life as follows: "Life is that property of matter that results in the coupled cycling of bioelements in aqueous solution, ultimately driven by radiant energy to attain maximum complexity."[28] Morowitz's approach has recently been described as being "from the perspective of complex systems dynamics," a label also used for the work of Stuart Kauffman and others affiliated with the Santa Fe Institute, where Morowitz has been on the board of directors for over a decade.[29] His study of the metabolic pathways common to all organisms led him to believe the earliest cell was probably a photosynthetic autotroph, in contrast to the reasoning of Oparin, Haldane, and VanNiel. By 1988 he, Bettina Heinz, and David Deamer had developed the theory and experimentally modeled the formation of simplest protocells—that is, spontaneously forming vesicles, self-enclosed by a bilayer of amphiphilic lipid molecules. They described how such a system could function to capture energy and nutrients. Revealing the recent influence of Peter Mitchell's chemiosmotic theory, they concluded: "if some of the amphiphiles are primitive pigment molecules asymmetrically oriented in the bilayer, light energy can be captured in the form of electrochemical ion gradients . . . thereby providing an initial photosynthetic growth process."[30] In contrast to "gene-first" scenarios, they argued, as Morowitz had since at least 1981,[31] that a membrane-enclosed structure or vesicle was a far more likely first step. (Hence the title of one of his many, highly readable popular science books, *Mayonnaise and the Origin of Life*.)[32] Such a lipid vesicle provided the basic separation of a compartment in which important biomolecules could be concentrated, and, in line with the understanding Mitchell had provided, it allowed for an energy-generating mechanism by the creation of ion gradients (e.g., proton gradients) across the membrane. Only in such an enclosed, energized space was it possible to imagine conditions in which large biopolymers, such as polynucleotides, could be synthesized and protected from chemical degradation. Morowitz's 1992 book *The Beginnings of Cellular Life* develops the story further; it is an elegant, clear exercise in the logic of what the most basic constituents of the last common ancestor surely had to include.[33] Morowitz teases out the strands of the metabolic pathways shared by all extant organisms and argues persuasively that this amounts to a portrait of the last common ancestor's metabolic capabilities.[34]

NASA support for Harold Morowitz's work has produced much more, however, than the *Mycoplasma* story and the list of requirements for a minimal cell, impressive as they are. By 1962 Morowitz, Pollard, and George Jacobs of NASA had formed what they called the Committee for Theoretical Biology. They had met at the NAS Space Science Board's study group in Iowa City in the summer of 1962 and, together with several other colleagues, had agreed that support was needed for the development of theoretical biology as a viable and vibrant discipline. Through Morowitz and Jacobs's efforts NASA Exobiology funding was obtained to support several month-long summer courses in the subject. "It was a real shot in the arm for theoretical biology. Theoretical biology was not well regarded in those days," according to Morowitz.[35] The group first convened to plan strategy on 30 October 1962 at the Nassau Inn in Princeton, New Jersey. Present at the meeting were Pollard, Morowitz, and Jacobs but, in addition, James Danielli of SUNY-Buffalo, Henry Quastler of Brookhaven National Lab, and Joseph Engelberg of the University of Kentucky.[36]

The participants felt that a ten-week summer institute should be convened—more than a month of regular lectures, discussions, and social activities undertaken together—as that would be important to stimulate the growth of a robust theoretical biology. Many theoretical problems seemed ripe for development, the group thought—not least, in the words of Engelberg, that "we should look for biological invariants to see what things are constant in a whole hypothetical population of many earths. It was felt that this would be related to life on Mars." In addition to conceiving of a summer institute, the committee concluded that "1) There is a developing area of theoretical biology; 2) It has promise of real power of interpretation; 3) To find out more," a larger group, from six to thirty-five people, should meet for four days; and "4) There should be a program to support sabbatical leaves, research associates and post doctorals."[37] Other members listed on the committee but not present at this first meeting were: Hans Bremermann of the math department at the University of California–Berkeley, John Gregg of zoology at Duke University, Herbert Jehle of physics at George Washington University, Edwin Taylor of the biophysics department at the University of Chicago, William Taylor of biophysics at Penn State (where Pollard had moved in 1960 and now chaired the department), and Martynas Yčas of the microbiology department at SUNY Upstate Medical Center in Syracuse, New York. A slightly later list also included Howard Pattee of biophysics at Stanford and Frederick Williams of zoology at the University of Minnesota. Many of them were engaged in work relevant to exobiology and origin of life studies. Pollard received an exobiology grant for $194,000 in 1964 to continue his work on "physics of cellular synthesis, growth and division."[38] The title of a paper from this period, invoking "artificial synthesis" of a bacterial cell, reflects the optimism for sweeping theoretical synthesis which was developing.[39] Yčas had pioneered the "metabolism-first" idea, that some kind of complex chemical processes or cycles could have begun before any organism existed on the primitive Earth; later, when they became enclosed by membranes, one could begin to speak of them as living systems.[40]

Many of the same personnel were collected by Orr Reynolds and George Jacobs of NASA, to form a Planetary Biology Advisory Subcommittee of the Space Sciences and Applications Steering Committee of NASA. This group, concerned in an even more focused way with exobiology matters, first convened on 22 November 1963, just as the news broke of President John F. Kennedy's assassination. The group felt that JFK, with his enthusiasm for NASA's mission, would have wanted their meeting to proceed, so they did. This group initially included Pollard, Morowitz, Jacobs, Quastler, but also Albert Szent-Györgyi; it continued to meet through the early 1970s.

This was the birth of an initiative that bore much fruit over the next several decades, such that today a vibrant field of theoretical biology exists and is considerably more respected within the life sciences than it was in the early

1960s. NASA money was indeed forthcoming, channeled through the American Institute of Biological Sciences (AIBS), for summer theoretical biology institutes in 1965, 1966, and 1968. The first and third were organized by Morowitz and his wife, Lucille, in Fort Collins, Colorado, and Traverse City, Michigan, respectively.[41] The 1966 meeting was organized by James Danielli. According to Morowitz, these institutes were quite an important stimulus, bringing together as they did a whole new generation of talents, most of whom became the key voices in theoretical biology today. Walter Elsasser was one of them. He was a Manhattan Project physicist, trained in the Copenhagen school, who became interested in biology. Elsasser wrote several books on theoretical biology, the most recent being *Reflections on a Theory of Organisms*.[42] He had been unable to get physicists even to listen to him prior to that time. But at these meetings he found a peer group that, although they criticized his ideas a lot, found them very interesting and was eager to talk to him. Another was Herbert Jehle of George Washington University. He was German-born and had spent World War II in a concentration camp for being a conscientious objector. He became one of the physicists who turned their attention to problems in biology in the years after the war.

Thus, yet another broad and important stimulus to life sciences, establishing the careers of many of the brightest of the current generation of stars, was supplied via the catalyst of NASA Exobiology. The roster of faculty recruited to teach at the three institutes reads like a who's who of theoretical biology today: at the first workshop (1965) were Brian Goodwin, Robert Rosen, Edwin Taylor, and Ernest Pollard. "Various people with an interest in theoretical biology heard about the workshop and showed up: Herbert Jehle, Walter Elsasser, Ross Ashby." The third workshop (1968), on Thermodynamics and Statistical Mechanics in Biology, "brought out a group of young people who were the future of Theoretical Biology: George Oster, Art Winfree, Charles Delisi, Jonathan Roughgarten, Byron Goldstein. On the faculty were Bruno Zimm, Peter Curran and Ernie Pollard, and Donald Carothers."[43]

Life at High Temperatures

Around Thanksgiving in 1967 a paper appeared in *Science* which was widely noticed in the exobiology community, though its author, Indiana University microbiologist Thomas Brock, had not been an exobiology regular. The paper reported that bacteria of numerous kinds had been isolated and grown in culture, from hot springs near boiling temperature in Yellowstone National Park. And Brock, stimulated by Elso Barghoorn and Stanley Tyler's as well as Preston Cloud's 1965 papers in *Science*,[44] closed with a note on "Thermal Biology and the Origin and Evolution of Life."[45] He observed: "It has been hypothesized that the microorganisms of hot springs are relicts of primordial forms of life. Such a speculation does not seem unreasonable when we consider that evidence of hot spring activity dates back to the Precambrian, and that certain rock formations (for example the Gunflint chert, 2 billion years old), which probably have been formed in hot spring deposits teem with fossil microorganisms which resemble the Flexibacteria so common in thermal waters today. If organic matter, macromolecules, and primordial organisms arose at high temperatures, low-temperature forms might be derived from them by mutation and selection."[46]

Shortly after the paper came out, Brock "was contacted by several people from Ames Research Lab, and at one stage an Ames researcher spent a week in [his] lab, collecting samples for lipid analyses. (The organic geochemists [including Kvenvolden and John Hayes] liked lipids as markers in fossils.)." According to Brock: "This was about the time of the moon launchings [i.e., early 1969]. I was invited to Ames to give a seminar, and Cyril Ponnamperuma was quite interested in my work. He invited me to spend a sabbatical there and I almost did it in 1969, but a medical problem kept me from coming. Later, Cyril and some friends organized a two-week trip to Iceland to which I was invited as the biology expert. This was funded by NASA through Boston College."[47]

On that trip to Iceland NASA Exobiology personnel, including Dick Young and Cyril Ponnamperuma, were interested in studying thermophilic bacteria, along the lines of Brock's suggestion. But they were also interested in seizing upon a unique opportunity to study an extreme, presumably abiotic environment. A brand new island, Surtsey, had begun forming near Iceland in 1967 because of an undersea volcanic eruption. This seemed to Young and Ponnamperuma an excellent opportunity to study a piece of newly created land as it was first being colonized by life; the life forms that first moved in must be capable of living in extreme environments like that of the early prebiotic Earth. Their study was completed on Surtsey; while examining the hot springs of the Icelandic mainland in the early spring of 1970 for bacteria such as Brock had found at Yellowstone, however, Ponnamperuma slipped and his leg went into one of the boiling pools.[48] He was hospitalized for weeks, mostly at Stanford University Medical Center, after being flown back to California. This put him frustratedly out of action during a crucial phase of analysis of the Murchison meteorite, as will be described later.

Numerous origin of life workers visited Brock's research site at Yellowstone, including Preston Cloud and J. William Schopf, trying better to understand the kind of environment in which the microfossils of the Gunflint chert lived and then were preserved. Australian specialist in stromatolites, Malcolm Walter, also visited. He discovered that many kinds of filamentous microorganisms, including some in the hot springs of Yellowstone, formed layered stromatolite structures by trapping sediment; previously, it had been thought that stromatolites found as fossils must almost certainly be formed by cyanobacteria at moderate temperatures such as those seen today in Shark Bay in western Australia.[49] Walter "wrote the textbook" on stromatolites soon afterward.[50] By the late 1980s and early 1990s enormous numbers of fossil stromatolites were known from Archean era rocks 2.5 to 2.8 billion years old. Some, preserved in chert, have also been found dating back to 3.4 or 3.5 billion years old; they are morphologically similar to later ones, but it is not absolutely certain that they were biotically formed. At least by the mid-Archean era (2.8 billion years and younger) most of the stromatolite organisms were clearly photosynthetic cyanobacteria, which seem to be responsible for the process of oxygenating the Earth's atmosphere, although it took hundreds of millions of years before sediments were sufficiently oxidized to allow any of the gas to build up free in the air.[51]

In 1987 Walter was invited by NASA Ames researcher David DesMarais to bring his stromatolite and hot spring experience to a NASA conference on planning ahead for Mars exploration. There was much brainstorming about how to know what kinds of environments to look for as possible places likely for life. In the wake of Woese's discoveries about thermophilic Archaea and the revelations of life at undersea hydrothermal vents (see chap. 5), the work that had been done on thermophilic microorganisms now seemed to NASA more relevant than ever to exobiology. Walter had never thought of NASA as a source of primary research funding prior to that time, he says (though he had been a member of Schopf's 1979–1980 NASA-funded Precambrian Paleobiology Research Group [to be discussed in chap. 5]). But by 1989 he wrote to DesMarais inquiring about NASA support, got connected with the Exobiology Program, and has been receiving some degree of NASA funding ever since.[52] Indeed, the 1987 brainstorming led to a follow-up Mars-oriented workshop on hydrothermal ecosystems, partially sponsored by CIBA Corporation, in 1995.[53]

The Chicken and Egg Problem

Origins of life (OOL) research was dramatically expanding during these years, above and beyond NASA's influence; the third international conference in France in 1970 was the largest yet. But the more researchers learned, the more they were faced with dilemmas to which there was no obvious solution. As we saw in the debates between Fox and Miller in the last chapter, by the late 1950s there had already emerged the central catch–22 of origin of life research: if DNA and RNA contain the information required to make the proteins crucial for metabolism, yet DNA and RNA cannot be synthesized and cannot function without the help of numerous indispensable protein enzymes, how can such a chicken-egg system have ever come about to begin with? This dispute has become more heated in the years since, with groups polarized into "metabolism first" and "replication first."[54] A discussion of two recent works on this problem can help outline the development of ideas in origin of life thinking from the late 1950s onward.[55]

In his book *Origins of Life* Freeman Dyson suggests a set of intermediate steps which he calls the "dual origin hypothesis"—that is, that metabolizing enzymes enclosed within a membrane, by far the simpler component of living systems, probably developed first; then later the much more highly constrained and improbable process of high-fidelity replication arose. Replicating molecules could have arisen separately or, more likely, within the membrane-enclosed metabolizing systems, as Morowitz emphasizes.[56] In either case, Dyson argues, the development of a symbiotic relationship between the two would then produce systems that could *begin*, over a long time, to approach the last common ancestor of all organisms alive today. Dyson is careful to point out that the dual-origin hypothesis is one he finds persuasive on philosophical grounds, not because it is supported by any conclusive piece of evidence. He finds the possible analogy with Lynn Margulis's theory of symbiotic origin of eukaryotic cells very compelling, for example.[57] It is of considerable interest to see a scientist so frankly admit to his philosophical preconceptions and offer them for our scrutiny. The contrast is so refreshing given the bulk of scientific writing that attempts to disguise these motivating wellsprings and to construct, instead, accounts of rational, stepwise logical processes of "blank slate," objective discoveries.

Being a physicist allows Dyson to see the extent to which a lot of biologists' thinking is predisposed by their own philosophical assumptions—for instance, why such an overwhelming majority of life scientists trained since Watson and Crick believe information-carrying molecules are more fundamental to life than biochemical metabolism.[58] This, despite the fact that, ever since researchers have seen the origin of life to be predicated upon the origin of DNA, RNA, or some other more primitive information-carrying molecule, the result has been the chicken-egg problem described earlier. He is less aware, or at least does not comment on, the degree to which his own reasoning is being guided just as forcefully by notions about "hardware" and "software" inherited from the culture of computer technology.[59] This is not to imply that use of these analogies in thinking about living systems is *necessarily* faulty but, rather, that, just as the dominance of machines in industrial, scientific cultures cannot be said to be historically unrelated to the growth of the mechanistic view of life from 1850 to 1950, these researchers ought at least to note that the ideas of hardware and software are not merely disconnected intellectual "ideas" floating around but also fundamentally *cultural* resources, being drawn upon here by scientists. Thus, it is worth asking the question: do these ideas come into the scientific arena freighted with any other interesting cultural or philosophical baggage?

Dyson opens with a gracious acknowledgment that he has *not* represented the ideas of some of the more prominent thinkers in the field, among them J. B. S. Haldane, J. D. Bernal, Sidney Fox, Hyman Hartman, Pier Luisi, Julian Hiscox, Lee Smolin, and Stuart Kauffman. That being said, however, Dyson has left out a bit too much in some places. Because the book "outlines a theory which explains how life began, and in fact scientifically defines what life itself is," it surely needs to credit those workers, at least in passing, when Dyson makes central ideas for which those others were primarily responsible. For example, Dyson emphasizes the need to distinguish between replication and reproduction in order to break the logical catch–22 deadlock that results when one considers DNA- or RNA-centered systems to be the sine qua non of life.[60] Dyson gives John Von Neumann credit for emphasizing the distinction between replication and metabolism. This is the most significant distinction Dyson rightly emphasizes in his book. But one can only wonder why he lauds Erwin Schrödinger and Von Neumann's early and vague approaches to this distinction, as in Schrödinger's influential 1944 book *What Is Life?* while so studiously avoiding mention of Sidney Fox and his school—those who first made the issue of "proteins first" versus the "nucleic acid monopoly" central in the origin of life debate. Of course, the two big-name physicists have become revered in science (and Dyson himself is a physicist), while Fox was a protein chemist who eventually became marginalized by the mainstream origin of life community.[61] So, if one is constructing a "forerunners" pedigree for one's most important idea, perhaps the temptation is overwhelming to attribute that idea to winners and silently pass over losers, especially if one at the outset intends to write a highly condensed narrative that disclaims any attempt at comprehensiveness. From a historian's point of view this practice is in itself an object of study.

Dyson points out that Schrödinger saw biology "through [Max] Delbrück's eyes," and historians have elaborated at some length on the construction of a master narrative of the history of molecular biology which emphasizes only the line from Schrödinger, Delbrück, and Salvador Luria to Watson and Crick. Dyson says that thus Delbrück's focus on replication (and later on nucleic acids) as the central feature of the origin of life gained undue prominence in the field and came to dominate the mind-set of most researchers. Here again, however, Fox (and his son Ronald) anticipated Dyson, stating this insight in terms of "paradigms" and their control of thinking in the field repeatedly over the last twenty-five years.[62] Thus, Dyson's failure to cite them, at least in passing, stands out.

Maynard Smith and Szathmáry's *The Origins of Life* sets out to describe and explain what they plausibly argue are the eight major qualitative transitions that have occurred in the history of life since the origin of replicating molecules.[63] The book is an eloquent and very illuminating analysis of these transitions and of some very important parallel trends among them.[64] But, as a result of such breadth of conceptual reach, it manages to survey only somewhat superficially the origin of life per se. The major transitions they address are: replicating molecules→populations of molecules in compartments; independent replicators→chromosomes; RNA as gene and enzyme→DNA and protein; prokaryote→eukaryote; asexual clones→sexual populations; protists→animals, plants, and fungi; solitary individuals→colonies; primate societies→human societies and the origin of language.

In Maynard Smith and Szathmáry's *The Origins of Life,* from its first page, the focus is on *information.* The question of metabolism being of equal importance, let alone first in time (as in Dyson), is very briefly raised,[65] only to be dismissed or minimized: their overall usage betrays a strong bias toward an "information-first" view of life. Their approach clearly assumes that life is synonymous with replication.[66]

There is no more *historical* a phenomenon in modern biology than the dialectically related rise of information theory and computers and the simultaneous importation of such analysis into biological thinking, beginning no later than

Schrödinger's 1944 work *What Is Life?*[67] Maynard Smith and Szathmáry tackle this strikingly parallel development of concepts right away. It would seem strange or incomprehensible to Darwin, they say, that template reproduction allows transmission of instructions in a homogeneous-looking, as yet unformed egg or zygote. The idea is much less strange to us because "we are familiar with the idea that patterns of magnetism on a magnetic tape can carry the instructions for producing a symphony."[68] Indeed, they close their book with a tantalizing guess that the move to transmitting information in electronic form may be potentially a transition on the scale of the other major transitions around which the book is framed. It is astute of the authors to recognize how much our cultural experience *enables* our view, especially on questions of such fundamental importance as "what is life?"

Being more or less complete advocates of the information-first approach to conceptualizing life, however, they seem to miss the other implication of the power of historical context. If our cultural experience enables our view, it also simultaneously *constrains* it. The primacy of computers and electronic information in our lives makes images of "programming" of instincts and "hardwiring" of certain traits highly compelling metaphors for how we think about "life" in late-twentieth- and early-twenty-first century, high-tech Western society. But these metaphors tend to channel one's thinking strongly, above and beyond the actual experimental evidence, as in the nature-nurture debate, in which "master molecule" and "inborn hard-wiring" metaphors have boosted the stock of biological determinism far above even the rapidly growing knowledge base of molecular genetics. We may well reflect on the dominance of such models when they say, "a living being resembles a computer, rather than just a program, although it has its own program as subsystem." The irony of the back-and-forth relations between culture and nature is never more provocative than in this passage,[69] in which computers and computer "viruses" are used as the standard against which to evaluate whether *biological* viruses should be thought of as truly alive. Is this not putting the cart before the horse in some fundamental ontological sense?

As we shall see in origin of life debates, the possibly crucial question that gets drowned out by talk of the primacy of information (and thus of nucleic acids) is: can there be any other central characteristic of living systems as fundamental as, or perhaps even more fundamental than, information? Granted, Maynard Smith and Szathmáry give a brilliant and powerful analysis of events since the evolution of information-carrying molecules. But their bias leaves us with the chicken-egg problem: if metabolism is dominated mostly by proteins but is a prerequisite for the functioning of nucleic acid information molecules, how can a system like our current living cell, even the simplest prokaryote, with each of these two parts totally dependent upon the other, ever have evolved in the first place? This is the issue upon which Dyson's book is so helpful.

That is not to say that Dyson is the first to raise this issue. As John Farley makes clear, ever since Leonard Troland's 1914 paper emphasizing autocatalytic enzymes and Muller's 1926 gene-first response, as well as Oparin's 1924 emphasis on metabolism, this tension has been a central focus of debate and discussion in the origin of life literature.[70] Noted advocates toward Muller's end of the spectrum have included Norman Horowitz and Carl Sagan. Toward the opposite end have been A. I. Oparin, J. D. Bernal, N. W. Pirie, and Sidney Fox. The boom of interest in an "RNA world," beginning with Altman and Cech's 1982 discovery of catalytic RNA molecules ("ribozymes"), was precisely because it was hoped this phenomenon would finally offer a way out of the impasse that dominated much of twentieth-century discussion. If the simplest nucleic acid information molecules can also simultaneously perform the enzyme role, previously thought only to be a property of proteins, then catalytic RNA molecules could be the "missing link" bridging the gap between these two now separate but interdependent functions. Maynard Smith and Szathmáry clearly hope ribozymes offered the solution to the catch–22.[71] But this now seems to have been excessively optimistic.[72] For, although RNA does seem to have the dual capabilities to bridge the gap, its monomers are so difficult to form spontaneously and are so short-lived under primitive Earth conditions, that the question of how to get from an abiotic world to the RNA world is not much easier to solve than before the RNA world transitional stage was known (see chap. 5 for more discussion).[73]

"Gemischers" versus "Analytikers"

A related distinction of long standing between origin of life researchers was whether they pursued a "synthetic," or "constructionist" approach, as Fox called his work, or an analytic one. One of the things Dick Young supported in Fox's work was the basic approach of combining substances (in the style of Oparin's coacervate mixtures or the "plasmogeny" of Alfonso Herrera, both active in the 1920s and 1930s).[74] Miller and Horowitz were almost as dismissive of Herrera's work as they were of Fox's, although they thought the creation of "simulata" (what had in the 1930s been called "cell model experiments") an interesting curiosity. Miller wrote to tell Horowitz about Herrera:

> Oró, [Robert] Sanchez and I were in Mexico City in early May at a symposium honoring Alfonso Herrera, who from about 1900 to 1940, conducted thousands of experiments trying to make "organized elements" from inorganic or organic materials. Some of the results are impressive (e.g. mitotic spindles) but of course this has nothing to do with the origin of life. Herrera's "organized elements" make Fox's microspheres look sick by comparison. Orgel, Oró and I have been talking (I don't know whether it will progress beyond this stage) about translating Herrera's book and perhaps including previous work in this area as well as more modern efforts (Fox) in this direction. We were even talking about borrowing your expression and calling the book "Simulata."[75]

Horowitz replied: "Funny, I never heard of Herrera. It just goes to show you what making a lot of noise will do for a man. [i.e., Fox]. Fox gets written up in every other issue of *C & E News,* while Herrera, whose work was similar, is unknown. Incidentally, I checked the word 'simulata' in the dictionary, and it seems to be non-existent. The correct word is 'simulacra.' Of course, if you prefer my invention, you are welcome to it."[76]

Much of Oparin's work on coacervates was of this kind (and thus similarly suspect in the eyes of Horowitz and Miller). The experiments of Krishna Bahadur, chemistry professor at the University of Allahabad in India, could also be seen as in this tradition. Bahadur's structures, called "jeewanu," are similar in size to Fox's microspheres, though they are complex mineral-organic structures. They have also been shown to have photosynthetic and nitrogen-fixing activity and thus belong to the "autotrophs first" approach rather than the Oparin-Haldane "hetrotrophs first" school of thought.[77] Some experiments by Adolph Smith and Gary Steinman can also be considered within the synthetic approach to origin of life studies; these experiments involving formaldehyde and ammonium thiocyanate are based on the work of A. L. Herrera.[78] Carl Woese's and Leslie Orgel's work, by contrast, each trying to work out the origins of the genetic code, were more in the analytic tradition.[79] So was John Oró's work on "organic cosmochemistry," including his first prominent discovery, of the formation of adenine from ammonium cyanide.[80]

In a 1973 review Lynn Margulis used similar constructionist/analytical categories to describe current research in the origin of life; she evidently thought both approaches had potential, as she called them, the "gimish" [*sic*] (more commonly *gemisch,* a Yiddish word for "mixture") and the "microanalytic" approaches:

> In both, those gases, liquids and substrata thought to be reasonably abundant are brought together under . . . conditions thought to be reasonably plausible for the early Earth: Energy is supplied . . . and after some period of time the materials produced are analyzed. At the end of the experiment the gimishers ask: "what has been made?" The analytikers prefer to carefully control each of the inputs . . . and ask at each step: "what exactly is produced, which is the most abundant product, how can the conditions be altered to yield more of some familiar biological molecules?" The results of many experiments of these sorts have been impressive to some of us.[81]

Clays

In reviewing a book by A. Graham Cairns-Smith, a physical chemist at the University of Glasgow in Scotland, Margulis noted that neither approach impressed him.[82] Cairns-Smith saw early on the impossibility of assuming a sudden, chance appearance of the whole nucleic acid–based replication system as we know it.[83] He sought a way out of the chicken-egg dilemma by following up on a suggestion made in 1949 by J. D. Bernal, that charged clay surfaces could have served as binding places in the prebiotic environment, attracting organic monomers and holding them in close proximity, thus greatly facilitating their combining to form larger, more complex organic polymers.[84] Cairns-Smith's suggestion was that clay or crystalline minerals could have served a considerably larger role: because of repeating patterns of charges in their structure, he suggested those patterns could act as primitive heredity mechanisms, making the prototype for life "clay genes," as it were. Then at some later stage, when a more complex organic heredity molecule had finally appeared, there could be a "genetic takeover" by that more efficient, sophisticated information molecule. Margulis found the theory provocative and highly suggestive. Cairns-Smith presented increasingly detailed and complex versions of his theory, first at the Roussel UCLAF origin of life conference in Paris in 1973, at a 1974 symposium at the Royal Society of London (at which James Lovelock also presented a version of the Gaia hypothesis),[85] then in a 1982 book.[86]

Interest in the theory has grown steadily, but only when Hyman Hartman joined forces and applied with him did Cairns-Smith first obtain any NASA funding. In 1970 Paecht-Horowitz, Berger, and Katchalsky at Israel's Weizmann Institute demonstrated that montmorillonite clays promote polymerization of protein-like polypeptide chains from amino acid adenylates (esters formed from amino acids and adenosine monophosphate [AMP]).[87] By the late 1970s Cairns-Smith's ideas had sparked a fair amount of interest at Ames Research Center, according to Hartman: "It was the Israelis, Amos Banin, Noam Lahav and co-workers who brought an interest in clays to Moffett Field [Ames]. James Lawless, Sherwood Chang and David White began to use clays to polymerize amino acids, etc. Banin interpreted the Mars data from Viking as due to iron-rich clays."[88] And by 1982 interest was sufficiently great that NASA supplied funds for Cairns-Smith and Hartman to organize a conference on "Clay Minerals and the Origin of Life" at Glasgow University (fig. 3.1).[89] While some research groups such as Stanley Miller's have remained highly skeptical, the clay theory has received a fair amount of publicity, if not a lot of NASA funding.[90] It was NRC/NASA Ames postdoc money, for the most part, which brought the Israelis to Ames to work on clays.[91] And Leslie Orgel used some of his NASA exobiology money over the years, particularly in the 1990s, to investigate the role clay minerals might play in helping to catalyze polymerization of nucleotides into oligonucleotides.

Moon Rock Analysis and the Murchison Meteorite

One of the chief tasks for which exobiology scientists saw the need to prepare was the scientific lode of samples that *Apollo* would be returning from the Moon, by mid–1969 if the ambitious program schedule was kept. (In fact, after several weeks in quarantine, the first samples, from *Apollo 11*, were divided

FIGURE 3.1. Conference on Clays and the Origin of Life, University of Glasgow, Scotland, 18–24 July 1983. This conference was convened by Graham Cairns-Smith and Hyman Hartman, with NASA funding assistance. *Left to right, front row:* T. J. Pinnavia, Hyman Hartman, Harmke Kamminga, *(behind)* Gustaf Arrhenius, Krishna Bahadur, H. Van Olphen, Sherwood Chang, M. M. Mortland, unidentified woman, G. S. Odin, S. W. Bailey, A. L. Mackay, W. D. Keller. *Second row:* R. C. Reynolds Jr., A. G. Cairns-Smith, Everett Shock. *Third row:* D. D. Eberl, Adam Cairns-Smith, H. Harder, P. L. Hall, (*right of globe*) Armin Weiss, James Lawless. *Fourth row:* W. J. McHardy, P. S. Braterman, N. W. Pirie, S. F. Mason, Noam Lahav. *Back row:* R. F. Giese, J. M. Adams, D. P. Bloch, D. S. Snell, Mme Odin and children. Not in photo: T. Baird, Amos Banin, L. D. Barron, P. J. Boston, R. C. Mackenzie, R. Mohan, P. Smart. (Courtesy G. Cairns-Smith.)

up among the labs waiting for them by the early fall of that year.) High-purity reagents, ultra-clean glassware, and sterile containments with glove boxes and other facilities had been prepared at a number of locations; among them John Oró's lab at the University of Houston, Preston Cloud's lab at the University of California–Santa Barbara, Warren Meinschein's lab at Indiana University, and Keith Kvenvolden's lab, in Cyril Ponnamperuma's Chemical Evolution Branch at Ames, had all been developed with substantial NASA funding.

Cloud was a well-known geologist, a veteran of NASA meetings, and member of the NAS. He had looked into geochemistry in addition to his work on stromatolites and Precambrian paleobiology generally. Oró was a biochemist who had followed up the Miller-Urey experiment with work on pathways for the prebiotic synthesis of adenine and other nucleotides from very simple starting molecules common in interstellar space.[92] Meinschein had been a geochemist for the petroleum industry; interest in organic compounds on the

FIGURE 3.2. The NASA Ames team responsible for the initial chemical analysis of the Murchison meteorite organics in 1970. *Left to right:* Etta Peterson, Jose Flores, Katherine Pering, Cyril Ponnamperuma, James Lawless, Keith Kvenvolden. Kvenvolden directed the analysis that found that the amino acids were racemic and thus of extraterrestrial origin. Pering, working directly for Ponnamperuma, analyzed the meteorite hydrocarbons. (NASA photo, courtesy of K. Pering.)

Orgueil meteorite had inspired him to move into academia to work full-time on extraterrestrial materials.[93] Kvenvolden had also been a noted geochemist in the oil industry before being hired by Ponnamperuma to, as he saw it, engage in the scientific adventure of a lifetime, preparing for the geochemical analysis of the first rocks ever to be studied from the Moon.[94]

By the time samples began arriving, the Ames group consisted of Ponnamperuma, Kvenvolden, mass spectroscopist James Lawless, organic geochemist Katherine Pering, and technicians Jose "Jesse" Flores and Etta Peterson (fig. 3.2). At first some groups thought they had detected native amino acids[95] and porphyrins[96] in the lunar samples, but upon careful control studies and analyses rerun by several labs, including those of the highest cleanliness standards, these claims did not pan out. Other than carbide from solar wind, the only carbon on the Moon seemed to be from a tiny amount of cosmic dust.[97] The Moon had no native organics, no prebiotic synthesis, going on. (Or, if it was occurring, the intense bombardment with solar radiation was destroying such compounds as fast as they could form.) The labs did, however, acquire truly "blank" organic standards this way, which could be compared with any other extraterrestrial sample that might come along.

It is truly fortunate that the Moon did not contain living organisms, from a back contamination point of view. On the first *Apollo* sample return mission, *Apollo 11*, it was only realized a few weeks before launch that the recovery ship scheduled to pluck the sealed capsule from the ocean did not have a crane strong enough to lift the entire capsule up onto the deck of the ship. An elaborate plan had been devised by the Planetary Biology Subcommittee, a scientific panel convened by NASA, whereby the capsule would be lifted while still sealed onto the deck, bolted directly to the portable quarantine facility on the ship by an airlock, and only then would the astronauts open the hatch and transfer themselves and the samples in sterile fashion into the portable quarantine chamber. The subcommittee was presented by NASA officials in April 1969 with the fait accompli that the necessary crane could only be fitted on a recovery ship after several months (in time for the *Apollo 12* mission); the procedure would thus be fatally compromised on the *Apollo 11* mission by lifting the astronauts aboard separately, after they opened the hatch of the potentially contaminated spacecraft floating in the ocean, exposing both air and sea to any potential contaminant organisms from the Moon. The subcommittee met on 3 June 1969 and drafted a letter of protest, which was sent to NASA administrator Thomas Paine, but its members were given to understand that nobody less than President Richard Nixon himself could authorize a postponement of the *Apollo 11* mission, and there was no evidence he would do so.[98] The scientists did not seriously believe that any life existed on the Moon, but they were aggravated at being asked to create a scientifically sound containment protocol, only to have it ignored at the last minute because of apparently political concerns. They felt this set a very bad precedent for future cases, such as Mars, where the chance of native life was felt to be considerably greater than on the Moon.[99]

In a fascinating case of historical contingency, a carbonaceous chondrite (a class of meteorites containing a significant amount of carbon) fell near Murchison, Australia, on 28 September 1969, just as the lunar sample labs were geared up and ready for unprecedentedly clean analysis of extraterrestrial material. Local officials, including a postmaster in that rural area, collected fragments and a great many were purchased by American collections. The Field Museum in Chicago obtained quite a lot of material, and some went to the meteorite collection in the Geology Department at Arizona State University, under the curatorship of Carleton Moore. The research group there included George Yuen and John Cronin, biochemists who first turned their attention to meteorite organics only after the fall of the Murchison rock (fig. 3.3). Moore realized what a unique opportunity was available, given the preparedness of the clean labs at Ames and other places. Past claims of organic compounds in meteorites had always been compromised by a high probability of contamination. Chemist Paul B. Hamilton of DuPont had put it thus: "what appears to be the pitter patter of heavenly feet is probably instead the print of an earthly thumb."[100] Now labs existed with truly "clean blank" standards, personnel who had trained intensively for several years to seek and eliminate all possible sources of contamination from

FIGURE 3.3. George Yuen (*left*) and John Cronin (*right*) in the meteorite biochemical analysis lab at Arizona State University, c. 1986. (Courtesy J. Cronin, ASU Research News.)

their reagents, and state-of-the-art gas chromatographs and mass spectrometers. So, Moore sent a sample of the Murchison meteorite to Ponnamperuma at Ames in late 1969 or early 1970.[101] Ponnamperuma gave most of it to Kvenvolden and told him to put his team to work on analyzing it for any organic biomolecules such as amino acids.[102] He gave a small subsample to geochemist Katherine Pering and assigned her to analyze the hydrocarbons, then he went on the expedition to Iceland in which his leg was badly burned in one of the hot springs there.[103]

Once the analyses were run, Kvenvolden visited Ponnamperuma in the hospital at nearby Stanford Medical Center and told him some extremely exciting news; not only did the meteorite definitely contain several different amino acids, but the amino acids occurred in racemic mixtures as well. This was a crucial new finding: earthly contaminants would be entirely the L-form of amino acids, since that is the only form Earth life makes or consumes. A racemic mixture was what one would expect for extraterrestrial synthesis by purely chemical means, that is, Miller-Urey style. In fact, the range of organic compounds in the meteorite was very similar to the range of compounds that had been found in Miller-Urey type synthesis experiments.

At this point, however, Kvenvolden says he got a very rude shock. Ponnamperuma angrily told him: "You are no longer responsible for this project. And don't tell anyone about these results." After a long and heated argument, in which Kvenvolden went over Ponnamperuma's head to Chuck Klein and eventually to Hans Mark, head of the Ames Research Center, Ponnamperuma relented, and the paper was published with the entire team as authors and Kvenvolden as lead author.[104] To be fair, it must be noted that at least one of the other participants does not agree with certain parts of Kvenvolden's account and thinks it totally uncharacteristic of Ponnamperuma to act in such a petty manner.[105] It is fortunate that, in the end, one of the more spectacular results produced by exobiology work up until that time was not tainted by whatever personal difficulties may have existed among some of the researchers.

On balance it should be said that Ponnamperuma's contributions were many: his experiments, bringing scientists from all over the world (including Oparin) to NASA Ames Research Center, and his roles as journal editor and as an administrator. These were an important part of why so much happened in exobiology in these years. Kvenvolden, for example, says "we have to give Cyril credit, he was the one that made the contact with Carlton Moore at ASU—and we got the [Murchison] sample and it was pristine. . . . Then we began to get these great results."[106] It is an age-old question in science: does the credit go to the person who puts the sample in the analytic machine, or does it go to the person who gets the sample, gets the funding, organizes the enterprise, gets the staff, and so forth, who had the vision and made it happen?

Further analysis on the Murchison in years since, especially in the lab of John Cronin and Sandra Pizzarello at Arizona State University, has found dozens of amino acids and many other organic compounds present, all of reliably extraterrestrial origin.[107] These are some of the more firm data that exobiology still has to stand upon. And they agree remarkably well with the detected organic molecules found in giant molecular clouds in interstellar space. When Cronin and Pizzarello announced on Valentine's Day 1997 that they had found enantiomeric excesses of some of the amino acids that were certainly of extraterrestrial origin, in some cases as much as a 56:44 ratio of L:D rather than the expected 50:50 racemic mixture, it was further exciting news.[108] For the first time it became possible to say with certainty that the preference for L–amino acids in earthly life forms *might* have been based on a bias that already existed in the organic molecules being delivered to Earth from space at the time life first arose. The question of how the stereospecific preferences of living things got started had been a mystery from the time that Louis Pasteur first discovered such preferences in 1848. Now, in trying to eventually solve that mystery, the new field of exobiology had contributed some solid pieces of data for the first time since. Even faced with a mystery on the scale of how life originated, exobiology had won some firm handholds.[109]

CHAPTER 4

Vikings to Mars

A major milestone in the history of exobiology was the 1976 landings on Mars by two NASA *Viking* spacecraft. By the time of their launch in 1975 there had been no more ambitious planetary exploration mission than the *Viking 1* and *2* spacecraft. Each carried fourteen experiments on the lander section of the spacecraft alone and more on the orbiting platform from which the lander was detached.[1] The mission cost a billion dollars, of which $59 million was for the biology instrument package (fig. 4.1). Another experiment onboard the lander, the gas chromatograph–mass spectrometer (GCMS), also cost $41 million and was interpreted in conjunction with the biology experiments. This was more money than was ever spent, before or since, for a single exobiology project or mission.

Viking did not detect unambiguous signs of life on Mars. The overwhelming consensus of the research community at the time was that the experiments proved Mars was lifeless, indeed, too hostile for life or organic molecules even to exist, at least in the top one meter or so of regolith (soil). Yet the mission provided enormous amounts of data relevant to exobiology, not least of which was the relative isotopic proportions of gases in the Martian atmosphere. This data was crucial to the recognition that one class of meteorites found on Earth, designated "Shergottite-Nakhlite-Chassignite" (SNC), are almost certainly of Martian origin (see chap. 8). In addition, the *Viking* results were striking confirmation of Lovelock and Margulis's predictions, based on their Gaia hypothesis, that Mars would be lifeless because of what was already known about its atmospheric gases from Earth-based observations. Norman Horowitz, not without his poetic or humanitarian moments, found inspiration from the very lack of life on Mars, as he interpreted the findings. Coming to a conclusion that sounds more like what one would expect from Carl Sagan (but from the opposite direction), Horowitz summed up the mood thus: "The failure to find life on Mars was a disappointment, but it was also a revelation. . . . it is now virtually certain that the earth is the only life-bearing planet in our region of the galaxy. We have awakened from a dream. We are alone, we and the other species, with whom we share the earth. If the explorations of the solar system in our time bring home to us a realization of the uniqueness of our small planet and thereby

FIGURE 4.1. Harold "Chuck" Klein, showing the Viking Biology instrument to the team that helped in its design. *Left to right:* Klein, Vance Oyama, Genelle Deverall, Glenn Carle, Richard Johnson, Gary Bowman, Bill Ashley, Fritz Woeller, Dwight Moody, Bill Chun. *Seated, at end of table:* Bill Berry, Bonnie Dalton, Marjorie Lehwalt, Bonnie Berdahl. (Courtesy H. Klein.)

increase our resolve to avoid self-destruction, they will have contributed more than just science to the human future."[2]

The process of thinking about how to define life was profoundly shaped in the exobiology community by brainstorming to design experiments capable of detecting life.[3] Thus, for Carl Sagan, Joshua Lederberg, and others, the *Viking* data could not be ignored and led to rethinking their basic assumptions. Those, like Sagan, who still held irrepressible hopes of finding life in the cosmos, were chastened by the *Viking* results; nonetheless, they did not give up their quest, turning more of their attention, for example, to comets and to SETI.[4] Indeed, by the 1996 discovery of putative fossil microorganisms in Martian meteorite ALH84001, their hopes were given new life, even on Mars, though the evidence appeared to support at best only ancient life there, billions of years ago.

A very small minority of scientists, most important among them Gilbert Levin, continued to believe the *Viking* results had indeed shown life on Mars. For them the revival caused by the Mars meteorite in 1996 felt like even more of a vindication. For James Lovelock the *Viking* project was the cradle of his

Gaia hypothesis for precisely the opposite reason: because of his certainty that there would not be life on Mars, at least not at present.

Viking, then, represented an important moment of redefinition and refocusing in the history of exobiology, even if the results could be read in very different ways. With this in mind, let us turn to a close look at the history of this mission and of its home at the Jet Propulsion Laboratory (JPL).

Lovelock, Horowitz, and the Jet Propulsion Laboratory

As early as 1959, Richard Davies and Max Gumpel at the Jet Propulsion Laboratory near Pasadena, California, were already at work on planetary exobiology. They got an early NASA grant to investigate ideas for an infrared (IR) Mars probe for detecting extraterrestrial life. Davies and Gumpel gave a preliminary report on this work as a talk at the 11–15 January 1960 first COSPAR meeting.[5] JPL took an early lead in exobiology work, and, because of its continuous role in planning and design of the spacecraft that would explore the moon and planets, it has always been a focus of much of that side of exobiology. By contrast, until recently NASA Ames was more focused on the origin of life.[6] Indeed, in 1959 a JPL report already called for development of new, larger rocket boosters that could carry a new generation of automated lunar and planetary probes. Then, after Soviet attempts at Mars launches in October 1960, a successful Venus probe launch in February 1961, Yuri Gagarin's flight in April 1961, and the Bay of Pigs debacle that came so quickly on its heels, the JPL became much more active in developing planetary exploration missions and the hardware to support them. Nothing less than recovering national prestige was at stake, in addition to ongoing scientific interests.[7]

NASA was moving quickly to recruit the best talent in instrumentation and basic science from all over the world. One man who combined both was research chemist and biologist James Lovelock, who in 1957 had also developed a highly sensitive new device, the electron capture detector (ECD) for gas chromatography. This device allowed detection of trace organic molecules in the atmosphere down to the parts per trillion range for the first time.[8] On 9 May 1961 NASA official Abraham Silverstein wrote to Lovelock, inviting him to come to the United States to work on development of the gas chromatograph (GC) for the lunar *Surveyor* spacecraft at JPL.[9] Lovelock eagerly agreed. His first NASA grant, for $30,100, was awarded before year's end and was channeled through the University of Houston,[10] where a tenured professorship for Lovelock at Baylor College of Medicine was arranged, "with a dream salary of $20,000 per annum." He was to live in Houston with his family for two and a half years and commute regularly to JPL for much of the next eleven years; he continued to visit JPL periodically as a consultant until just before the launch of the *Vikings* in 1975.[11] Because of ideas that he first developed on physical life detection experiments, in March 1965 Lovelock was also put to work on an early Mars probe design, called *Voyager,* among other things to develop the GC as a life detection instrument.[12] His description of the discussions between scientists and engineers is highly evocative of the heady sense of mission at JPL during the 1960s, as designing and launching probes to the Moon and then the planets became a reality.[13] "As one whose childhood was illuminated by the writings of Jules Verne and Olaf Stapledon I was delighted to have the chance of discussing at first hand the plans for investigating Mars," he recalled some fifteen years later.[14]

As Lovelock describes it, the early meetings at JPL on life detection strategies for Mars probes had quickly settled into a rut. The strategies all sought to detect Earth-like microorganisms by immersing them in liquid culture broths and then looking for their metabolic by-products.[15] This was true of Vishniac's Wolf Trap, of Levin's Gulliver, and of Vance Oyama's early ideas. Lovelock thought it was far too limiting to make such narrow, "Earthcentric" assumptions about potential Mars organisms. Challenged to come up with a more robust strategy to look for evidence of life, he argued that one ought to look for entropy-reduction phenomena.[16] After a few days of thinking it over, he suggested the most obvious activity of living things which offsets entropy was that they keep the gas composition of a planetary atmosphere far from chemical equilibrium. For example, if a planet's atmosphere contained significant amounts of both methane and oxygen simultaneously, for any length of time, Lovelock argued, this is so far from the equilibrium condition that it is strong presumptive evidence of life. Living things must be constantly replenishing two such reactive gases or their levels would not remain high for long.

By September 1965 geneticist Norman Horowitz had become the new head of the Biology Division at JPL, a position he held until 1970 (while still working part-time on the faculty of nearby CalTech). As such, Horowitz came to oversee much of the planning of life detection experiments. Although Congress was not looking favorably at the *Voyager* mission (the project was postponed so much by a vote of 22 December 1965 as to effectively kill it),[17] Lovelock had published a first paper on his thinking and was on the verge of attaining a powerful new insight.[18] He realized that the gases that living organisms most actively affect, especially carbon dioxide, methane, oxygen, and water vapor, are just those gases that most dramatically shape the climate of the planet. He claims to have had a flash of insight one September day at JPL, in which he first wondered if living organisms might actively control the climate of a planet, via feedback mechanisms, to keep the conditions there favorable for their own survival and growth. Immediately blurting out his insight in discussions with Horowitz, Carl Sagan, and Dian Hitchcock, he found them skeptical but sufficiently intrigued to encourage him in his thinking.[19] Indeed, Hitchcock, a philosopher by training, had been collaborating on Lovelock's ideas about physical life detection for some months already; the two would eventually publish together in Sagan's journal *Icarus*.[20]

Horowitz, according to Lovelock, "was open-minded": "although he disagreed with my views about the Earth and its atmosphere, he thought, as the good scientist he was, that they should be heard." Horowitz arranged for Lovelock to give a paper on his ideas to the American Astronautical Society[21], and he invited Lovelock to the second NASA conference on the origins of life, to be held at Princeton in May 1968, where Lovelock first met Lynn Margulis.[22] Lovelock found the reception of his ideas cool at the NASA meeting, with the exception of the Swedish specialist in chemistry of the oceans Lars Gunnar Sillén.[23] He recalled that most of the older scientists at the meeting, especially Preston Cloud, were unsympathetic to his concepts.[24] Nonetheless, he worked steadily at the ideas, especially after 1970, when Lynn Margulis began to collaborate with him on the Gaia hypothesis. All the while, he continued as a consultant at JPL, largely designing other scientists' instruments.

His and Horowitz's concerns notwithstanding, work on the latest versions of Wolf Trap, Gulliver, and Oyama's experiment (now called the "gas exchange" experiment, or GEx) all went ahead on continued NASA funding. So did the development, by Klaus Biemann, Juan Oró, Leslie Orgel, and their team, of a gas chromatograph and mass spectrometer to be sent to Mars to analyze organic compounds present in the regolith. Lovelock came up with the crucial means for hermetically linking the gas chromatograph to a mass spectrometer when those instruments eventually were sent to Mars on the *Viking* spacecraft, the next iteration of design after Congress finally definitively canceled *Voyager* in the wake of the summer 1967 race riots in many U.S. eastern cities.

Lovelock called the new field spawned by the Gaia hypothesis "geophysiology." He later described its origins thus:

> It arose during attempts to design experiments to detect life on other planets, particularly Mars. For the most part these experiments were geocentric and based on the notion of landing an automated biological or biochemical laboratory on the planet. . . . Lovelock took the opposing view that not only were such experiments likely to fail because of their egocentricity, but also that there was a more certain way of detecting planetary life, whatever its form might be. This alternative approach to life detection came from a systems view of planetary life. In particular, it suggests that if life can be taken to constitute a global entity, its presence would be revealed by a change in the chemical composition of the planet's atmosphere. . . . The reasoning behind this idea was that the planetary biota would be obliged to use any mobile medium available to them as a source of essential nutrients and as a sink for the disposal of the products of their metabolism. Such activity would render a planet with life as recognizably different from a lifeless one. At that time there was a fairly detailed compositional analysis by infrared astronomy of the Mars and Venus atmospheres, and it revealed both planets to have atmospheres not far from chemical equilibrium. Therefore, they were probably lifeless.[25]

Because of the state of chemical equilibrium in the atmospheres of both

Venus and Mars, Lovelock predicted from the first Gaia insight in 1965 that both planets were lifeless. Consequently, he was skeptical about the large expenditures on the *Viking* biology instruments, above and beyond his earlier skepticism about the conceptual basis of the instruments, now thinking the money could be much better spent on other measurements on Mars.

Yet now an additional, much deeper insight dawned upon Lovelock. Given the so-called faint young sun paradox, the fact that the biota was so actively shaping the chemical environment of the biosphere (including the atmosphere) took on new explanatory power. The sun had been cooler, as much as 30 percent cooler, at the time when life first originated on Earth. Yet during the entire 3.5 billion years or so since life had appeared, it seemed clear that the Earth's surface temperature could not have varied by nearly as much as 30 percent from present values: living things could not have survived and proliferated if the Earth had been that much cooler than at present. Either the Earth had been warmer than it should have been at the origin of life, relative to now, or, more likely, living things were regulating the temperature, so that modern temperatures were cooler, relative to how much the sun had warmed, than they would be on a lifeless planet. Because the main means of regulating the Earth's surface temperature known at the time was the so-called greenhouse effect, dependent upon gases given off and consumed by living organisms (CO_2, methane, water vapor, among others), it did not seem impossible that the biota could regulate planetary temperature, decreasing the greenhouse effect slowly over eons, to compensate for the increasing heat of the sun. (Later, it turned out, the biota also regulates cloud formation and thus dramatically alters the amount of incoming solar energy reflected back to space as another powerful way of regulating temperature.)[26]

Perhaps, Lovelock began to think, the biota acted as a cybernetic system that regulated temperature, pH, oxygen level, and other parameters in just such a way as to maintain conditions on Earth suitable for the survival of life. As mentioned earlier, Lovelock's idea was at first received quite coolly by the scientific community, even at a 1968 NASA-sponsored origin of life meeting where interdisciplinary thinking was the norm.[27]

Although he was not a fan of the Gaia hypothesis, Norman Horowitz agreed with a number of Lovelock's views. Lovelock shared Horowitz's feeling that sterilizing Martian landers was unnecessary: "The concept of contaminating a virginal Mars with Earth-life seemed the stuff of fanatics, not scientists, and the act of sterilization hazarded the delicate and intricate instruments we wanted to send to Mars."[28] In a more piquant passage, Lovelock described his view of life detection experiments as follows:

> the engineering and physical sciences of the NASA institutions was often so competent as to achieve an exquisite beauty of its own. By contrast with some very notable exceptions, the quality of the life sciences was primitive and steeped in ignorance. It was almost as if a group of the finest engineers were asked to design an automatic roving vehicle

which could cross the Sahara Desert. When they had done this, they were then required to design an automatic fishing rod and line to mount on the vehicle to catch the fish that swam among the sand dunes. These patient engineers were also expected to design their vehicle so as to withstand the temperatures needed to sterilize it for otherwise the dunes might be infected with fish-destroying microorganisms.[29]

Yet Horowitz also felt that the Wolf Trap, Gulliver, and other designs shared the basic flaw of assuming that Martian microbes, if they did exist, would do well in a wet environment, since all those designs involved saturating Martian regolith with a liquid broth of nutrients. In Horowitz's way of thinking this produced conditions wildly unlike those of Mars; he thought so still more after July 1965, when the *Mariner 4* space probe showed Mars to be a cratered, dry planet. (Even President Lyndon Johnson, after looking at the *Mariner 4* photos, concluded that "life as we know it with its humanity is more unique than many have thought."[30] *Mariner 4* led Carl Sagan, in his enthusiasm for the possibility of life, to observe that satellite photographs taken from six thousand miles above Earth also showed no signs of life.)[31] Measurements the spacecraft made of the Martian atmosphere found it to be much thinner than previously supposed. The pressure of the air was too low for liquid water to exist on the planet's surface. "CO_2 was its major component, with only a trace of water vapor," recalled Horowitz. "That discovery gave me and my collaborators, George Hobby and Jerry Hubbard, the impetus to design an instrument that would search for life on a dry planet. That instrument was the pyrolytic release experiment. . . . I never applied [to NASA] for funding to develop the experiment, since the funds were provided by JPL."[32]

Because of the *Mariner 4* results, Horowitz was among those who proposed that Antarctica, specifically the very coldest, driest desert valleys there, was a better analog for Mars than most other sites on Earth, yet even they, he said, were overwhelmingly hospitable places for life compared to the Martian environment.[33] Horowitz and his collaborators, Roy Cameron and Jerry Hubbard, began to study the microbiology of the driest, most inhospitable parts of Antarctica to understand whether life could survive there at all.[34] They later claimed to have found some of the only naturally sterile soils on Earth (14 percent of their samples) from these valleys, claiming this made life on Mars still less probable than previously thought and proving that sterilizing spacecraft to be sent to Mars was pointless because conditions there were so much harsher than those sufficient to render some Antarctic soils totally sterile.[35] Cameron and Richard Davies also launched a similar expedition in 1966 to the Atacama Desert of northern Chile.[36]

In response to these findings both Levin and Vishniac began to test their own life detection devices on the soils from the Antarctic Dry Valleys. In 1972 Vishniac's Wolf Trap was able to detect organisms in some of the samples that Horowitz, Cameron, and Hubbard had found sterile, rendering a more optimistic view of the possibility of life on Mars.[37] His studies of the microbiology of these valleys was to make Vishniac the first fatality in the field of exobiology, when he slipped and fell to his death from an Antarctic cliff on a sampling expedition in December 1973.

In general, the preparations for *Viking* gave a big boost to research on microbial life in extreme environments. Thomas Brock, the expert in thermophilic microorganisms, for example, was invited back to Ames: "In the early 1970s, I was invited to Langley Field for a large NASA meeting, which was focused on the Viking project. My talk was focused on life in extreme environments and basically dealt with the question of what were the environmental requirements for life. Carl Sagan seemed to be running this meeting."[38] Sagan and others were prompted to try to define living systems more than ever, not merely as a theoretical matter for origin of life studies; now the need was great to define what one should look for, what would count as life. From the *Viking* era date Sagan's jocular speculations about the possibility of finding "squamous purple ovoids" or "macrobes," large, visible life forms that justified the need for a television camera to be mounted on *Viking* as one "life detection experiment."[39] He gave a more sober assessment in the article on the subject "Life" which he wrote for the 1974 edition of the *Encyclopedia Britannica*.[40]

In a discussion of the search for life on Mars soon after the *Mariner 4* results, Horowitz spoke of life's extreme adaptability, even to harsh desert conditions. He described in some detail, for example, the remarkable water-conserving adaptations of the kangaroo rat of the Arizona/California Mojave Desert. But he concluded on a more skeptical note, "even Southern California is not as dry as Mars, and I am not suggesting that Mars is inhabited by kangaroo rats and that the first life-detection device on Mars should be a mousetrap."[41]

All this is not to say that Horowitz thought life on Mars impossible. American culture was influenced strongly in a similar direction by Frank Herbert's science fiction novel *Dune*. Released in mass paperback just at the time of the *Mariner 4* results from Mars and positing an entire complex culture exquisitely adapted to the conditions of a desert planet, the book went on to become far more than a cult classic. (Herbert has also been credited with inspiring the nascent environmental movement; he constructed an entire ecology from immense sandworms to microscopic organisms crucial to the desert ecosystem's stability and to the plot.)[42] Summarizing his own thinking in a paper in *Science,* Horowitz wrote that the *Mariner 4* data were "very depressing news for biologists, but if I have learned anything during 6 years of association with the space program, it is that people with manic depressive tendencies should stay out of it. . . . The fact is that nothing we have learned about Mars—in contrast to Venus—excludes it as a possible abode of life."[43] Although he concluded, "it is certainly true that no terrestrial species could survive under average Martian conditions as we know them, except in a dormant state," Horowitz nonetheless kept open the possibility. He reasoned (and the later discovery of dry water channels from a time of flooding in Mars's distant past confirm his thinking): "But if we admit the possibility that Mars once had a more favorable climate which was gradually transformed to the severe one we find there today, and if we accept the possibility that life arose on the planet during this earlier epoch, then we cannot exclude the possibility that Martian life succeeded in adapting itself to the changing conditions and survives there still."[44]

Horowitz was highly skeptical but not so much that it prevented him from accepting the logical possibilities of the problem. "It is not optimism about the outcome that gives impetus to the search for extraterrestrial life," he said; "rather, it is the immense importance that a positive result would have." When one multiplied the probability of success by the importance of the problem, he concluded, "the value so obtained is high." *Mariner 4* did not conclusively answer the question, Horowitz argued, but it did prove that we now had, or very soon would have, the technology capable of doing so.[45]

In the same paper in February 1966 Horowitz described the current state of the Gulliver experiment (fig. 4.2), after he had been on board as scientific advisor to Gilbert Levin for three and a half years.[46] Urey-Miller type chemistry led to the assumption that some organic products would be common throughout the solar system, and these compounds were the ones that should be selected for the radioactively labeled substrates in the nutrient broth. Formate, lactate, and glutamate were good choices on these grounds and were readily metabolized to CO_2, he said. (Apparently, none of the biologists designing or reviewing the experiment were aware or remembered that formate was capable of reacting in a purely chemical way, with peroxides for instance, to produce CO_2 as well.)

Yet ever since the *Mariner 4* results, as mentioned earlier, Horowitz decreased and soon dropped his involvement with Gulliver (soon to be renamed the Labeled Release, or LR, experiment) and began working on a life detection device that would not require organisms to grow in liquid water. This was the beginning of what came to be known as the Pyrolytic Release, or PR, experiment, one of those actually chosen in 1969 to fly to Mars on *Viking*. "In a way, it was Levin's machine turned upside down."[47] Horowitz discussed the concept briefly in the February 1966 *Science* paper: one could use radioactively labeled carbon dioxide to test for photosynthesis in a sample of Martian regolith because, "if there is life on the planet there must be at least one photosynthetic species."[48] Regardless of whether water or some other substance was used by organisms as the reducing agent, the carbon fixed would thus show up as radioactively labeled organic compounds. This could be volatilized by heating (pyrolyzing) the organic matter in an oven after a suitable incubation time and after first flushing all of the original labeled CO_2 from the system. Then the organic carbon would be converted back to labeled CO_2 and could be measured by a Geiger counter, just as in the LR experiment. As he noted, Horowitz got all his funding for the PR device through JPL; he never needed to apply for money from the Washington headquarters Exobiology Program as Vishniac and Levin did.

FIGURE 4.2. Gilbert Levin field testing the "Gulliver" Mars life detection device (later called the Labeled Release, or LR, experiment) in the California desert, summer 1965. (Courtesy G. Levin.)

Building and Launching *Viking*

By 1968 the canceled *Voyager* had been replaced by the planned *Viking* Mars mission, and NASA advertised a competition among all submitted life detection schemes, to decide which four experiments would be chosen to get built and sent to the Martian surface on the *Viking* lander. In December 1969, from over fifty submissions, the four experiments chosen were Horowitz's PR, Levin's

LR, Oyama's GEx, and Vishniac's Wolf Trap.[49] A planning committee was created to oversee design and construction of the Biology Instrument package; the contractor, TRW of Redondo Beach, California, had the lowest bid and got the contract to build it. The committee consisted of the four experimenters, with Wolf Vishniac as the initial chair, plus Joshua Lederberg and Alex Rich, scientists who it was felt could be more objective since they did not have experiments of their own at stake. Vishniac, it soon turned out, was too laid-back and willing to allow everybody their say; with the mix of strong egos on the committee nothing would get decided, and things did not move forward. Each experimenter thought his own approach by far the most important, yet all the experiments had to function in a common environment inside the same experiment package (see fig. 4.1). As just one example, Horowitz argued that the temperature inside the package should be kept as low as possible. Having designed a dry experiment, he had no qualms about making uncomfortable those who insisted on such non-Mars-like wet experiments, as he told Lederberg: "There is to be an important meeting at TRW this Friday to make decisions regarding the thermal environment of the biology package. I intend to press for as low a temperature as possible—0°C rather than the 15°C agreed on before the decision was made to land mission B at a high latitude. I would be glad to go even lower if I thought there was a chance it would be acceptable to the wet experimenters. I hope I will have your support if it should turn out to be necessary to poll the team."[50]

Data from *Mariners 6, 7,* and *9* in the years since 1965 had confirmed that Mars had a thin atmosphere and was a cold, rocky, desert planet. *Mariner 9* in 1971 had arrived in the middle of a planet-wide dust storm with greater than 100 mph winds that lasted for months. Moving piles of dust, sorted by grain size and thus having different shades of gray, now appeared ever more certain to be the explanation of the changing colored surface features that had tempted observers since Percival Lowell to imagine vegetation zones shifting with the seasons.

Before long Harold "Chuck" Klein was invited to join the committee as the new chair. He brought the same capable administrative talents that he had brought to directing the Ames Exobiology Program and then all of Life Sciences at Ames. Klein's managerial style worked, and though the Viking Biology Committee was noted by many as one of the most contentious groups of people ever assigned to work jointly, he managed to keep the group together and the project moving forward, if notoriously behind schedule. Said Klein, "I think NASA was really looking for a 'moderator'—not necessarily a 'leader'— and I suppose they came to me because I ostensibly had a reputation for being pragmatic, able to deal with people, and experienced at formulating compromise solutions in difficult situations. (I had the nickname, 'Rabbi,' among some of my associates.)"[51] Klein's level-headed calm would turn out to be most important of all in the days and weeks after *Viking* landed on Mars and after results from the experiments began to come in. Oversight by Gerald Soffen at JPL

FIGURE 4.3. "On Mars": posing beside a full-scale replica of the *Viking* lander at Jet Propulsion Laboratory in Pasadena. Biochemist Leslie Orgel (*far right*), with his wife, Alice, and son, and Gerald Soffen, senior Viking Project scientist, c. 1975. (Courtesy L. Orgel.)

(fig. 4.3), Klein's superior as overall director of all twenty-seven Viking science experiments, was equally important.

Another *Viking* experiment crucial to exobiology was being designed by Klaus Biemann of MIT, the world's most renowned specialist in mass spectrometry; he had been working with Mars in mind since the 1964–1965 NAS Mars meetings.[52] Now he headed a team including Salk Institute biochemist Leslie Orgel (fig. 4.3) and John Oró of the University of Houston, specialist in (and founder of) the new field of organic cosmochemistry. They were attempting to build a miniaturized gas chromatograph (GC), mated to a mass spectrometer (MS), such that organic compounds separated by GC could then be fed one by one into the attached MS, where they could be identified by molecular weight. In the words of its designers, finding the structures and abundances of organic molecules on the Martian surface

> seemed important because we hoped that the nature of Martian organic molecules would provide a sensitive indicator of the chemical and physical environment in which they were formed. Furthermore, we hoped

that the details of their structures would indicate which of many possible biotic and abiotic syntheses are occurring on Mars. . . .

Since much is known about the degradation of organic compounds under the influence of high temperature, pressure, irradiation, etc., the absence of organic compounds above a certain limit of detection might eliminate certain sets of conditions that otherwise could be postulated to exist or to have existed at the surface.[53]

It was thought by many, including Horowitz, that the GC-MS data would be the most useful of all in telling something about the possibility for life on Mars. It could report on the identity and quantity of organic molecules necessary to build living cells (or possibly left over from no longer living cells). Thus, it did not depend upon the chance of encountering still-living cells to give information relevant to past or present life; even if the biology experiments all yielded negative results, finding organics relevant to life would still be highly suggestive. At the very least even if life had never evolved on Mars, many thought that prebiotic organic molecules must surely have formed there, Miller-Urey style. If prebiotic chemistry on Mars had been frozen by changes in the planet's climate and atmosphere in an intermediate stage before life emerged, to many exobiologists a survey of those compounds seemed just as great a scientific treasure trove as finding extant life: it was like having a snapshot of the development of a terrestrial planet in an earlier stage, perhaps similar to what Earth had passed through.

As development of the *Viking* instruments progressed, Horowitz and his team discovered that Miller-Urey synthesis on Mars was more than just a theoretical matter. In test runs of their pyrolytic release (PR) device they exposed simple inorganic gases in a simulated Martian atmosphere to light from a xenon arc lamp and found that Miller-Urey type organic compounds were being synthesized.[54] They determined that it was the ultraviolet wavelengths that were catalyzing the synthesis from carbon monoxide as carbon source. Because this process of carbon fixation would mimic the living response that the PR instrument was designed to detect, they had to shield the lamp with an ultraviolet filter in their design, lest the experiment give a false positive. In June 1972 the group had found that a similar reaction could occur with methane as the carbon source; as Horowitz described it to Miller: "Ellis Golub, a post-doc who is working with Hubbard at JPL, finds that methane is converted to organics (formaldehyde?) when it is irradiated with long-wavelength UV (longer than 2500Å) in the presence of Vycor. The identification of HCHO is not certain yet, and I am hoping he will finish that before he leaves in July. . . . The reaction is different in some ways from that involving CO, as might be expected, since one is an oxidation and the other a reduction. There are still plenty of mysteries left."[55]

Thus, as the mission approached, Horowitz opined that the GCMS experiment would probably be of even greater importance than the biology package, which was constantly plagued with delays. As he expressed it to Leslie Orgel of the GCMS team: "The Viking biology package is experiencing severe difficulties, as you have probably heard. I am happy to hear that GCMS is in good shape, however. I consider it the most important instrument on Viking."[56]

Indeed, problems with the biology instrument were not limited merely to the difficulties of getting the team to work together. Fearing the complexities of getting all four experiments to function problem-free in a single instrument, NASA's Viking Project manager James Martin issued a directive on 1 July 1971 declaring: "It is project policy that no single malfunction shall cause the loss of data return from more than one scientific investigation."[57] In November and December 1971 TRW and NASA Ames personnel under Chuck Klein worked to simplify the biology instrument. It simply had too much going on in the space allotted. In 0.027 cubic meters—a box about the size of a gallon milk carton—were 40,000 parts, half of them transistors.[58] Several items were eliminated, including a Martian gas pump, an onboard carbon dioxide gas system, and one control chamber each for the GEx (Oyama's) and light scattering (Wolf Trap) experiments.[59] By 24–25 January 1972 Walt Jakobowski and Richard Young from NASA headquarters met with people from the Viking Project Office, Martin Marietta Corporation, and TRW "to discuss ways to remedy the problems, especially cost, which had escalated to $33 million."[60] Alas, by the end of the month James Martin had concluded that one of the four biology experiments would have to go. Klein, Lederberg, and Rich, the Biology Team members who did not have a stake in any one of the experiments, met to discuss priorities; shortly afterward, by 13 March 1972, NASA headquarters had decided that Vishniac's light-scattering experiment was based on the least Mars-like conditions and therefore it should be the one to be sacrificed. The lot fell to administrator John Naugle to convey the bad news to Vishniac.[61] The entire Viking Biology Team met immediately and showed rare cohesiveness in criticizing the decision at headquarters to drop the Wolf Trap. "Young & Soffen were on the hot seat" to defend the priorities of headquarters. "While stopping short of mutiny—and still promising to work hard—Klein said that the team wanted a better explanation of why Wolf Trap was dropped."[62]

Vishniac was, as one might expect, the most upset of all. But all protests were in vain; the decision of headquarters was final. This put Vishniac in an almost untenable position with regard to funding, bringing the full brunt of "NASA envy" upon him in his exposed and vulnerable position.[63] Vishniac managed to continue his studies on microorganisms in Antarctica; he also began collecting samples for another microbiologist, E. Imre Friedmann, who specialized in endolithic microbes (those that live entirely within rocks) and sought them in Antarctic rocks, postulating that they might be analogs of Martian organisms.[64] But for Vishniac that, too, came to an end with his accidental death when he fell from a cliff there while collecting samples, on 10 December 1973.[65]

Was the headquarters decision justified? In retrospect the Viking Biology package costs continued to escalate; even without inclusion of the Vishniac experiment the total came in at $59 million, surely one of the most expensive space experiments by far and one of the big high-budget space missions that triggered, in response, the "faster, better, cheaper" approach that Dan Goldin later brought as NASA administrator. The Biology Team members felt this would have happened anyway, even with all four original experiments still included; in hindsight, however, it is hard to believe that at least some reduction in cost overruns was not achieved by the tough decision.

It should be noted that the relevance of Antarctic ecosystems as models for exobiology research remained very much alive after Vishniac's death and is seen by many as one of his legacies to the field. E. Imre Friedmann has been a leader in this field and started an entire research group at Florida State University to investigate the endolithic microorganisms. His work was funded by NASA Exobiology:

> My first NASA grant started in 1977 and since then I have been supported without interruption. I remember how difficult it was at the first time to find the proper channel where to apply. As a recent immigrant to the US I was relatively inexperienced in these matters and I found the vast organization that is NASA frighteningly complex and impenetrable (this was the time before the Internet, where instant information is at hand). It took me more than a year (after mailing an application to the wrong address and waiting for an answer, while missing the deadline) until I found the Exobiology program and Dick Young, who was very helpful from the beginning.[66]

Michael Meyer, who later became the head of the NASA Exobiology (now Planetary Biology) Program upon John Rummel's departure in 1992, was a postdoc in Friedmann's lab beginning in 1985, after he had completed his Ph.D. degree on cryopreservation of marine diatoms. Friedmann ran two important workshops on the relevance of cryptoendolithic organisms to the biological evolution of Mars, on 11–13 October 1985 and on 26–28 October 1990. Chris McKay contributed a paper to Friedmann's monumental 1993 *Antarctic Microbiology* volume in which he argues for the continued relevance of Antarctic ecosystems, not only for Mars but for exobiology research on Europa and other locales as well.[67] Friedmann remained active in Antarctic research, funded by both NASA and NSF, until 1997 near his retirement, when bureaucratic red tape at NSF made getting continued funding from that agency too difficult.[68]

*B*y the time the *Viking 1* and *2* spacecraft launched from Cape Kennedy, on 20 August and 9 September 1975, respectively, the team had written a description of the experiments for *Nature*.[69] A special issue of the journal *Origins of Life* was also in preparation, describing the experiments in much greater detail. A feeling comes through from those involved of a sense of the historic nature of their enterprise, but they were also aware of how complex the experiments were and how limited was their ability from Earth to check up on ambiguous results or run additional controls. Richard Young wrote a history of the mission to date.[70] Harold Klein penned an overview of the biology package and its development.[71] Knowing the sensational nature of the mission, Klein seemed to feel more than most the responsibility to educate the press and the public about keeping a cautious, scientific attitude toward the experiments. The special issue of *Origins of Life* also contained detailed descriptions of each of the three remaining biology experiments, authored by members of each team. Levin's coauthor was his chief co-experimenter on the LR, Patricia Straat.[72] Jerry Hubbard wrote for the Horowitz PR team.[73] And Oyama, Bonnie Berdahl, and the rest of the team described the GEx experiment.[74] This set a pattern that was repeated in special *Viking* experiment issues of several journals at different phases of the data collection, interpretation, and disputation of the results: Klein would write a general overview, expressing a broad consensus of the outcome, then each of the experiment teams would write up separately their individual results and opinions.[75]

The Bicentennial Anticlimax: What *Viking* Found and What It Did Not Find

Viking 1 landed in a basin, the plain of Chryse, on 20 July 1976, seven years to the day after *Apollo 11* had landed on the Moon. (Initially, a 4 July landing to celebrate the U.S. Bicentennial had been hoped for; in the end calm heads prevailed, as extra time was needed to assess the safety of the possible landing sites more carefully. Too rocky a site might cause the descending lander to tip over upon touchdown.) *Viking 2* landed a few weeks later, 3 September 1976, on the plain of Utopia, halfway around the planet and considerably closer to the North Polar cap, in an area that had the highest measured levels of atmospheric water vapor. After a short time of stabilizing systems, *Viking 1* began to transmit a television image of the Martian surface, and on 28 July a mechanical arm with a scoop dug a trench about five centimeters deep in the Martian regolith and delivered samples to the hoppers from which the biology instruments and the GCMS drew. When the first television images came in, "a new reality was created." Science experiments manager Gerald Soffen said: "Mars had become a place. It went from a word, an abstract thought, to a real place."[76] No longer the stuff of fantasy novels, open to the full span of what different people's imaginations could envision, now there was a real landscape to engage with mentally. Little did any of the researchers yet suspect how multifaceted, even enigmatic, this new place with the pink sky and dusty, rocky red landscape could prove to be.

First, the inorganic analysis team led by Benton Clark of Martin Marietta Corporation, using an x-ray fluorescence spectrometer, discovered remarkably high levels of sulfur, in the form of inorganic sulfate, in the Martian regolith. Phosphorus was also thought to be present (it is found in the Martian atmosphere).[77] Next, when the scoop delivered the sample to the GCMS, the indicator said the hopper was still empty—that is, that no sample had been delivered.

Eventually, it was thought that most likely the indicator was malfunctioning, but the glitch introduced a cloud of uncertainty into interpreting the GCMS results when they finally began to come in. The biology experiments, meanwhile, had plenty of surprises of their own to offer.

Every one of the biology experiments yielded evidence of activity from the very first run. The pyrolytic release experiment gave one reading consistent with production of organic matter (e.g., by photosynthesis), and the reading was high enough compared to his prestated requirements that even Horowitz was briefly shaken about his doubts over the existence of life on Mars. But this result was not repeatable. When wetted in the gas exchange experiment, the "soil" (regolith) released oxygen "in amounts ranging from 70 to 770 moles per cubic cm. Heating the sample to 145°C for 3.5 hours reduced the amount of O_2 released by about 50%. There was a slow evolution of CO_2 when nutrient was added to the soil."[78] By three days into the first run (1 August 1976) the gas production had decreased considerably, leading some to suspect that the reaction was chemical rather than biological. That is, it may have been produced by a potent reactant present in the sample which was used up via chemical combination with the water or nutrients.[79]

Levin's Labeled Release experiment showed the most potent reaction of all three. Recall that the nutrient solution added to the sample contained a mixture of "the following acids: formic, glycine, glycolic, D-lactic, L-lactic, D-alanine and L-alanine, each . . . uniformly labeled with $_{14}C$. The volume of nutrient delivered to the sample contains approximately 260,000 cpm, each of the 17 carbons of the added substrates thus contributing about 15,000 cpm."[80] There was an immediate peak of labeled CO_2 release in the first minutes after the nutrient solution was added, followed by a slow, continued release over the many days during which measurements continued. The amount of CO_2 released amounted to approximately 15,000 cpm, or the amount as if a single carbon atom had been cleaved at the same spot from the entire pool of a single substrate.[81] The plot of data looked somewhat like a bacterial growth curve (though it lacked an initial lag phase); furthermore, if the soil was first heated to 160°C for three hours the activity was completely destroyed.[82] The effect was partially destroyed by incubating the soil at 40–60°C, and the activity was "relatively stable for short periods at 18°C," but lost after long term storage at 18°C. All of these data seemed to Levin to be almost completely consistent with what one would expect from a biological reaction. He was tentative at first, but the subsequent controls convinced him that the best explanation of the LR results could well be the existence of microbial life on Mars.[83]

Horowitz was as puzzled as any by the results but determined not to abandon his earlier caution. Given that results were being released to the press on practically a daily basis, the nation, indeed the world, was getting the chance to observe science in process in a new way. *Viking* officials, especially Klein, worked hard to explain the slow, deliberate process by which the experiments had to be checked, different kinds of controls tried, and so forth. But the results were simply too unexpected; at each new trial that should have brought clarity in choosing between a chemical or biological explanation of the results, the ambiguity stubbornly persisted. Unused to doing science with an audience looking in at every step in the process, on 7 August Horowitz told the press: "We hope by the end of this mission to have excluded all but one of the explanations, whichever way that may be. I want to emphasize that if this were normal science, we wouldn't even be here [i.e., at a press conference]—we'd be working in our laboratories for three more months—you wouldn't even know what was going on and at the end of that time we would come out and tell you the answer. Having to work in a fishbowl like this is an experience that none of us is used to."[84]

As many, including Horowitz, had thought, the GCMS results were beginning to look as though they would be awfully useful in sorting out the ambiguous results from the biology experiments. "As one observer noted, the gas chromatograph–mass spectrometer was the court of appeals in the event that the biological experiments did not present a clear verdict."[85] But perhaps the greatest surprise of all came from the GCMS, once analysis had been run. The GCMS team decided, given the need to clarify the confusion developing around the biology results, to gamble that the device actually had received a sample in the first scoops (the remote control arm had jammed after that, so it was quite a while before another sample might be delivered to the instrument); they ran the first analysis on 6 August 1976, after heating the sample to only 200°C (which was not expected to volatilize any organics if they were present). The instrument worked well and behaved as though a sample had indeed been present. So, a follow-up analysis was run on 12 August with the remainder of the sample to look specifically at the organics. If life were responsible for the biology experiment results, organics should certainly be present (though their presence did not *necessarily* mean those results must be biological).

To Biemann's surprise and everyone else's there were no organic compounds at all, down to the level of a few parts per billion that the instrument could detect.[86] This was a great shock. Like most of the *Viking* scientists, Gerald Soffen, "once he assimilated the fact that the GCMS had found no organic materials, walked away from where the data were being analyzed"; all he could think was: "That's the ball game. No organics on Mars, no life on Mars." Soffen "confessed that it took him some time to believe the results were conclusive. At first, he argued . . . that there must have been no sample present in the GCMS because there had to be organics of some sort on the planet. . . . To his dismay, the data [from the second sample] indicated that there was a sample in the instrument and that the sample was devoid of organics."[87] On subsequent repeat runs the results were the same.

Later investigators, like those present at JPL in August and September 1976 (with the noteworthy exception of Gilbert Levin), have been forced to conclude that "since the infall of meteorites and interplanetary dust should be carrying organics to Mars at a rate of over 100,000 kg per year, the absence of organics suggests that they are being actively destroyed. The destruction . . . could be due solely to . . . solar UV."[88]

Juan Oró of the molecular analysis team called an ad hoc meeting of *Viking* scientists: he had a theory about the source of gas production in the biology experiments. Oró recalled from some of his earlier biochemical work that formate, one of the carbon sources in the LR nutrient mixture, could, in the presence of a catalyst, be easily cleaved by hydrogen peroxide (H_2O_2) or other peroxides to form CO_2 and water. Oró thought the iron oxides on the Martian surface could be excellent catalysts and that the peroxides would be formed by photolytic chemistry in the atmosphere and on the surface of Mars because of the high levels of solar UV.[89] Thus, the same UV exposure might explain both the lack of organics in the top five cm of the "soil" and the sudden, rapid CO_2 production when a sample of the surface material, containing UV-produced peroxides, was first brought in contact with the LR nutrient solution. The rapid production of oxygen gas in the GEx experiment when the soil was first wetted, he thought, might be from the same peroxides splitting the water to release oxygen gas.

According to Levin, Oró was highly concerned with receiving priority for this idea and made any scientist who stayed to hear his theory sign a paper saying he would not publish on it before Oró did. Levin thought this attitude suspect.[90] Shortly afterward, in a press conference in which the GCMS results were announced, Klein also told the press about Oró's new theory. In Oró's account, "Chuck Klein was very correct in saying, now, you're going to be presented with observations that according to Levin indicate the possibility of life on Mars. But one member of the molecular analysis team has a relatively simple chemical explanation, so the press was divided in two groups [of opinion]. And the basic theory was published the next day in the *Los Angeles Times*."[91]

Oró and many others carried out simulations of the effect of UV on organics: "the ultraviolet light gets to the surface, producing H_2O_2 and oxidizing any organic compounds. We did some experiments in the laboratory simulating Martian conditions and the half-life of any organic compound is at most two months."[92] Cyril Ponnamperuma and a team at the University of Maryland added peroxide to a sample of Levin's nutrient mixture, which Klein sent them; they found a very similar response and amount of CO_2 evolution to what was seen in the Mars LR experiment.[93] Oyama and several of his coworkers eagerly embraced the chemical oxidation theory as the most likely explanation of their GEx results. They proposed, after some lab work, that γFe_2O_3 was the most likely oxidant.[94]

Levin thought everybody jumping on the peroxide/chemical explanation bandwagon was being just as nonobjective as if one staunchly insisted on a biological explanation. He pointed out that the control run of the LR, on a sample heated to 160°C, had completely killed the response; why should peroxides act that way? he asked. Proponents of the chemical theory replied that 160°C might have been enough to destroy a peroxide. Levin recalls that he collected the six Biology Team members as well as Leslie Orgel and asked them to write on a slip of paper a temperature they would agree would clearly differentiate between a chemical and a biological reaction. There was remarkable unanimity among the seven independent "secret ballots": they all picked 50°C. That is, a reaction that was active below 50°C but ceased fairly sharply at that temperature was probably biological, they thought. Levin asked Fred Brown, the LR instrument contractor, whether he could program the LR device aboard *Viking* so that it would heat a sample to only 50°C rather than to 160°C. Brown was able to run the LR at 51°C using only some of the heaters, and the activity was almost totally eliminated.[95]

As a control, Levin suggested trying the 50°C heating again. The second time the instrument ran at 46°C. The response was a 70 percent reduction in the reaction. This was extremely suggestive to Levin, whose research experience with distinguishing fecal coliforms bacteria from other coliforms had impressed upon him that a fairly small temperature difference, from 37 to 44°C, was enough to completely suppress the growth, of all but fecal organisms. But the results also seemed to be in striking accord with the prediction each of the seven scientists had made. Levin pressed the other scientists to admit that a temperature difference of 46 to 51°C could not possibly affect the chemical reaction and must therefore be biological. But the six others immediately retracted their commitment to the 50°C number, Levin says, and they insisted that it could still be due to a chemical reaction.[96] Their caution may be partly ascribed to the fact that they had only a single pair of data points, with replication difficult or impossible to achieve on an instrument that was so far away.

The data were as confused and ambiguous as ever, having some "chemical" and some "biological" features. But with so much at stake—not only life on Mars but the possibility of seeming impetuous, unscientific, or insufficiently cautious before a world audience—the underlying bedrock epistemological assumptions of the experimenters were thrown into sharp relief. This can be viewed as a giant artifact caused by the abnormal fish bowl conditions under which the science was being carried out, or, alternately, as a unique opportunity *because* of the abnormal conditions (analogous to the fortuitous timing of the impact and analysis of the Murchison meteorite) to obtain a window into parts of the process of doing science which would normally be hidden from view. Perhaps in the spirit of Schrödinger and Heisenberg, we must entertain both views simultaneously to gain a full picture of the nature of science, at least of exobiology. Since the "big science" of the post–Second World War period, and particularly in the case of exobiology, to speak of the science artificially extracted from the public relations context that served as such nourishing soil for its development would be arbitrary indeed.

Levin and Straat continued to make the case that the interpretation of the biology results from *Viking,* at least the results from their LR experiment, were still open. By 1979, however, almost all other scientists concluded that the chemical explanation was more likely.[97] In that context Levin and Straat were viewed as being intransigent; they were rapidly marginalized. By 1988 they wrote that the balance of the evidence now seemed to them to have tipped *in favor* of a biological interpretation.[98] By the 1990s Straat was no longer writing on the subject, but Levin became still more convinced after the 1997 *Mars Pathfinder* results that water might exist in significant quantities not far below the surface of Mars; thus, life was more likely. Similarly, he considered that the August 1996 announcement of the discovery of putative microfossils in a Martian meteorite gave broad support for the case for Martian biology, even if those possible organisms were from over three billion years in the past.[99] Like Carl Sagan, Levin raised the possibility that Earth biota could have been seeded by Mars meteorites long ago when Mars was still habitable, or vice versa, now that it was recognized that meteorites were in fact moving at least in the Mars to Earth direction.[100]

In 1997 a popular book appeared, championing Levin's cause and presenting him as a scientific genius suppressed by the establishment.[101] Levin's former *Viking* colleagues and the new generation of exobiology researchers had largely ignored Levin's writings for the past fifteen years; however, the new book by Barry DiGregorio caused Harold Klein sufficient irritation that he felt compelled to respond, hoping to silence the argument once and for all.[102]

Klein pointed out that Levin's argument consisted of two main propositions; only one of them had been properly and directly addressed, he said. "The two main arguments . . . are, first that the responses seen on Mars are virtually indistinguishable from those shown by a variety of terrestrial organisms and second, that laboratory attempts to reproduce the LR results, based on non-biological mechanisms, cannot account for the results."[103] Klein said all rebuttals had concentrated on the second argument, while little attention had been paid to the first. He went on to outline a number of characteristics that the presumed Martian microbe or microbes must have, in order to fit with the data. First, they needed to live in an anaerobic environment devoid of liquid water at temperatures averaging (even at a sheltered depth of 5 cm below the surface) between -33 and $-73°C$.

Second, the organisms must survive after being brought from that ambient environment and placed in a storage container at an average of 15 to 18°C within the *Viking* lander. The samples were held at that temperature for eight days, at which time they were placed in an incubation chamber at 10 to 13°C. Two days later, ten days after being scooped up and dumped into the spacecraft, the sample had 0.115 mL of an aqueous solution of the organic carbon sources added. After being put through these changes, the microbial species (or spp.) must *immediately* release gas (within the first four minutes, as the first measurement showed substantial gas already released by that time, continuing straight up to 1100 cpm released within the first hour); Klein emphasized that the reaction took off immediately without the lag phase characteristic of most microbial growth curves. Then it leveled off after about twenty-four hours and ceased when carbon "approximately equivalent to one of the added carbon atoms [was] released, and over 90% of the added nutrients remain[ed] unaffected."[104] Klein noted the further improbability for a living organism to have done all of these things: next, when the sample was treated with a second dose of nutrient solution, no further release of radioactive gas was seen.

Finally, while fully active after ten days of storage at 15°C, these organisms must "lose their ability to metabolise when the nutrient mixture is [first] added after 84 days of storage at this temperature."[105] Klein argued, "it is possible that examples can be found in which a single species, or group of organisms, can duplicate one of these elements, and that another . . . group of organisms can duplicate a different one. But the likelihood that any single species, or group of terrestrial organisms, can reproduce the aggregate of observations made under conditions similar to those experienced during the Viking LR experiments is infinitesimal. . . . To claim that terrestrial organisms could reproduce all aspects of the LR data, is unsubstantiated."[106]

Carl Sagan, in his mature reflections about Mars, was skeptical. But in 1993 he still held out the prospect for counterintuitive local variations, saying:

> Within the emerging exobiology community [in the early 1960s] there was, as there is today, a spectrum of beliefs about the likelihood of extraterrestrial life. There were those who, like Philip Abelson, for example, argued that the environment of Mars, particularly the low water activity, was a demonstration that the planet is lifeless. And, of course, in retrospect, you've got to doff your hat to Abelson. He was right. But we argued that you could not be sure, that for the first time examining a planet in which there had been at least smoke, if not fire, about extraterrestrial life, you had to be careful. Lederberg and I wrote a paper on oases, that is, microenvironments, that conditions deviated from the norm, and there certainly today seem to be such microenvironments [on Mars]. So that's one area of debate. . . .
>
> Some people thought life was more likely than other people thought, but I think what bound us together was the importance of the question, including the importance of negative answers.[107]

Not long after Klein's rebuttal, Levin and his case for a revised LR experiment that would resolve the ambiguities of the *Viking* results received front-page coverage in the *Washington Post*. While occasionally tongue-in-cheek, the piece did give Levin a considerably more sympathetic forum than he had found among the scientific community.[108] The case for life on Mars perked up with a prominent article in the *Proceedings of the National Academy of Sciences,* which argued that the *Viking* GCMS would have been unable to detect some of the most likely organic compounds delivered to the Martian surface by meteorites.[109] In retrospect some have argued that the GCMS was too insensitive to detect organic matter in amounts found in the number of cells suggested by Levin's interpretation of the LR data; it had been assumed in the instrument's design that, if cells were able to grow, higher levels of organics must be present all around them. Further discoveries of subsurface water ice by *Mars Odyssey* in February and March 2002 have continued to reveal, much like the observations of *Mariner 4* did in 1965, that Mars is a sufficiently complex place to repeatedly overturn past scientific certainties. Levin has been vindicated on a number of points. (The case of the meteorite ALH84001, discussed in chap. 8, illustrates this point further.) We still have a very small set of locations from which surface samples have been taken and samples only to a depth of five centimeters. Perhaps at a depth of a meter, ten meters, or more sufficient shielding from UV and sufficient frozen water, possibly even liquid water, are available to make organic compounds viable. Perhaps even life. Some might argue that the stunning discoveries at hydrothermal vents, of the "third kingdom" of Archaea (see chap. 5), or of the endosymbiotic behavior of bacteria that later turned into mitochondria, chloroplasts, and other cell organelles should make researchers more cautious than Klein in predicting what microbes might and might not be capable of. At bottom this turns upon a basic attitude toward the degree of adaptability of living organisms; what is more unlikely, life on a harsh planet such as Mars or Europa or life (even complex multicellular animals) at many atmospheres of pressure and temperatures approaching 150 to 200°C near undersea hydrothermal vents? (In such a situation the "micro environment oases" invoked for Mars in 1962 by Sagan and Lederberg are also extremely relevant.) Those on different sides of that divide will tend to disagree about the meaning of a great many kinds of evidence. They will conduct different, often complementary kinds of research.

Levin argues that even Lovelock's test for life on Mars has been met because of the amount of carbon dioxide in the atmosphere. Since the proposed oxidants have never been conclusively proven to exist, Levin argues, living organisms are the likely source that recycles CO into CO_2.

Carl Sagan died in December 1996, Gerald Soffen in November 2000, and Harold Klein in July 2001; they will not see the outcome of the story. Perhaps Levin could yet get his follow-up LR experiment on a future Mars mission, as he hopes. A planned automated sample return mission in the decade after 2010 could answer many questions as well. Mars can wait, it seems. After showing up human intellectual foibles for well over a century now, Mars has all the time in the world.

Part III

Broadened Horizons, 1976–2000

CHAPTER 5

The Post-Viking Revolutions

The years from 1976 to the 1990s were a time of even greater ferment in exobiology than the 1950s to 1975. Several new seriously stultifying factors to origin of life research appeared, about which consensus emerged almost simultaneously around 1980. In the wake of *Viking* and these new realizations, massive reconceptualization was required. This was true for the origin of life problem itself and for almost all that was known about conditions on the primitive Earth. Iris Fry has described how Creationists jumped on the new quandaries and reconceptualizations to claim that origin of life work had reached a "crisis" that science cannot resolve: "They also revel in data indicating that the time available for the emergence of the first living systems was much shorter than previously thought. The natural emergence of complex biological organization already evident in the simplest cell, they claim, is even less likely within such a short geological time frame. They conclude that the need for a designer is strongly supported by the new findings."[1] Scientists, however, have viewed the situation from a fundamentally different philosophical point of view. Instead of seeing a disproof of the scientific approach, they have seen a crisis that called for creative thinking and innovation. Exobiology science has responded dramatically, across the board, with new research agendas and reformulation of many of its most basic assumptions.

Having been incubated at JPL in the years leading up to *Viking,* the Gaia hypothesis, as a scientific theory as well as a broadly influential social metaphor, came to maturity during this period. By the late 1980s scientists began to realize that it had made significant contributions to what later would be called "Earth System Science."

A flood of new data poured in during these years as well, about the existence of hitherto unknown but nonetheless complex communities of life forms living around hydrothermal vents at the bottom of the deep oceans, about more and more ancient microfossils narrowing the time window in which life must have originated, about the Archaea, about comets, the impact of extraterrestrial bodies with Earth, the relationship of such impacts to climate and to mass extinction, the lunar and Martian origin of many meteorites that had landed on Earth, and, finally, new laboratory data on membranes and on the ability of RNA to act as an enzyme. This information catalyzed new lines of thinking in laboratory work, but, more important, it turned the attention in exobiology more sharply than ever toward the heavens—not just to other planets but also to comets, asteroids, and meteorites as objects of extreme interest for thinking about the origin of life on Earth.[2]

Perhaps most important of all, debates over "punctuated equilibrium" theory in evolutionary biology, the recognition (beginning in 1974 but not widely accepted until 1984) that the Moon probably formed from a violent catastrophic collision between Earth and a Mars-sized body,[3] then, in June 1980, that the dinosaurs were in all likelihood extinguished by an asteroid impact on Earth sixty-five million years ago combined to startle astronomers, geologists, biologists, and even exobiologists into recognizing that they had been wearing rather dogmatic "gradualist" blinders, inherited from Darwin and his mentor, Charles Lyell.[4] As if to underscore the point for any still dozing, six weeks after the first publication of the dinosaur-asteroid impact theory, a dozing Mt. St. Helens took the world by surprise and erupted in one of the most violently explosive displays in recorded history. Although Gould's punctuated equilibrium theory remains controversial, more rapid change in evolution, cosmic as well as terrestrial, became less unthinkable. The renewed "catastrophist" astronomy, geology, and evolutionary biology since 1980, as well as the discovery of the "third kingdom" of Archaea and the firm establishment of Lovelock's ideas, owe much to the field of exobiology and to NASA funding. So do thriving new fields of research on the "RNA world" and on possible hydrothermal settings for the origin of life.

Hydrothermal Vents, Archaea

In January 1977 scientists exploring the hydrothermal vents in the pitch blackness at the Galápagos rift, 2.5 km deep in the Pacific Ocean, got the surprise of their lives. Entire ecological communities of life were thriving profusely in the pitch blackness, where no photosynthesis was possible for primary production. Not just microorganisms but complex tubeworms several feet long, crabs, and many other creatures grew quite happily at temperatures and pressures previously thought impossible and, it was soon discovered, were supplied nutrition entirely from chemosynthetic primary production by sulfur-oxidizing bacteria and other chemolithotrophs (bacteria that can obtain energy purely from oxidation of inorganic compounds).[5] John B. Corliss, at the University of Oregon, and Holger Jannasch, marine microbiologist at Woods Hole Oceanographic Institute, were among the first biologists to study these new life forms and ecosystems.[6]

In October and November 1977 Carl Woese and his research group at the University of Illinois, working on projects funded by NASA Exobiology since 1975, announced one of the most remarkable discoveries of twentieth-century biology (fig. 5.1). Studying the 16s ribosomal RNA of many different microor-

FIGURE 5.1. Carl Woese, at work in his lab at the University of Illinois, Urbana, examining by hand some of the voluminous 16s rRNA data that led to recognition of the Archaea as a "third domain," 1976. (Courtesy C. Woese.)

ganisms, the researchers found that methanogens (methane-producers), halophiles (microbes that can tolerate high salinity), thermophiles, and hyperthermophiles (microbes that can live at high and ultra-high temperatures), all of which had previously been classified as bacteria, were as different from them as the bacteria were from eukaryotes (all plants, animals, and fungi are eukaryotes). Woese and his colleagues called this new "third kingdom" of organisms the Archaebacteria (later Archaea), and they argued that nature really contained three discrete divisions of life: the Archaea, the Eubacteria, and the Eukarya. This difference was more fundamental, they argued, than the older division into prokaryotes and eukaryotes (essentially, bacteria vs. everything else including humans). Furthermore, the Woese group suggested that the Archaea were on the oldest part of the tree of life, closest to the "root," or last common ancestor of all forms living today.[7] Thus, as soon as the hyperthermophilic organisms of the undersea vents were recognized and determined to be Archaea, many others besides Woese's group began to speculate about the relevance of the Archaea for the origin of life (OOL), given their lineage and their capabilities for living under harsh conditions. (It is worth noting that, by 1998, with new data Woese came to believe that the last common ancestor was actually a heterogeneous population of cells with considerable horizontal gene transfer, rather than a discrete single entity.)[8]

What made the intellectual breakthrough of seeing a fundamentally tripartite division in living nature so difficult? Woese himself thinks it is a classic case of what Thomas Kuhn called a "paradigm shift."[9] There is some reason, however, to suspect that Woese's training may have caused microbiologists to regard his initial claims with skepticism. He earned his Ph.D. degree in Ernest Pollard's unusual new Biophysics Program at Yale University in 1953. Among burgeoning new "biophysics" departments of the immediate postwar period, Pollard's Yale department was something of an unusual beast, and Pollard's personality was a source of friction with many who even thought of themselves as allies.[10] Even if one accepts that, as with his younger colleagues Morowitz and Woese, Pollard was "ahead of his time" in his sweepingly interdisciplinary approach to biophysics, it is nonetheless clear that this would create disciplinary rivalries and bad blood, sufficient to serve as barriers to the easy acceptance of revolutionary new ways of seeing "theoretical biology," above and beyond the paradigm-breaking nature of the ideas themselves. Robert MacNab, also of Yale Biophysics, said that as late as 1974 his work on bacterial flagella was still regarded with deep and basic suspicion, even dismissal, by microbiologists such as Raymond Doetsch, primarily because he was not trained as a microbiologist and therefore "did not know the first thing about bacteria; for example, that one simply cannot *see* flagella in unstained living preparations by light microscopy."[11]

In response to this alternate interpretation, Woese's own perception is that "in my case it was a paradigmatic issue primarily, the fact that I wasn't a microbiologist was secondary. The prokaryote-eukaryote dichotomy, since Stanier and VanNiel's 1962 paper, had been absolute dogma in microbiology. And, of course, biologists in general also had traditionally accepted it lock, stock, and barrel."[12]

Nor did the resistance, at least in many circles in evolutionary biology, end with the broad general acceptance of Woese's three-kingdom doctrine in the 1980s. Ernst Mayr at Harvard, for example, put up a strong argument against a three-kingdom view of life.[13] And he attempted to recruit others, such as Lynn Margulis, to his cause.[14]

Influenced by Woese's discoveries about archaebacteria and his belief that such "extremophiles" and chemolithotrophs were probably the most ancient life forms, Benton Clark, a veteran of the *Viking* mission, began to reason that hydrothermal vents would have been common in the early history of the Earth and suggested an origin of life based on sulfur compounds as the key energy sources. In addition, John Corliss, John Baross, a specialist in microbial life in extreme environments from the University of Washington, and others argued that, because life was able to thrive at such temperatures and pressures, with conditions more stable than the vicissitudes of the ocean-atmosphere interface, the vent environment was a more likely place for the origin of life. They suggested a high-temperature origin, probably first of Woese's "archaeal" life.[15] Hyperthermophiles quickly became a "hot topic" in origin of life research,[16] and headlines began to appear speculating on "life's first scalding steps" and other similar titles.[17]

The Primitive Atmosphere

There is another important part of the intellectual context that made a high-temperature origin of life attractive at this time. The problem was twofold: geochemists had finally begun, after many years, to convince most of the research community that the Earth's early atmosphere was probably *not* chemically reducing (hydrogen-rich) but, rather, neutral. Second, as older and older microfossils were found, the time window available for the origin of life process was drastically narrowing. We will look at each of these in turn.

From nearly the beginning of modern scientific work on the origin of life, some prominent geologists and geochemists argued that the composition of the Earth's early atmosphere might not have been chemically reducing, despite how central this point was for Oparin and for the 1953 Miller-Urey experiment. William Rubey, a geologist who wrote papers in the 1950s and was a contemporary of Urey, pointed out as early as 1951 that CO_2 and H_2O, not CH_4 and NH_3, were the main gases coming out of volcanoes.[18] According to biochemist John Cronin, "Urey's reduced atmosphere, although influential, was kind of an anomaly that flourished for awhile until modern ideas of planetary formation and evolution made it untenable. Much of the early work didn't assume a reduced atmosphere, e.g., the Chamberlins in 1908 and Haldane in 1929." With Harrison Brown and Hans Suess's 1949 work on terrestrial atmospheric noble gases, it became clear that the Earth's atmosphere was not derived from some primordial H_2-rich primary atmosphere, and with William Rubey's 1951 ideas about a secondary atmosphere arising from degassing of the earth's interior and H. D. Holland's 1962 work on the redox state of the mantle, says Cronin, "Urey's atmosphere began to lose favor pretty early with geochemists and atmospheric scientists, although due to Miller's work and its hold on the 'popular' imagination it continued to hold sway in the wider OOL community for some time. Since it is not possible to absolutely rule it out for some brief period and/or in specialized locales in the early Archaean period, it still has its adherents."[19]

Penn State University geoscientist James Kasting states, "I would say . . . however, that it was really Jim [James C. G.] Walker who did the most to change our ideas about the nature of the early atmosphere. His 1977 book, *Evolution of the Atmosphere,* laid the foundation for the weakly reduced, CO_2-H_2O-N_2 atmosphere that is currently favored. Dick [H. D.] Holland also played a role in all of this, although his 1962 model was a multi-stage one that started off strongly reduced and then became weakly reduced later on." He adds a recent afterthought:

> I should point out that during the last few years, I have come to realize that there should have been significant abiotic sources of CH_4 on the early Earth from submarine outgassing. There is some discussion of this in my chapter in Andre Brack's 1998 book, *The Molecular Origins of Life*. However, even that discussion is now somewhat out of date. Most of the methane probably comes from serpentinization of ultramafic rocks and perhaps from impact catalyzed reduction of CO_2. My latest thoughts have not yet been formally written up. I don't think that early Earth had a highly reduced CH_4-NH_3 atmosphere, but I do think it had substantial amounts (100 ppm or more) of CH_4, in addition to CO_2, H_2O, N_2, and traces of CO and H_2.[20]

Keith Kvenvolden's 1974 book *Geochemistry and the Origin of Life* reprinted several of the original papers from the 1949–1962 period, gaining wider attention for the view that the early atmosphere might not have been reducing in nature. Geophysicist (and editor of *Science*) Phil Abelson also made the case for carbon monoxide as the primary form of carbon, rather than methane, in a 1966 paper that both Stanley Miller and Norman Horowitz took immediate notice of.[21]

By 1980 science journalist Richard Kerr wrote in *Science* that the consensus of the research community (Miller was still a prominent exception) was now leaning toward a nonreducing atmosphere at the time life first began on Earth. And, because Miller's latest experiments with CO, H_2O, and other less-reduced gases showed drastically reduced yields of organic compounds produced in a Miller-Urey apparatus, the apparent lesson was that synthesis of organic building blocks for life was more difficult than had been believed. One of the cornerstones of the optimistic OOL research paradigm of the generation since 1953 now seemed very shaky at best.[22]

The Narrowing Time Window

In 1954 Stanley Tyler and Elso Barghoorn reported on the first Precambrian microfossils, nearly two billion years old from the Gunflint chert on the northern shore of Lake Superior.[23] In 1965 Barghoorn, his graduate student J. William Schopf, and longtime stromatolite expert Preston Cloud announced a new round of such discoveries, continuing into 1967.[24] From 1967 to 1969

Barghoorn and Schopf received thirty-five thousand dollars per year in NASA Exobiology funds; in 1969, with newly minted Ph.D., Schopf set up a lab at UCLA, with fifty thousand dollars per year in NASA Exobiology money to expand the search.[25] Through the late 1960s a rapid string of discoveries of microfossils piled up; by 1977 Barghoorn and his new student Andy Knoll found convincing microfossils as old as 3.4 billion years in the Fig Tree series of South Africa.[26] Very few older rocks were known, and most of them had been so metamorphosed that there was little hope of finding convincing microfossils any older than those already found. By February 1978 Stephen Jay Gould wrote in his widely read column in the magazine *Natural History:* "If prokaryotes were well established 3.4 billion years ago, how much further back shall we seek the origin of life?"[27] He pointed out that conditions on Earth had only been suitable for life for at most a few hundred million years prior to the Fig Tree organisms, which were eubacteria. Yet, citing Woese's November 1977 discovery that the common ancestry of Archaea and Eubacteria must lie even further back, Gould concluded that the origin of life must have occurred very rapidly and almost immediately after conditions for it permitted. The contrast of this new conclusion with the "long, drawn-out" scenario so deeply ingrained in the OOL community of the 1950s and 1960s, made obvious what a deeply rooted prejudice the "long, slow process" model had been since Darwin.

In reality, in the community itself the realization of the shortening time window had been dawning rather more steadily and earlier than Gould's essay seemed to suggest. As early as 1968 exobiologist Alan Schwartz (now head of his own research group [fig. 5.2]), for example, had been "struck by the rapidly decreasing 'window' for the origin of life which fossil discoveries was generating and wrote a short manuscript on the subject. I sent it to a geochemical colleague for criticism. His response was that the realization of the shortness of the time available was pretty much common knowledge," so Schwartz, chagrined, never submitted the manuscript.[28] Nonetheless, many researchers were only just coming to this realization, so Gould's basic point was valid: between the late 1960s and about the time of Woese's announcement of the three kingdoms, the OOL community did slowly come to a new view. The process of life's origin either could not be long and drawn-out, or else a lot of the early stages (the formation of organic building blocks) had to take place in extraterrestrial settings. The strong possibility of a nonreducing atmosphere seemed to confirm this conclusion and to press home the other major intellectual shift to which Gould was pointing. Origin of life chemistry, it now seemed clear, could not have been a matter of chance, random bumping together of molecules requiring endless billions of years, as George Wald posited in an influential summary of the field written shortly after the Miller-Urey experiment.[29] The chemistry must have been constrained by some natural limits to lead spontaneously in the direction of living systems fairly directly and rapidly—perhaps as little as ten million years to go from abiotic conditions to cyanobacteria, according to one 1994 estimate.[30] Thus, amid the intellectual disorientation and reorientation of this

FIGURE 5.2. Alan Schwartz and his research group at the University of Nijmegen, the Netherlands, 1987. (Courtesy A. Schwartz).

period, even if nobody was really sure at first how such chemistry must work, it seemed the news was not all bad for origin of life work.

In early February 1977, less than a month after the new undersea vent discoveries, UCLA paleontologist J. William Schopf and Indiana University geochemist John Hayes had a conversation with NASA Exobiology's Dick Young to try out a new idea on him. In the wake of the OPEC oil embargo and the economic slump that followed in the United States, post-*Apollo* NASA budgets shrank even faster than they had before. The fat times of the 1960s and early 1970s were only a memory now. Still, Schopf had been thinking for some time that Precambrian paleobiology needed a concentrated period of intense close group effort by leading researchers in the field and in related disciplines such as geochemistry, prebiotic chemistry, microbiology, climatology, and atmospheric chemistry. Schopf dreamed of a fourteen-month-long Precambrian Paleobiology Study Group (PPRG) centered at UCLA. Dick Young thought the idea a good one and said that "'in principle' his program 'might possibly' be interested in supporting such a project."[31] Encouraged, in March 1977 Schopf contacted those he hoped would form the nucleus of such a group, to begin putting together a detailed, formal grant proposal. Included were Hayes, Hans Hofmann, Ian Kaplan, David Raup, and Malcolm Walter. Almost immediately, Schopf got a windfall: he received word in April that he had been selected to receive a

$150,000 Alan T. Waterman Award from the National Science Foundation, enough, he thought, to cover perhaps half the cost of his PPRG dream project. Thus, he applied to NASA Exobiology in January 1978 for only the same amount in matching funds. By June an expanded fourteen-member group met at UCLA for a planning session; by November Dick Young notified the group that the funds had been approved. In late May 1979 Hayes, Hoffman, and Walter set off on a four-week field trip to Australia, Africa, and Canada to fill in gaps in a complete geological sample collection representing the entire Archaean and Proterozoic eras. A total group of twenty-four scientists then convened in July to begin studying the entire collection, regular meetings, and the preparation of reports. About half the group was in residence at UCLA for the entire fourteen-month period of the PPRG; some were there for periods of weeks or months; the remainder worked solely at their own institutions, save for the final group meeting in August 1980. The group produced *Earth's Earliest Biosphere,* a massive compendium volume of everything known to date on Precambrian paleobiology and much of what was known in many related areas such as prebiotic organic synthesis and the evolution of the Earth's environment in the period after life appeared.[32]

A very similar effort was organized by Schopf nine years later, also with help from NASA Exobiology funds, to focus more intensively on the slightly later Proterozoic period and to take into account the explosion of new research in the intervening decade. This resulted in 1992 in a second volume, *The Proterozoic Biosphere,* which has become as much a standard encyclopedia of the field as the first book was.[33] In 1993 Schopf announced new microfossil discoveries from the Apex chert formation of western Australia that pushed the oldest known microfossils, which Schopf suggested bore strong resemblance to existing cyanobacteria, back to 3.45 billion years ago.[34] Schopf states that the two crucial, intensive, synthetic PPRG research groups did so much to consolidate and catalyze work in Precambrian paleobiology, and in generally relevant exobiological topics, that he sees NASA funding as crucial to the spectacular progress this field has made in the past thirty-five years. On the initial 1978 PPRG application he planned to staff the project with the best relatively young scientists available, rather than well-established luminaries in the field. As a result, the proposal was strongly criticized by two senior reviewers, probably Barghoorn and Cloud, Schopf speculates. Despite these negative reviews, Schopf says,

> Dick Young, then Exobiology officer, . . . funded us. (He, in my opinion, was the great hero in the matter.) His faith bore fruit. The product of our work (*Earth's Earliest Biosphere,* . . .) was judged the 1983 "Outstanding Volume in the Physical Sciences" by the Association of American Publishers. Years later, again with NASA funding, I set up a second PPRG, . . . the product of which (*The Proterozoic Biosphere*) was judged the 1992 "Outstanding Volume in Geography and Earth Science" by the

Association of American Publishers. As far as I am aware, receipt of two such national awards is unprecedented—and both were based partly or wholly on NASA funding. . . . the two great PPRG volumes have, I believe, both set the standard and charted the course of the field of Precambrian paleobiology for every interested scientist, worldwide. Without NASA's backing, I can't imagine how this would have happened.[35]

NASA Exobiology strikes again. Twice in the same spot.

As we shall see in chapter 8, work has not always been so completely free of criticism for Schopf and his UCLA group (most recently, the 3.45 billion year old Apex chert microfossils have been questioned as possibly artifacts), but there can be no doubt that they have indeed contributed much to setting the standard for research in Precambrian paleobiology. They (and Schopf in particular) have become a powerful force to be reckoned with in exobiology, so much so that one recent book referred to Bill Schopf as the "dean of the early fossil record."[36]

The Gaia Hypothesis

A major exobiology meeting convened at NASA Ames Research Center on 19–20 June 1979. With all the new data pouring in, John Billingham of Ames saw a need to reconsider the big questions, both in origin of life research, what was known of conditions relevant to life on other planets, and SETI; as a result, he arranged the "Conference on Life in the Universe."[37] It was here that Benton Clark proposed the model cited earlier for OOL based on sulfur biochemistry. Soon after this, Dick Young retired as the head of Exobiology (now called Planetary Biology) at NASA headquarters in Washington, D.C. Donald DeVincenzi, his deputy, who had trained under Young for a year as well as in administrative positions at NASA Ames, became the new Exobiology head in August 1979.

Since the *Viking* results had so strikingly borne out Lovelock's prediction that Mars would be lifeless based on its atmospheric chemistry, Lovelock and Margulis (fig. 5.3) and their Gaia hypothesis got a prominent place on the agenda of Billingham's conference. This was a crucial turning point for the theory. Not only was it being given a high-profile podium just at the time Lovelock's first book on Gaia came out; perhaps just as important was that Stephen Schneider, a leading atmospheric researcher from the National Center for Atmospheric Research (NCAR) in Boulder, Colorado, was at the meeting and was much impressed by the potential power of the Gaia hypothesis. It was Schneider who critically addressed the idea and its promise in a 1984 mass-market book, *The CoEvolution of Climate and Life,* and in a television documentary produced in 1985 by the BBC's "Horizon" and the American "NOVA" series.[38] In addition Schneider, along with Penelope Boston, organized the first major conference to evaluate the scientific merit of the Gaia hypothesis, under the auspices of the

FIGURE 5.3. James Lovelock and Lynn Margulis, codevelopers of the Gaia hypothesis, and Spanish microbiologist Ricardo Guerrero, c. 1990. (Courtesy J. Lovelock).

American Geophysical Union, in March 1988.[39] And in a series of meetings at Ames in 1981–1982 on the evolution of complex and higher organisms, convened by Billingham and David Raup, the participants reached the following major conclusion: "Of special interest, is the controversial Gaia hypothesis, which proposes that living things have prevented drastic climatic changes on the Earth throughout most of its history. This view, regarded as highly speculative and tentative by many workers, has yet to be rigorously examined. If it proves to be correct, and if climatic stabilization can be shown to be a likely consequence of the activities of life on other worlds as well, then we may expect that extraterrestrial life is abundant throughout the universe. An effort should be made therefore, to *determine whether the Gaia hypothesis is valid.*"[40]

Given the potential fruitfulness of the Gaia hypothesis, recognized no later than this time by many in the exobiology community, it is a fascinating phenomenon worthy of study just how much resistance Gaia generated in the geology, atmospheric science, climatology, and evolutionary biology communities.

Charles Darwin had some good rhetorical reasons for clinging so tenaciously to his term *natural selection,* despite intense criticism that, to many, it implied an anthropomorphic, voluntaristic "selector" in nature.[41] And in a story with some interesting parallels James Lovelock's term *Gaia* was attacked from the beginning; the same charges were brought: it's anthropomorphic (no matter how many times he said, "I meant it as a metaphor"), you're assigning *agency* to a natural process and therefore secretly slipping a supernatural Creator back in through the back door, and so forth. Ironically, this time it was the hard-line natural selectionists (W. Ford Doolittle, Richard Dawkins, John Maynard Smith, and William Hamilton) who attacked the metaphor for having voluntarist overtones, having themselves worked hard to press the "selfish gene" metaphor to supplement the natural selection of their revered forefather Darwin.[42]

From the beginning the key technical criticism was how behavior by a microorganism that benefited the biosphere as a whole but not itself (and might even sometimes be detrimental to its own survival, such as the first release of oxygen by anaerobes) could ever evolve and persist by natural selection. And Lovelock acknowledges that the early versions of the theory, up through his 1979 book *Gaia: A New Look at Life on Earth,* suffered from an inadequate consideration of this question.[43] He developed the "Daisyworld" mathematical model, in collaboration with Andrew Watson of Reading University, to answer these objections.[44] The 1981–1982 NASA ECHO Workshop participants, who found the hypothesis intriguing said: "Although many of us are skeptical, we agree that the Gaia mechanism approaches one extreme of a spectrum of possibilities (ranging from total control of a planet's environment by its organisms to total lack of control) and that much further study is needed to determine the causes of large-scale environmental stability and change. . . . The Gaia hypothesis in particular could be investigated by seeking to identify evolutionary mechanisms (if any such exist) that are capable of selecting organisms whose activities promote global environmental stability."[45]

A key intellectual barrier was the idea in geology, evolutionary biology, and environmental science that the environment changes and affects organisms but that organisms themselves were mostly passive recipients of such selective forces. For most of these researchers it required a deep reconceptualization to see living organisms as potent forces, shaping conditions on Earth just as powerfully (or perhaps more so) as they were being shaped by those external conditions. But in addition the name *Gaia* drew a great deal of fire for suggesting, via the image of the ancient Greek Earth goddess, everything from vague New Age mysticism to teleology reimported into biology after a 150-year struggle by evolutionary biology to banish it. In the ensuing "take no prisoners" firefight, Lovelock has modified his theory to reflect the valid points his critics have driven home.[46]

Exobiology (and, more recently, astrobiology) after the disappointment of *Viking* has fully incorporated Lovelock's insight (usually without attribution) that life detection strategies need, insofar as possible, to be "non-Earthcentric."[47]

After the modifications of the theory as presented in Lovelock's second book in 1988, more researchers in the exobiology community found Lovelock's theory acceptable. Harold Morowitz wrote, for instance, that origin of life researchers now needed to understand that "in [Lovelock's] sense, life is a property of planets rather than of individual organisms." This view was complementary, rather than contradictory, with the traditional biology view that sought to define life by comparing what all living organisms have in common.[48] Indeed, under the name *Earth system science* the core of the modified Gaia theory is now mainstream science, but, say the critics, "never under the name Gaia."

Lovelock, however, tenaciously defends *Gaia* and insists that "names are important."[49] Describing one striking episode, he says:

> I stuck with the name Gaia because my Green friends and quite a few scientists regarded a change of name as a betrayal and so do I. I did try the neologism "geophysiology" for scientists and it worked for a while until the snarling dogs realized it was just another name for Gaia. I overheard a distinguished geophysicist at NCAR say to a young scientist, "I will not have you use the word geophysiology—it's just closet Gaia." [In] Mary Midgley's new book *Science and Poetry* . . . she deals in full with the name Gaia and why it was rejected by so many scientists. . . . A great deal of the fuss over Gaia is because I work as an independent and only rarely go to meetings of scientists. It is hard to appreciate the work of someone you do not know.[50]

Thus, as with Woese, even those whose ideas got off the ground in the intense interdisciplinary environment of NASA Exobiology in the 1960s could run into trouble because of plain old disciplinary turf defense, if the main body of the discipline, such as geology or climatology, was still outside of the exobiology context. Lovelock has written at some length on this problem, making it difficult if not impossible for a scientist to operate outside academia as an "independent."[51] He himself barely managed it, even with a long track record of training and research in prestigious British government science establishments prior to transitioning to independent status as an inventor and a consultant to NASA and to industry groups.

Lovelock believes that since the late 1990s or so the climate has improved to some extent. But still not enough that many of the neo-Darwinians with whom the vitriolic public conflict occurred will ever openly credit the term *Gaia,* even if they accept most of what is now called Earth System Science. Says Lovelock:

> The grandees over here are ready to admit, even at small meetings, that they were wrong to ridicule Gaia, but apart from Bill Hamilton no one will go public. John Maynard Smith used his powerful influence to have Tim Lenton's article "Gaia and Natural Selection" published by *Nature* as the lead article. Richard Dawkins, at a closed meeting in Oxford of about 25 scientists, said after I had spoken on Gaia and evolution, "Jim

has his disciples and I have mine, they both get it wrong." John Lawton, now head of the UK Research Council, NERC, had an editorial in *Science* on Earth System science, which generously acknowledged the Gaian contribution. It could be much worse.[52]

John Lawton's acknowledgment of Lovelock and Gaia is certainly more than many scientists who face such opposition ever see in their own lifetime: "Physicists have long understood the 'Goldilocks effect'—why, in general terms, Earth's natural blanket of atmospheric CO_2 and distance from the sun make the planet 'just right' for life, neither too hot (like Venus) nor too cold (like Mars). James Lovelock's penetrating insights that a planet with abundant life will have an atmosphere shifted into extreme thermodynamic disequilibrium, and that Earth is habitable because of complex linkages and feedbacks between the atmosphere, oceans, land and biosphere, were major stepping stones in the emergence of this new science [Earth System Science]."[53] Lovelock sees an interesting parallel between the opposition to the new "catastrophism" that broke through during this period and the opposition to Gaia theory. (Kuhn's *Structure of Scientific Revolutions* seems to be widely read among exobiology scientists, especially those who perceive themselves as outsiders.)[54] Both, he claims, were so basically opposed to a powerful Kuhnian paradigm that intense opposition was inevitable:

> So powerful was this dogma [of Lyellian/Darwinian gradualism] that it persisted, in spite of abundant contrary evidence, until Alvarez and his colleagues produced almost unequivocal evidence for an impact catastrophe as the cause of the KT extinction. During the 150 years from 1830 to 1980, any mention of sudden evolutionary change was treated as if it were heresy and most geologists found it prudent never to speak of catastrophes. It took the hard evidence and the superior rank of the Nobel Laureate Alvarez, to break the ice. Even so, he was amazed by the fury and bad manners of those Earth scientists who still continued to attack his research. So I am indeed naive if I think that the even more heretical theory of Gaia will be recognized by the great Church of Science. Young scientists, who imagine that they have nothing to lose, occasionally break ranks, as in the *New York Times* article, but even then only obliquely.[55]

So, what is the Alvarez discovery to which Lovelock refers, and how did it come about? At least partly, the reader by now may not be surprised to hear, with help from NASA funding.

Of Asteroids, Mass Extinctions, Dust Storms, and Nuclear Winter

Physicist Luis W. Alvarez (winner of the 1968 Nobel Physics Prize) and his son Walter of the University of California–Berkeley Geology Department had noticed an anomalously high level of the rare metal iridium in the very thin clay layer at the boundary between the rocks of the late Cretaceous period and the early Tertiary (the K-T boundary). It occurred to them that iridium was almost exclusively known from extraterrestrial sources such as asteroids and meteorites. Thus, the Alvarezes began to examine samples of the K-T layer from different locations around the world to see whether the iridium anomaly was local or more widespread; they found it to be global in its occurrence. This immediately suggested the possibility of a large asteroid impact, the explosion from which was large enough to distribute extraterrestrial material all over the globe and which, not incidentally, might finally answer the age-old question of what had brought about the sudden end of the dinosaurs (and so many other species that this was called a mass extinction by paleontologists).[56] When their paper, with coworkers Frank Asaro and Helen Michel, was published in *Science* on 6 June 1980, it provoked both excitement and skepticism, as noted earlier. Walter Alvarez had been supported by NSF funds, the remainder of the team by Department of Energy funds, and Luis Alvarez additionally received NASA money for the work.[57] Subsequently, the Alvarez team was funded by NASA Exobiology to continue its research.[58] By October 1981 a meeting had been convened in Snowbird, Utah, of paleontologists, specialists in asteroid impacts, iridium spikes, and so forth, to evaluate the Alvarez theory. The consensus was strongly in favor of the Alvarez team's theory. Follow-up calculations indicated that an asteroid of about ten kilometers in diameter was necessary to produce the iridium levels measured. The search began for the geological remnant of what must be a very large crater, hundreds of kilometers in diameter, produced by the impact. By the late 1980s it appeared that the Chicxulub formation, on the bottom of the Gulf of Mexico, just east of the Yucatan Peninsula, was indeed the crater made by the K-T impact. Calculations soon showed that the amount of dust thrown into the atmosphere by such an enormous explosion would block out the sunlight for months or perhaps years, dropping photosynthesis levels and temperature so drastically that it could more than account for the mass extinctions, including the dinosaurs.

The investigation of mass extinctions under NASA auspices did not end with the Alvarez paper; it was only just beginning. David Raup, a well-known paleontologist from the University of Chicago and the Field Museum, had been a member of Schopf's PPRG in 1979–1980. His first direct contact with NASA Exobiology, however, came in July 1981, when, at the invitation of NASA Ames's John Billingham, he chaired the first of three workshops devoted to the "Evolution of Complex and Higher Organisms," the so-called ECHO workshops, held at Ames. The succeeding sessions were held in January and May 1982. Raup had studied in some depth the extinction of marine species in the geologic past. After the very first ECHO meeting, he and his younger colleague Joseph J. Sepkoski Jr. were stimulated to think further about how often these extinctions came in massive clusters.

By March 1982 Raup and Sepkoski published a paper in *Science* demonstrating that there had been no less than five major mass extinctions and launching a search for their (perhaps astronomical or astrophysical) cause.[59] Their work showed that the average "background" extinction rate was between 2.0 and 4.6 families per million years of geologic time. The mass extinction events stood out even more dramatically than had previously been realized: these episodes reached extinction rates of 19.3 families per million years. As Raup later put it, describing how important the ECHO meetings had been as a stimulus to this new line of research, "Largely as a result of interactions at the meetings, . . . Raup and Sepkoski launched a statistical analysis of data bearing on a proposition made earlier by another of the participants (Fischer) to the effect that biologic extinctions on Earth have had a periodic distribution in geologic time, and that the periodicity is driven by extraterrestrial forces."[60] The analysis was published in the *Proceedings of the National Academy of Sciences (PNAS)*.[61]

When they had completed their statistical analysis, Raup, in May 1984, wrote: "The publication of this new analysis . . . led, in turn, to the publication of no fewer than five papers by geologists and astrophysicists, proposing mechanisms for the extraterrestrial driving force. . . . Whereas this line of research is far from complete, it is clear that the ECHO meetings played an important role in catalyzing these new initiatives in space research, initiatives which may have far-reaching consequences for biology as well as for the space sciences."[62]

Raup and Sepkoski were subsequently funded by NASA Exobiology, from 1983 to 1994, when Raup retired. As Raup put it: "My own funding from NASA started, as you can see, shortly after the workshops. Not coincidental."[63] Summing up his experience with NASA for this research, as opposed to NSF, where competition and increasing paperwork requirements made funding steadily more complicated and unreliable, he continued: "John Billingham was the prime mover in the effort to extend the origin and early history of life studies to more recent evolutionary history. John and I worked closely to arrange the workshops, select participants, and get funding from Headquarters. The report speaks for itself. The group meetings were a wonderful experience in the mixing of disciplines and were responsible directly or indirectly for a variety of research collaborations and initiatives. . . . My motives for using NASA rather than NSF or other funding sources are obscure. I had been supported by NSF off and on for 20 years at that time but it was getting more and more difficult and time-consuming. Thus, the less formal, more personal, atmosphere of NASA was attractive. Also, the kind of synoptic work I did probably fit better with the NASA culture than that of NSF. It was a good experience all around."[64]

After their *PNAS* analysis convinced Raup and Sepkoski that a periodic mass extinction cycle needed much closer attention, and, well before the paper came out in print, astronomers did indeed begin hypothesizing many possible causes. "Through word of mouth, preprints, and particularly news stories in *Science* and *Science News* [in September 1983]; researchers who . . . think more about outer space than the fossil record heard about the proposed 26 million year periodicity. The rush was on."[65] When Luis Alvarez showed the preprint to astronomer Richard Muller at Lawrence Berkeley Labs, for example, Muller had postulated within an hour "that an unseen companion [star] circling the sun once every 26 million years could be responsible."[66] *Nature* published five papers by separate research groups, including one by Rich Muller and Walter Alvarez, coming to a similar conclusion in the same issue. Most concluded that the star must be a "brown dwarf" (a substellar object intermediate in mass between a star and planet) of low luminosity; otherwise, it would have been noticed already by astronomers. Raup and the Alvarez team immediately began organizing a conference, held on 3–4 March 1984 at Lawrence Berkeley Labs, on hypothetical multiple comet impacts and their effect on evolution. Alvarez recalls: "Almost everyone active in the field attended. Gene Shoemaker spent an entire afternoon telling us why no one should believe in 'Rich's star,'" which was soon dubbed "Nemesis." Still, at least at the time of his writing in 1986 or 1987, Luis Alvarez believed the case for Nemesis and periodic extinctions (on a 28.5 million–year cycle) was quite strong. It should be noted, however, that by 1990 the consensus of the scientific community leaned against periodicity being real, though the idea is still kicking around.[67] As Raup put it:

> If one were to poll miscellaneous geologists, paleontologists, and astronomers, I think you would find a strong consensus opposed to periodicity. The negative views would be based on some or all of the following arguments:
>
> * Statistical support for periodicity in the extinction record is weak or flawed.
> * The Nemesis orbit would be unstable.
> * None of the other proposed mechanisms is viable.
>
> On the other hand, the idea is still around and many people would jump on any new data that might confirm periodicity. I think Rich Muller is still confident of finding confirmation through dating of lunar impacts or by finding Nemesis in sky surveys. . . . For me, periodicity may or may not be real. Arguments on both sides are good ones and we can't do much more until a new and independent source of data appears. But the idea is certainly alive.[68]

The Alvarez asteroid theory was at least partly responsible for the convening of several important scientific meetings: the NASA Ames ECHO meetings as well as the October 1981 meeting in Snowbird, Utah, mentioned by William Hartmann at the opening of this chapter. But one of the first and politically most important fallouts from the Alvarez asteroid extinction theory was described by Luis Alvarez: "Soon after my colleagues and I published our impact hypothesis, a group of atmospheric experts at the NASA Ames Laboratory examined it in detail. They confirmed our general conclusions but thought that the dust cloud would fall out more quickly than we had predicted. A study that grew out of that work is the now-famous 'nuclear winter' paper that proposed that smoke from fires set by exploding nuclear weapons would similarly block out sunlight worldwide with consequences similarly dire.... The fact that neither of the two superpowers' nuclear-weapons establishments had thought about the possibility of a nuclear winter has sobered everyone concerned with fighting a nuclear war."[69]

The team at NASA Ames included Richard Turco, Owen Toon, Thomas Ackerman, James Pollack, as well as Pollack's former Ph.D. advisor, Cornell astronomer Carl Sagan. Sagan and Pollack had studied the planet-wide dust storms on Mars first clearly seen by *Mariner 9*. They had begun, along with Turco, Toon, and Ackerman, modeling the dust cloud after the Alvarez asteroid impact and soon realized a similar dust cloud might have similar or even worse effects after even a "limited" nuclear war. But they had overlooked the effects of smoke from forest fires and buildings ignited by nuclear explosions, as Sagan was soon to realize. While visiting Ames for the last ECHO meeting in May 1982, Sagan talked with Pollack and Toon about the recent article by Paul Crutzen and John Birks in the environmental science journal *Ambio* on climatic effects of smoke from nuclear war.[70] Pollack soon arranged to use Ames's Cray supercomputer to run climate simulations using both smoke *and* dust effects. On 6 April 1982 Richard Turco mentioned the Crutzen and Birks article at a NAS special meeting on climatic effects of nuclear war, where he presented the findings of the Ames team on dust effects. He said that results from the new model, including smoke and dust effects, should soon be forthcoming.

In the first year and a half of the Reagan Administration the new aggressive nuclear policies of the United States government caused great worry among many citizens. The anti-nuclear movement dramatically picked up steam, including the nationwide Nuclear Freeze movement, from 1981 to 1982. Jonathan Schell wrote a powerful and very influential series of articles in the *New Yorker*, published in 1982 as the book *The Fate of the Earth*. In the politically polarized climate surrounding the administration's decision to put forward-based Pershing II nuclear missiles in NATO countries in Western Europe, dramatically shortening the Soviet Union's perceived response time window, the Reagan Administration perceived much anti-nuclear activism as disloyal. Thus, when members of the Ames team, most of whom were federal civil servants as employees of NASA, began to publicize their results, pressure was exerted from the top down, through the NASA administration, to put a stop to the work. In the fall of 1982, at an American Geophysical Union meeting in San Francisco, Jim Pollack was scheduled to report on latest results of the Ames study on smoke and dust from nuclear war. He was pressured by both the director and assistant director of NASA Ames the day before the meeting to cancel the talk.

Pollock and Sagan decided, instead, to plan a peer review meeting of their findings for 22–26 April 1983 at Harvard.[71] Their idea was to hold a scientific peer review meeting, closed to the public and press, to make clear that the study (now known as TTAPS from the initials of its authors) was not motivated politically and was being judged by the scientific community based entirely upon its scientific credibility. The meeting produced much productive scientific criticism and fine-tuning but basically affirmed the conclusions of the TTAPS study.[72] The revised manuscript was submitted to the journal *Science* on 4 August 1983 and published there on 23 December.[73] Their basic conclusions were that, under almost all imaginable scenarios of nuclear exchange above a few hundred detonations, the smoke and dust would be sufficient to block out almost all sunlight for months, years, or even decades. The "nuclear winter" resulting would be sufficient to cause the extinction of most life forms on Earth, certainly of all human life. The only way to prevent such an irreversible tragedy, many concluded, was to cease any thought of war-planning scenarios in which either side hoped to "prevail" over the other. A large segment of the public was convinced that both sides must reduce their nuclear arsenals to fewer than a thousand warheads as soon as possible; otherwise, even an accidentally escalating nuclear exchange could very quickly pass the threshold above which the nuclear winter result was inevitably triggered.

Meanwhile, in September the Soviet Union shot down Korean Air flight 007, killing hundreds of innocent civilians, when the commercial passenger plane accidentally strayed into Soviet air space. Cold War rhetoric was turned up to even a higher level; in response to the deteriorating political climate, the TTAPS group scheduled a public presentation of their results early, at a conference on the "World after Nuclear War," in October 1983, at the Washington, D.C., Sheraton Hotel. That same month the made-for-TV film *The Day After* aired on nationwide television, with a panel discussion afterward on nuclear policy and the effects of nuclear weapons, including Sagan, Elie Wiesel, and Henry Kissinger. (The film was very frightening, yet it did not take into account at all the compounding effects of nuclear winter.) In all the years in which NASA Exobiology funds produced scientific findings with high-profile public relations dimensions, few moments, surely, matched this one for historical drama, political impact, and direct implications for the human future. A week after the TTAPS paper appeared in *Science,* on New Year's Eve, Carl Sagan gave a high-profile "lay sermon" to thousands of people packed into the Cathedral of St. John the Divine in New York City, imploring humanity to respond to the nuclear winter findings by raising its consciousness and adopting whatever activism was necessary to prevent such a tragedy from occurring. Gone was the lighthearted, wisecracking Sagan of the "Johnny Carson Show" in the years leading up to *Viking*. In his new incarnation Sagan still had an ego that could provoke his opponents, but the seriousness of the consequences of his science had produced a change; emerging was a spokesman for science who would soon advise the Pope and the Soviet Central Committee on the scientific and policy implications of the nuclear winter study.

In an article from this time, summarizing the past efforts of the NASA Exobiology Program and describing the changes in emphasis that had occurred since *Viking,* the new Exobiology head, Donald DeVincenzi, listed the currently supported research agenda (table 5.1). One can see the influence of both the

TABLE 5.1 *Donald DeVincenzi, 1984 Summary of Exobiology Scientific Goals*

These goals include the study of:

1. Biogenic elements (including studies of abundance of CHONPS[a] in the universe, including in interstellar molecular clouds)
2. Chemical evolution (including Miller-Urey type simulations, organic compounds on meteorites [Cronin],[b] possible role of clays in synthesis of oligomers [Cairns-Smith and Hartman])[c]
3. Origin of life (including sequence-specific templating [Orgel],[e] origin of genetic code [Woese],[d] studies on microspheres [Fox] and similar structures, origin of metabolic systems)
4. Organic geochemistry (including search for microfossils [Schopf, Knoll], diagenesis of organic matter, modeling of ancient climates [Pollack, Kasting][f] for correlation with properties in the geologic record)
5. Evolution of higher life forms (including Alvarez asteroid extinction work, Raup and Sepkoski on periodicity of mass extinctions and possible cause)
6. Solar system exploration and SETI (detection of life and life-related organics beyond the Earth [Biemann], instruments, especially GC, to send to Titan and to comets, SETI program)

[a] *CHONPS* stands for carbon, hydrogen, oxygen, nitrogen, phosphorous, sulfur.
[b] John Cronin first received NASA funding, $45 thousand per year, in 1975; it increased steadily every year, reaching $115 thousand by 2000. Cronin to Strick, personal communication, 6 December 2000.
[c] A. Graham Cairns-Smith to Strick, personal communication, 28 December 2001 and 8 January 2002; Hyman Hartman to Strick, personal communication, 3 February 2002. Hartman was funded from 1980 to 1987 at $40–50 thousand. In addition, Cairns-Smith and Hartman received funds to organize a July 1983 meeting in Glasgow on Clay Minerals and the Origin of Life.
[d] Leslie Orgel to Strick, personal communication, 11 January 2002; Orgel received funding for this work steadily from 1969 to 2001, totaling $4,652,528. In addition, he had a contract for $56,896 from 1969 to 1977 related to the *Viking* GCMS project.
[e] Carl Woese to Strick, personal communication, 14 January 2002. Woese's funding rose steadily through these years; in 1977 he received $73 thousand and by the early 1990s $100 thousand or more per year.
[f] James Kasting and James Pollack were at first co-PIs on this grant; by the late 1980s Kasting had taken it over and has been funded continually "on the order of $60–80K per year since that time." Kasting to Strick, personal communication, 19 December 2001.

"Life in the Universe" conference as well as the ECHO meetings; DeVincenzi prominently included "evolution of higher life forms," stating that this research was being pursued through "projects dealing with the possible influence of solar and galactic events on this process. These include further characterization of rock samples showing an anomalously high iridium content at the Cretaceous-Tertiary boundary. Current efforts are also being focused on examining the relationship between the proposed impact events (which may have caused these anomalies) and biological extinctions. They include developing models of atmospheric dust dispersion, which may have caused profound changes in light intensity and temperatures, and also a more careful examination of the extinction record itself."[74]

Using cautious scientific language DeVincenzi only hinted obliquely at the controversial nature of the impact theory debate and the periodic extinctions discussion; the work being supported was the Alvarez group, the ECHO meetings, and Raup and Sepkoski. He was hinting even more obliquely at the highly politically charged studies of "atmospheric dust dispersion," sharply reducing "light intensity and temperatures;" NASA was still supporting Turco, Toon, Ackerman, and Pollack in their modeling studies on these topics, despite the Reagan Administration's profound distaste for the resultant nuclear winter theory.[75] Pollack had begun, in 1981, to collaborate as well with James Kasting on modeling climates on the ancient Earth, using many of the same techniques developed for analysis of the K-T asteroid impact and the nuclear winter scenario.

Scientific as well as political attacks were directed against the nuclear winter theory. The debate pushed along dramatically the development of complex computer modeling of climate. By 1990 the TTAPS group published a follow-up paper that responded to many of the technical critiques.[76] Their results showed a somewhat less severe climate scenario than in the 1983 study; they argued, however, the basic phenomenon of nuclear winter remained an inescapable consequence. The collapse of communism in Eastern Europe in 1989 and in Russia in 1991 and the less aggressive nuclear stance of the first Bush Administration moved the issue out of the headlines. Some might argue (Sagan for one, Turco for another) that it was the danger of nuclear winter which was one important factor starting the process of moving U.S. government policy away from that of the early Reagan years.[77]

In his summary of NASA Exobiology's goals DeVincenzi seemed to have internalized quite a bit of the logic of the Gaia theory, stating, for example, that "there is a clear relation between the processes which are believed to have occurred on the primitive Earth with those that are occurring today, where the Earth's biota is, in effect, acting as a modulator of processes occurring on a global scale. It is just this relationship which is becoming more and more prominent as a major new NASA thrust for the future. . . . It is the clarification of this relationship which will lead to the most fundamental breakthroughs in understanding . . . the origin of life."[78]

Exogenous Delivery of Organic Compounds

In August 1986 the Space Sciences Board of the NAS held a meeting in Snowmass, Colorado, which began a series of meetings through 1988, leading to the 1990 publication of *The Search for Life's Origins*.[79] The Planetary Biology and Chemical Evolution Committee was chaired by Chuck Klein and included Hyman Hartman, John Cronin, George E. Fox, Andrew Knoll, John Oró, Toby Owen, Norman Pace, David Raup, Norman Sleep, Jill Tarter, David Usher, and Robert Woodmansee (with Sherwood Chang, Mitchell Sogin, and Carl

Woese as consultants), the majority of them NASA Exobiology grantees. The report expressly set out to reconceptualize exobiology in light of new findings from 1986 spacecraft to Halley's comet, new consensus that the primitive atmosphere was probably not reducing (pp. 80–81), theories that hydrothermal vents could serve as good sites for prebiotic synthesis (p. 81), the possibility of clays as initial genetic systems/sites of synthesis (pp. 85–86), findings of the ECHO Report (including the K-T asteroid theory, pp. 100–101), RNA world issues, among other things. The authors concluded that "at the very least, research on the possible effects of large-body impacts has sensitized the scientific community to think more in terms of cosmic influences on Earth systems."[80] A very similar note was struck by Chris Chyba in 1992: "Missions to Halley's comet [turned exobiology thinking outward from Earth, but] perhaps just as important was the psychological effect of the suggestion made in 1980, that a large impact played a role in the extinction of the dinosaurs. After this provocation impacts' possible role throughout Earth's history began to be examined in earnest."[81]

Indeed, it was in July 1986, just as the NAS SSB Committee was beginning this reassessment process, that Carl Sagan proposed to his new grad student Chris Chyba that Chyba "attempt a quantitative analysis of the role of infalling organic compounds from comets, meteorites, and cosmic dust in the origin of life." This became Chyba's doctoral thesis.[82] Chyba quickly joined the stable of up-and-coming talent funded by Exobiology, now under the direction of John Rummel, who took over from DeVincenzi in 1986. According to Rummel, Chyba's work was strongly attacked by Stanley Miller and his former student Jeff Bada. But the "shouting matches" between Miller and new approaches, in Rummel's view, could often be scientifically fruitful. He cited both Chyba's work on exogenous delivery of extraterrestrial organics and Everett Shock's work on the possibility of prebiotic organic synthesis at hydrothermal vents: "Chris had some very good results about how much cosmic dust had been raining down on the planet for a long time and the potential for that to bring in organics. Stanley was of the opinion that anything that brought in organics that wasn't the Miller-Urey experiment was somehow disrespectful. . . . It was funny to hear Stanley tell you about how anything brought in from outer space would be destroyed by deep sea vents anyway and so why should we bother with that sort of thing and of course so was all the stuff that was produced in the atmosphere. . . . Jeff Bada and Stanley to some degree, their disagreements with Everett Shock about the potential for hydrothermal vent systems to generate organic compounds has always been an interesting one. That's more of the same."[83]

Thus, as new approaches developed in Exobiology under Rummel's watch and as Miller-Urey type experiments seemed less relevant or out-of-date to much of the new younger generation of researchers, the Miller school, centered at the University of California San Diego (UCSD) and nearby Scripps Institute of Oceanography, fought back to maintain a prominent place in the field. By 1992 its members had organized a large research group with five main principal investigators (PIs) and their twenty students and had negotiated with NASA to create

TABLE 5.2 *Exobiology Budget History (in Thousands of Dollars)*

Program Component	Fiscal Year						
	1986	1987	1988	1989	1990	1991	1992
Exobiology baseline R&A	4,340	4,705	4,908	(5,050) 4,742[a]	5,076	5,423	6,294
Exobiology NSCORT	—	—	—	—	—	—	925
Exobiology flight (SSEX, GGSF)	0	434	550	760	657	1,100	2,760
SETI Microwave Observing Project	1,574	2,175	2,403	2,260	4,233	11,500	12,250
Total	5,914	7,314	7,861	7,762[a]	9,966	18,023	22,229

[a] After "Appropriations Integrity."

a new entity called NSCORT (NASA Specialized Center of Research and Training).[84] Table 5.2 shows steady growth in expenditures during the years of Rummel's tenure as Exobiology chief, including the first year of NSCORT funding.[85]

The principal investigators in the Exobiology NSCORT group are Stanley Miller at UCSD, Leslie Orgel at Salk Institute for Biological Studies, Gustaf Arrhenius and Jeffrey Bada at Scripps Institute of Oceanography, and Gerald Joyce at Scripps Research Institute. From its creation it has continued to be funded in the one million–dollar per year ballpark, under the aegis of Michael Meyer, Rummel's 1992 replacement as the fourth Exobiology chief in the "Dynasty." NSCORT was designated a "virtual center," with the purpose of encouraging more collaboration among the five senior researchers and twenty students spread over four separate institutions. In this sense it pioneered the "virtual center" idea that NASA expanded so dramatically with the creation in 1997 of the virtual Astrobiology Institute, linking research groups all over the country. J. William Schopf at UCLA is a supportive reviewer of the NSCORT group (Miller was a member of his 1979–1980 PPRG). One of its most central functioning institutions has been a biweekly journal club for the twenty students, to which the senior PIs "specifically are 'disinvited.'"[86]

Many of the Miller/Bada points of view, such as their profound skepticism about "ventists" having anything relevant to say about origin of life, are staked out clearly in the book coauthored in 2000 by Bada (with Christopher Wills), *The Spark of Life: Darwin and the Primeval Soup*. Here Bada also defends the possibility of a reducing atmosphere on the primitive Earth to a degree not supported as enthusiastically anywhere outside San Diego.[87] The NSCORT group is fairly negative in its attitude toward Cairns-Smith's "clay genes" origin scenario; however, its members think plausible J. D. Bernal's earlier, more modest suggestion that clays may act as catalysts upon which the first organic polymers may have been built up from their monomers.[88] Although Bada allows more credit for these approaches than Miller (he says some kind of "genetic takeover" scenario was probably likely, even if not from clay genes), essentially, they are still the "analytikers" that Lynn Margulis labeled them in 1973; less precise, controlled approaches still smack to them of the messy "gemischer" approach.

The RNA World

In the fall of 1982 a paper was published announcing the discovery that certain small RNA molecules in the protist *Tetrahymena* were capable of acting as enzymes, not just information-carrying molecules.[89] One of the authors, Tom Cech of the University of Colorado, was soon contacted by Cliff Brunk of the UCLA biology department, a member of Schopf's research group. The Schopf group wanted Cech to come down to UCLA and give a talk on the "ribozymes," as the catalytic RNA molecules had been dubbed, because of the discovery's extremely suggestive implications for the origin of life. Cech gave the talk on 16 November 1983; according to him, "I didn't even know what origin of life research was at the time! I was unfamiliar with the key work of Leslie Orgel, also of Manfred Eigen. The UCLA visit was an important learning experience for me, making me aware that there were these earlier ideas and I'd better know about them. Prior to then, I hadn't thought any farther back than 'a primordial organism.' . . . Is the work important for origin of life? The consensus is 'yes,' the truth is 'we don't really know.'"[90]

Cech and his work received an enthusiastic reception; soon word spread through the origin of life research community. There was cautious optimism that this might validate the "RNA World" scenario suggested by Leslie Orgel (see fig. 4.3) fifteen years previously, that is, that the chicken-egg paradox (of how to get a protein catalyst–DNA information system up and going, when both parts depend upon the other in order to be made and to function) could be resolved if a simpler molecule such as RNA could possibly be an earlier stage, if it could only be shown that RNA could act as an enzyme, in addition to its known information-carrying functions. Schopf recalls that attendees did not just walk out immediately seeing Orgel's RNA World had come into full bloom; rather, "Folks, I think, were a bit skeptical about the RNA World implications. Remember that in the origin-of-life business, 'seemingly good ideas' are plentiful; what takes the time and effort is to show that a 'good idea' has a counterpart in reality. For the RNA World, that came slowly, gradually, and somewhat later."[91]

Nonetheless, within a year or two, caution had been largely replaced by enthusiasm; there was a tremendous blossoming of research into the possibilities of the RNA World scenario.[92] By 1989, a remarkably short seven years after the first papers independently discovering ribozymes, Cech and Sidney Altman (leader of the other group, at Yale) were awarded the Nobel Prize in chemistry for the work. Perhaps the most significant reason the work was thought so important was its origin of life implications.

By 1991, however, hardly had the Nobel checks gone into the bank, when serious problems began to emerge, such as Cech hints at in his quote. Gerald Joyce of Scripps Research Institute had been a student in Orgel's lab in the late 1960s when Orgel first proposed the RNA World idea. Now he published an article explaining that the questions left unanswered about how to get to an RNA World were still so great that it was not any kind of answer to the original origin of life.[93] He began researching the pre-RNA world, or how to get to RNA to begin with and how protein synthesis could have evolved using RNA.

The problems with prebiotic synthesis of RNA were numerous. For one thing, the Miller group meticulously documented that the half-life of ribose, the key sugar needed, was very short under prebiotic conditions; it simply would not remain around long enough, even if formed, to react with other molecules to form nucleosides and nucleotides, let alone an RNA polymer. Leslie Orgel, in a more recent review, concluded there are still at least eight major difficulties in the chemical steps needed to form RNA.[94] These have been summarized by biochemist John Cronin as follows:

1. Ribose is only a minor product among many sugars produced by simple prebiotic reactions, e.g., the formose reaction.
2. Ribose is not very stable.
3. Phosphate is possible in only low concentrations in prebiotic oceans due to the insolubility of calcium phosphate.
4. There are apparently no good prebiotic routes to the pyrimidine nucleosides.
5. Positionally specific phosphorylation of nucleosides is difficult prebiotically.
6. How could nucleotides have been activated for polymerization? A thermodynamic problem.
7. A paradox: In ribozymes considerable chain length is required for replicative fidelity, but fidelity could only be realized in short chains by an error-prone primitive ribozyme.
8. The concerted effects of some or all of the above.[95]

According to Cronin: "The skepticism about an RNA world is not skepticism toward the possibility that in the course of its early evolution life went through a period in which RNA catalysis (ribozymes) was important or maybe even dominated biochemistry, but rather toward the idea that this biochemistry was primitive, i.e., represented first life. It is widely believed now that there were necessarily preRNA worlds."[96] Stanley Miller's group, for example, "has been interested in fashioning a pre-RNA that does not rely on the traditional pyrimidines and purines. . . . Another possible pre-RNA that the NSCORT researchers have been studying is peptide nucleic acid."[97] Woese's fruitful line of investigation, tracing back toward the last common ancestor and its very early form of 16s rRNA, also guarantees RNA study a prominent place in future studies.[98] Thus, an RNA World has now become a significant chapter in the story of the origin of life on Earth. The very first chapters in that story, however, remain unknown and the subject of speculation and differing camps of thought.

Although this survey of exobiology and origin of life ideas since *Viking* has not attempted to be comprehensive, it shows clearly that major reorientations have occurred during the past twenty-five years. The conceptual shifts are profound fundamental underpinnings of the new, more comprehensive discipline of astrobiology. In particular, the study of extraterrestrial bodies and the effects of their impact on Earth as well as the study of environmental conditions broadly and how they coevolve with living systems from the very first origin of those systems have both moved to the front burner as never before. There is now a prominent role for catastrophist impact thinking, for thinking about life at extraordinarily high temperatures and other extreme conditions, and for Earth System Science (or Gaia-type ideas, if one prefers) about the tightly linked evolution of living organisms and the planet on which they arise. All of these ideas seemed marginal or even heretical twenty-five years ago.

CHAPTER 6

The Search for Extraterrestrial Intelligence

From the beginning of the extraterrestrial life debate its most exciting and controversial aspect was the search for intelligence.[1] Unlike microbes, intelligence holds the potential for tapping into the experience and knowledge of other minds in answering the great questions of the universe. By the beginning of the Space Age the hypothesis of the American astronomer Percival Lowell that intelligent Martians had built canals on their dying planet, as well as the debate over unidentified flying objects (UFOs), had shown just how controversial the subject could be. Still, if a method could be found for confirming the existence of extraterrestrial intelligence, it would leapfrog theories of the origins of planets and life and go directly to the Holy Grail—minds similar to or different from ours but capable of contemplating the universe.

With the development of new techniques and detectors in radio astronomy, such a method became feasible just as the Space Age began. Although it was not part of NASA's early plans, the Search for Extraterrestrial Intelligence (SETI) was a logical extension of the search for microbial life and origins of life research. It was only a matter of a dozen years before this logic began inexorably to work its way into NASA thinking. Once it did, it proved so controversial that the idea saw a long phase of study, followed by a minimal and then considerable research and development program, only to be terminated by congressional politics with a tiny fraction of the proposed observational program completed. The story of SETI in NASA is a story of high ideals, internal and external politics, and ultimate disappointment. But it is a story that must be viewed in the larger context than NASA and even national politics and whose end has not yet been written, perhaps even within NASA.

Origins of NASA SETI: The Study Phase, 1969–1982

During the first decade of its existence NASA showed little interest in searching for interstellar communications. The space agency naturally had a greater interest in the immediate prospects for exobiology in our solar system, and, as we have seen, embraced the direct search for life in the solar system very early in its history. The paper "Searching for Interstellar Communications," published in *Nature* in 1959 by the physicists Giuseppe Cocconi and Philip Morrison one year after the founding of NASA, held little interest for an agency focused on planetary exploration. Even Frank Drake's first radio search for such communications in 1960, poetically known as "Project Ozma," passed virtually unnoticed at the space agency. A 1961 meeting on interstellar communication, sponsored by the National Academy of Sciences at Green Bank, West Virginia, did include two NASA employees, astronomers A. G. W. Cameron and Su-Shu Huang, both experts on planetary system formation. But their participation was based on individual interest and expertise, not NASA planning. Still, a meeting in 1963 on "Current Aspects of Exobiology," held at the Jet Propulsion Laboratory (a NASA-funded contractor administered by the California Institute of Technology) and devoted almost entirely to planetary exploration, included Drake's paper "The Radio Search for Intelligent Extraterrestrial Life." This signaled a potentially broader interpretation for exobiology; it was not, however, one that NASA was yet ready to incorporate into its programs.[2]

NASA's first publicly expressed interest in SETI came in 1970, not from planetary exobiologists but from an expert in space medicine, an area of responsibility at NASA's Ames Research Center in California. The person who would play a pivotal role in launching and sustaining a SETI program within NASA was John Billingham, a physician who had worked on the *Apollo* program space suits and now headed the Biotechnology Division at Ames. Billingham had obtained his medical degree from Oxford in 1954 and had spent six years at the Royal Air Force Institute of Aviation Medicine at Farnborough, where he researched physiological stresses imposed on aircrews under conditions of high speed and high altitude, especially heat stress. His work on aviation medicine brought him frequently to the United States, where he represented the Royal Air Force at scientific meetings and joint meetings with the U.S. Air Force. His interest in space medicine was spawned by *Sputnik,* which prompted him to submit to the British Interplanetary Society several papers on the control of cabin conditions for spacecraft and the protection of astronauts from the severe conditions on the Moon. These published papers brought him to the attention of NASA, and in 1963 he became chief of the Environmental Physiology Branch of the Crew Systems Division at Johnson Space Center in Houston. It was here that he tackled the physiological and medical problems associated with the *Mercury* and *Gemini* flights and played an early role in the design requirements for the *Apollo* spacesuits.[3]

After three years in Houston, Harold "Chuck" Klein invited Billingham to come to Ames as an assistant chief in the Biotechnology Division of Ames Life Sciences. Drawn by advanced research and development focus at Ames, as opposed to the more immediate operational duties in Houston, Billingham now worked in much the same area but with applications to future spaceflight. The Biotechnology Division was only one part of Life Sciences at Ames. On the top floor of the Life Sciences Building was the Exobiology Division, which Klein had headed before taking over as chief of all Life Sciences at Ames. Because they were located in the same building, Billingham ran into these "strange and interesting people" who were working on chemical evolution and the origin and evolution of life. Among the forty or fifty people working in the division at the time were Cyril Ponnamperuma, Sherwood Chang, and Richard S. Young.

Through these interactions Billingham became increasingly intrigued with extraterrestrial life and was led to the recent book by Joseph Shklovskii and Carl Sagan, *Intelligent Life in the Universe* (1966). "I read it from cover to cover, and it's one of those things that one remembers very vividly. I sat back and said 'Wow!'" This book in turn led him to the work of Frank Drake, Philip Morrison, the Green Bank conference, and a half-dozen others. "Then I sat down, and over a period of some months it began to dawn on me that nobody had asked a key question. And the key question was, if you were serious about conducting a search for other intelligent life, how would you do it? . . . how would you do it if you wanted to make it a really large-scale enterprise? I mean a very thorough enterprise, instead of a shoestring operation." Billingham had made a crucial realization: "In the back of my mind, I guess I also had this notion that, 'Gee, NASA is supposed to explore space, and here I am sitting in NASA and here are all these people on the top floor who are studying exobiology, only they're thinking about microbial life. If there's anything in this business of searching for intelligent life, maybe one should ask a second question, and that is, if indeed there is a way to put together a thoroughgoing approach, is it also possible that NASA at some future time may actually become interested in adding SETI to its existing base of scientific activity.'"[4]

Thus were the seeds for the NASA SETI program planted. Before proceeding any farther, Billingham took the prudent step of convincing Ames director Hans Mark that the subject might be worth pursuing at Ames. But Mark urged caution; before any major study, a mini-study of the problem of interstellar communications should be undertaken. This was done in the summer of 1970, concurrently with a more visible NASA-sponsored weekly lecture series on interstellar communication, also organized by Billingham. The speakers for the latter project included Carl Sagan on interstellar communication, A. G. W. Cameron on planetary systems, Cyril Ponnamperuma on chemical evolution, Ronald Bracewell on interstellar probes, and Frank Drake on the search strategy with radio telescopes. The results, published in 1974 under the title *Interstellar Communication,* documented for the first time in a public way NASA's early interest in the subject.[5]

A lecture series was one thing, a NASA program quite another. In this sense Billingham's mini-study took on importance beyond its inconspicuous beginnings. The study produced optimistic results and led to a decision to conduct a full-scale study the following summer as part of a summer faculty fellowship program in engineering systems design sponsored by NASA, Stanford, and the American Society of Engineering Education (ASEE). Billingham and his Stanford colleague James Adams had been running this fellowship program at Ames since the mid–1960s; the program would run for twenty years and would be one of NASA's important contributions to education.

For the interstellar communication summer study, with the advice of Hans Mark, Billingham and Adams now brought in Bernard Oliver. Oliver, an electrical engineer, vice president for Research and Development at Hewlett Packard and a participant in the famous Green Bank meeting in 1961 on interstellar communication, was Billingham's senior by fourteen years. He, too, would become crucial to NASA's SETI program. As early as grammar school in Aptos, California, Oliver was an avid science fiction reader. There, he recalled, "I certainly got the theme of a populated universe, and the concept of interstellar travel, of course, is what we all dreamed of in those days." He was a believer in extraterrestrial life, even though Sir James Jeans was at that time proposing the rarity of planets and life in the universe. Oliver obtained his degree in electrical engineering from CalTech and Stanford, and went east to work for Bell Labs on automatic tracking radar. It was during this work, as early as 1950, that he was astonished to learn by his own calculations that the ten kilowatt powers they worked with on radar could communicate anywhere in the solar system and with some further capability might even reach the nearest stars. Oliver left Bell Labs and went to Hewlett Packard in 1952, but he never forgot the implications of his calculation. After reading about Frank Drake's Project Ozma in a news magazine, he visited Green Bank and attended the first conference on interstellar communication there in 1961. As president of the Institute of Electrical and Electronic Engineers (IEEE) in the mid–1960s, Oliver traveled the country giving talks on interstellar communication and its feasibility, because it was, as he recalled "still hot on my mind." Already at this time he had the concept of using a large array of antennas for this purpose. He was one of those invited to Ames for the 1970 lecture series, where he spoke on "Technical Considerations on Interstellar Communications." His enthusiasm, combined with his technical expertise, was infectious: "Once our society becomes convinced of the existence of intelligent life elsewhere in the galaxy," he wrote, "we will embark on the greatest voyage of discovery in all our history."[6]

Thus, Bernard Oliver, "Barney" as his colleagues knew him, became the technical genius behind what came to be known as " Project Cyclops." Billingham and Oliver made sure that the twenty faculty they gathered from around the country in the summer of 1971 included those with expertise in details of antenna elements, receiver systems, and signal processing as well as more general problems about the probability of life in the universe and search strategies. By the end of the summer Oliver and his colleagues had produced plans for a detector consisting in its final stages of an "orchard" of perhaps one thousand one hundred–meter antennas covering a total area some ten kilometers in diameter. Cyclops was an ambitious project, but the system had the capability of starting out small and building more if the first few antennas detected no signals.

The Cyclops report is important for many reasons, ranging from the technical to the inspirational. It explicitly set forth the premises that by now were part of the "orthodox view" of extraterrestrial life proponents: that planetary systems were the rule, rather than the exception; that many planetary systems would contain at least one planet in the stellar "ecosphere," where temperatures are moderate enough to allow an oxidizing atmosphere and liquid water on the planetary surface; that organic precursors of life would form in abundance either from the primordial atmosphere or from material deposited by carbonaceous chondrites; that main sequence stars cooler than F5 spectral type would have lifetimes sufficiently long for biological evolution; and that intelligent life would evolve in these stellar systems. The report also suggested that we have no way of knowing the longevity of technological civilizations other than by making contact with them and that interstellar contact may greatly prolong the lifetime of races by "sharing an inconceivably vast pool of knowledge." Access to the "galactic heritage," Oliver wrote, "may well prove to be the salvation of any race whose technological prowess qualifies it."[7]

Among the fifteen conclusions of the Cyclops report were that signaling was vastly more efficient than interstellar travel; that the microwave region between one and three billion hertz (1–3 gigahertz) was the best place to search for such signals from the Earth's surface; and that the region between the spectral lines of hydrogen (1420 MHz) and the hydroxyl radical (1665 MHz) was a natural "waterhole" frequency for communication because there was less interference from natural radio waves. The report found it technologically feasible to build a phased array for interstellar communication across intergalactic distances and concluded that any directed beacon would most likely be circularly polarized and highly focused ("monochromatic") with spectral widths of one hertz or less. This last conclusion called for a high-resolution detector, and one of the major contributions of the Cyclops system was to propose a signal-processing system to analyze the two hundred–megahertz (MHz) bandwidth of the waterhole with a resolution not exceeding one hertz. Even concentrating on the waterhole, two hundred million channels would have to be searched. Rejecting scanning spectrum analyzers and the Fast Fourier Transform as too slow or too expensive, the report concluded that an optical spectrum analyzer would carry out the job most efficiently. This scheme, which made use of photographic film, an optical Fourier Transform, and a high-resolution vidicon tube, would still have required two hundred optical spectrum analyzers. The cost of the entire ambitious undertaking was six to ten billion dollars over ten to fifteen years. This cost estimate doomed Cyclops to any development effort in the real world. The fact that it could start out small and expand later was lost in the several billion-dollar price tag for the total project. Nevertheless, the Cyclops study marked a watershed in the application of technical expertise to the problem of interstellar communications. And, aside from its technical contributions, the Cyclops report came to an important administrative conclusion: that the search for extraterrestrial intelligence should be established "as an ongoing part of the total

NASA space program, with its own funding and budget." Toward this end, with the approval of Mark, in late 1972 Billingham began a Committee on Interstellar Communication.[8] By March 1973 the committee had produced "A Program for Interstellar Communication," Phase A of an Interstellar Communication Feasibility Study. By March 1974 it had a more comprehensive "Proposal for an Interstellar Communication Feasibility Study." The resulting documents remained unpublished, but briefings by both Oliver and Billingham to NASA administrator James Fletcher, chief scientist Homer Newell, and NASA's Office of Aeronautics and Space Technology (OAST) led to funding of $140, 000 from the latter in August 1974. Fletcher was supportive; the previous year he had written that "it is within the realms of possibility, in fact, likely that technically advanced civilizations may exist on the planets of distant stars. Communications with such far-off islands of intelligence may someday be begun, with effects on man's home planet that can now be only imperfectly imagined."[9]

With minimal funding in hand, at the beginning of 1975 Hans Mark formed an Interstellar Communications Study Group consisting of Billingham, astronomers Charles Seeger and Mark Stull, and Vera Buescher. Buescher was "the planet's first full-time interstellar secretary," as Billingham later put it, "the glue which held us all together." Others, including Oliver, David Black, and John Wolfe, remained closely associated with the group. The OAST funding was used primarily for a series of six SETI science workshops chaired by Philip Morrison, two further workshops on extrasolar planet detection, and one workshop on cultural evolution.[10] These workshops proved to be another landmark in SETI history and a critical stimulus to enlisting support by the wider scientific community (fig. 6.1). It was also during these workshops that the acronym *SETI* was adopted, "to differentiate our own efforts from those of the Soviet Union and to emphasize the search aspects of the proposed program." The Soviets had previously discussed communication with extraterrestrial intelligence, or CETI, but Billingham and his colleagues were sensitive to the fact that "communicating" was politically more explosive than merely searching. Sober scientists might undertake the search, but, if it came to communication, a much broader spectrum of society needed to participate. That was one issue that need not be addressed in an embryonic SETI program.

Having considered interstellar travel, robot probes, and electromagnetic signals, the Morrison report confirmed that radio signals were the optimum method for interstellar communication. It showed graphically the "free space microwave window" and the "terrestrial microwave window," indicating the best frequencies for interstellar communication, taking into account the Earth's atmosphere; these charts would appear repeatedly in SETI literature as justification for narrowing the frequency dimension of the search. The report also recognized that the search for signals had to be limited in direction or frequency or both. Although no consensus was reached on a search strategy, the report gave the first public discussion of a possible bimodal method for the search, a detailed look at selected target stars and a broad-brush all-sky survey, which became the

FIGURE 6.1. Members of the Science Workshops on Interstellar Communication, also known as the "Morrison workshops," 1975–1976, photographed in front of the Life Sciences Building at NASA Ames. *Front row:* Frank Drake, A. G. W. Cameron, Philip Morrison (chair, holding SETI license plate), Ron Bracewell, Bruce Murray. *Second row:* Bernard Oliver, Harrison Brown, Jesse Greenstein, Fred Haddock, Eugene Epstein, John Billingham. *Third row:* Bill Gilbreath, Yoji Kondo. *Fourth row:* Sam Gulkis, John Wolfe, Charles Seeger, Robert Edelson, Gerald Levy. *Back row:* Vera Buescher, Mark Stull, H. R. Brockett, Robert Machol. Their deliberations resulted in the landmark volume *The Search for Extraterrestrial Intelligence* (1977), known informally as the "blue book." (Courtesy SETI Institute.)

hallmark of the NASA program. As opposed to the optical spectrum analyzer of the Cyclops report, the Morrison report noted that large-scale integrated circuit technology had improved so much in the five years since Cyclops that "it now appears possible to build, at reasonable cost, solid state fast Fourier analyzers capable of resolving the instantaneous bandwidth into at least a million channels on a real time basis." This was to be a crucial point that would be the basis for the NASA SETI hardware. As we shall see in the next chapter, the studies of 1975–1976 also revived interest in the possible existence of extrasolar planetary systems and stimulated another NASA/ASEE summer study of a method for detecting them.[11]

Like the Cyclops report, the Morrison workshops reached a number of important administrative conclusions. The participants agreed that "it is both timely and feasible to begin a serious search for extraterrestrial intelligence." They also argued that the search fell under NASA's mandate:

> It is particularly appropriate for NASA to take the lead in the early activities of a SETI program. SETI is an exploration of the Cosmos, clearly within the intent of legislation that established NASA in 1958. SETI overlaps and is synergistic with long-term NASA programs in space astronomy, exobiology, deep space communication and planetary science. NASA is qualified technically, administratively, and practically to develop a national SETI strategy based on thoughtful interaction with both the scientific community and beyond to broader constituencies.

Accordingly, Hans Mark established a small but formally constituted SETI Program Office at Ames Research Center in 1976, within the Extraterrestrial Research Division formed in that year from the Exobiology Division. Headed by John Billingham, aided by John Wolfe, Mark Stull, Vera Buescher, and Mary Conners, and made possible by the continuing support of Hans Mark (director of the Ames Research Center) and Harold Klein (director of Life Sciences at Ames), this was the first institutionalization of SETI within NASA.

The mention of deep space communications and planetary science in the Morrison report and the discussion of a bimodal strategy signaled the interest of the Jet Propulsion Laboratory and the support of its prospective director, Bruce Murray, for SETI. Thus, at both JPL and Ames the innovative SETI programs stemmed from the personal interest and support of the new directors. The interest at JPL developed naturally, since JPL ran the Deep Space Communications Complex (part of the Deep Space Network) and had expertise in the radio astronomy needed for SETI. But the crucial ingredient was Murray, who, as professor of planetary science at CalTech, had participated in the Morrison Workshop on Interstellar Communication in April 1975 dealing with planet detection. After discussions with the JPL radio astronomy group about what role JPL might play in SETI, Murray championed a Sky Survey strategy against the skepticism of the Ames group, which pushed for a more traditional Targeted Search. In the fourth workshop in December 1975 Billingham and Seeger had presented a paper on "Ames-JPL Plans" for a detector. By 1977 JPL had a SETI office, headed by Robert Edelson, generating ideas about how JPL should contribute. Jill Tarter, who would later emerge as the project scientist, joined the SETI team from the University of California–Berkeley about this time. After some initial conflict an Ames-JPL partnership emerged that would become a major feature of NASA's formal SETI program.[12]

As with any project, funding was the perpetual problem constantly in the forefront if any progress were to be made. Thus began the selling of SETI. *Outlook for Space,* a report prepared in 1976 by contributors from all the NASA centers to guide NASA's thinking for the next twenty-five years, viewed investigations into the origin and existence of life, whether microbial or intelligent, as an important part of NASA's space objectives through the end of the century. Such statements appear in planning documents only after considerable lobbying by proponents. Again, Billingham had played an important role on this committee, resulting in SETI's first significant appearance in a formal NASA study at a high planning level. The possibility of increasing the scope of NASA's exobiology program from the search for microorganisms within the solar system to the search for extrasolar planetary systems and radio signals from extraterrestrial intelligence—from the confines of the solar system to the entire cosmos—was a breathtaking leap. But by 1976, the year of the *Viking* landers and the bicentennial of the United States, SETI was becoming respectable in NASA, if only in the smallest of ways.[13]

Propelled by the Morrison workshops and emboldened by *Outlook for Space,* Billingham and others sought to devise a program that might be funded. SETI would be significantly unlike most NASA endeavors. It would have no spacecraft, no launch risks, and no possibility of equipment failure in space. Political and economic realities and the revolution in digital electronics dictated that SETI would have no Cyclops system with a vast collecting area. Instead, the embryonic program would use existing radio telescopes to which would be attached specialized detectors and signal-processing apparatus whose construction would be the main objective of the funding. The proposed total cost of the SETI program as calculated in the late 1970s, including five years of research and development and ten years of operation, would be about one hundred million dollars, some 10 percent of the billion-dollar *Viking* project but roughly equal to the cost of *Viking*'s biological experiments.

In June 1979, with the possibility of significant funding on the horizon, NASA sponsored a landmark conference at the Ames Research Center on "Life in the Universe," the conference that also played an important role in furthering the Gaia concept. With the impetus provided by the Morrison workshops, NASA by this time had formally adopted a search strategy—the bimodal strategy that not only made sense scientifically but also satisfied the desire of both JPL and Ames to work on the project. Billingham and Wolfe at Ames and Edelson at JPL coauthored the paper given at the 1979 conference, the first to lay out the NASA program in detail. Referring to their "modest but wide ranging exploratory program," the authors described a ten-year effort "using existing radio telescopes and advanced electronic systems with the objective of trying to detect the presence of just one signal generated by another intelligent species, if such exists." Again the emphasis on detection was significant, since NASA was not prepared to communicate. JPL would undertake Murray's Sky Survey at frequencies from one to ten gigahertz (nine billion channels), while Ames would concentrate with more sensitivity on the Targeted Search among some seven hundred–plus stars within twenty-five parsecs (eighty light-years) of Earth.[14] Its one to three billion hertz encompassed two billion single-hertz channels.

In their joint paper the Ames-JPL authors characterized the concept of intelligent life as a hypothesis widely held in the scientific community. They

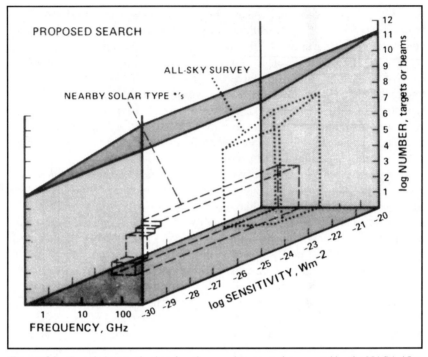

FIGURE 6.2. Cosmic haystack, showing the search space to be covered by the NASA / Jet Propulsion Laboratory Sky Survey and the NASA Ames Targeted Search. The Targeted Search was designed to have greater sensitivity, while the Sky Survey would observe in more directions and over a broader frequency range. Both were terminated in 1993, with parts of the Targeted Search continued by the Project Phoenix sponsored by the SETI Institute. (Courtesy NASA.)

viewed the hypothesis as resting on two postulates: that life is a natural consequence of physical laws acting in appropriate environments and that a physical process that occurs in one place (as on Earth) will occur elsewhere. As a practical matter, the group also adopted the assumption that some fraction of extraterrestrials would be "providing an electromagnetic signature we can recognize." They pointed out that, although many searches had been undertaken with comparatively primitive data-processing systems, the NASA system could achieve a ten million–fold increase in capability over the sum of all previous searches. And they recommended a major effort to develop the necessary equipment. The key instrument, known as the Multi-Channel Spectrum Analyzer (MCSA), and its software algorithms were the heart of the system, the means by which the "cosmic haystack" could be searched for its "needle." A three-dimensional graphical representation of the cosmic haystack in this article first dramatically depicted the magnitude of the task (fig. 6.2). Having examined several spectral analysis techniques, the group agreed with the Morrison study that "the digital approach is far superior in terms of capability, flexibility, reliability, and cost."

FIGURE 6.3. SETI Science Working Group, 1981. *Front row:* Sam Gulkis, Eric Chaisson, Frank Drake (chair), Jill Tarter, Don Beem, Peter Boyce. *Back row:* Woody Sullivan, Bernie Burke, Mike Davis, George Swenson, Ben Zuckerman, Jack Welch. (Courtesy SETI Institute.)

By 1979 the Ames-JPL group had a detailed idea for a coherent SETI program but not much money to carry it out.[15]

During the 1970s NASA had studied the SETI problem; during the 1980s, the Ames and JPL groups continued the push to implement the recommendations of the studies. Studies and refinements would continue, notably in meetings during 1980 and 1981 of a SETI Science Working Group (SSWG), composed of radio astronomers and engineers who could provide essential independent review and advice. Headed by John Wolfe of Ames and Sam Gulkis of JPL, this working group once again confirmed the microwave region as preferable, endorsed the bimodal strategy, envisaged a five-year R&D effort to design, develop, and test prototype instrumentation, and examined in more detail the instrumentation and strategies required (fig. 6.3).[16]

In the end, however, no amount of study would get the job done. To convert concepts and discussion into hardware and software required funding. And before funding was forthcoming NASA still had to overcome skepticism both from the scientific community and from Congress. In this effort they were not helped by broader events. Even as the Morrison workshops were under way in

1975, a broad challenge to the basic assumptions of SETI was launched. In particular, Michael Hart and David Viewing independently argued that, if interstellar travel is taken seriously and given the immense astronomical time scale available, the fact that there are no intelligent beings from outer space on Earth is an observational fact that argues strongly that extraterrestrials do not exist. Given the age of the universe and the time needed for intelligence to develop, Hart and Viewing proposed, extraterrestrials should have populated the galaxy. At a velocity of one-tenth the speed of light, Hart argued, this would have occurred in a mere one million years. Moreover, the argument required only one spacefaring extraterrestrial civilization. The existence of the thousands proposed by SETI proponents was implausible because it was unlikely that every advanced civilization had chosen not to engage in space travel or had destroyed itself in nuclear war. The bottom line, if this rationale held, was that "an extensive search for radio messages from other civilizations is probably a waste of time and money."[17]

The "where are they?" argument, minus Hart's conclusions, had been first casually raised in conversation by the physicist Enrico Fermi in 1950. Known as the "Fermi paradox," it gathered momentum during the 1970s in parallel with NASA's plans for a SETI program. By 1979 an entire conference was devoted to the question of "where are they?" centered on the Fermi paradox. The argument was elaborated and emphasized especially by physicist Frank Tipler, who took the extreme position that the logic was so compelling that it was a waste of taxpayers' money to undertake a search. In 1983 astronomer and science fiction writer David Brin termed the paradox the "Great Silence" and reviewed the scenarios that might account for it in terms of a modified Drake equation, taking into account a "contact cross-section" between extraterrestrials and contemporary human society.[18]

Meanwhile, skepticism in Congress was also proving a hindrance. In early 1978 the program unexpectedly received Senator William Proxmire's notorious Golden Fleece Award for "the biggest, most ironic, or most ridiculous example of wasteful spending." Proxmire, chairman of the Senate Appropriations Subcommittee, with jurisdiction over NASA funds, stated that NASA, "riding the wave of popular enthusiasm for 'Star Wars' and 'Close Encounters of the Third Kind,' is proposing to spend $14 to $15 million over the next seven years to try to find intelligent life in outer space. In my view, this project should be postponed for a few million light years." Proxmire noted that there was not a scintilla of evidence for life beyond the solar system, that even if living beings existed they were so distant that they would be dead and gone by the time we received a message, and that Earthlings had enough difficulty communicating with one another. He particularly objected to the costs associated with the JPL Sky Survey and suggested that "at a time when the country is faced with a 61 billion budget deficit, the attempt to detect radio waves from solar systems should be postponed until right after the federal budget is balanced and income and social security taxes are reduced to zero. After detailed congressional hearings in September 1978, the Subcommittee on Space Science and Applications of the House Committee on Science and Technology supported NASA's proposal to initiate a SETI program. The Golden Fleece had done its damage, however; the House and Senate Appropriations committees elected not to provide any money.[19]

NASA bridled at such criticisms in unusually stark terms. In their aftermath NASA administrator Robert Frosch wrote: "It is a time of the 'golden fleece' for SETI, and I presume it will be a time of golden fleeces for other things we try to do. The 'golden fleece' idea, the idea that searches, gropings for knowledge whose purpose we do not understand are silly and some kind of a ripoff, results from sheer lack of understanding, lack of imagination, and lack of perception of the meaning of the history of the human race."[20]

NASA continued to fund SETI at a subsistence level after 1979 until thwarted again by Senator Proxmire, who (this time being affected by Tipler's argument) on 30 July 1981 placed an amendment on the floor of the Senate which provided that no FY 1982 funds should be used to support SETI. "Three years ago, NASA requested $2 million for a program titled SETI. The idea was that they are going to try to find intelligence outside the solar system. Our best scientists say that intelligent life would have to be beyond our galaxy. I have always thought if they were going to look for intelligence, they ought to start right here in Washington." Proxmire was clearly peeved that the program had not been halted three years ago and offered the same arguments to terminate it finally now. "In this year of all years," he concluded, "we should not fritter away precious Federal dollars on a project that is almost guaranteed to fail." The amendment was unopposed, and, during the Joint House-Senate Conference on NASA's FY 1982 appropriations, Proxmire prevailed, effectively killing all funding for SETI for 1982. Frank Drake undoubtedly spoke for most SETI scientists when he wrote: "The ultimate irony is that while all of this has been taking place, Senator Proxmire has been frantically maneuvering to preserve excess subsidies to dairy farmers. Congress did not want this, but again he prevailed. The cost to the taxpayer for the excess subsidy, not the basic subsidy, is between $500,000 and $1,000,000 per day. Every two days enough funds to run SETI for a year are diverted to this end."[21]

Despite this setback, NASA boldly decided to return to Congress for full funding in FY 1983. The agency was supported by the "decadal review" of astronomy by the National Academy of Sciences, which recommended SETI as one of seven moderate programs that NASA should implement. Although the Hart-Viewing-Tipler arguments had precipitated a crisis in SETI thinking, proponents of the search had counterarguments that convinced many in the scientific community. Frank Drake and Barney Oliver argued that interstellar travel and colonization were too expensive and that radio communication was vastly more efficient across interstellar distances. Cornell astronomer Carl Sagan was among those who argued that, on some interstellar diffusion models, travel would be slower than Hart envisioned. And astronomer Michael Papagiannis argued that perhaps the extraterrestrials were in the vicinity of the solar system but undetected. Although uncertainties abounded in all of these arguments, the SETI proponents had one major characteristic of Western science on their side: empiricism. Philip Morrison expressed it as follows: "It is fine to argue about N [in the Drake equation]. After the argument, though, I think there remains one rock hard truth: whatever the theories, there is no easy substitute for a real search out there, among the ray directions and the wavebands, down into the noise. We owe the issue more than mere theorizing." This was a call repeated again and again as the NASA SETI groups sought funding from Congress.[22]

Back in that world of funding and politics, after activities that included a discussion between Sagan and Senator Proxmire which emphasized civilizations rather than science and again with the backing of Hans Mark (now deputy administrator of NASA), SETI funding was restored for FY 1983 at the level of 1.5 million dollars. Finally, NASA was ready to begin a sustained research and development program culminating in an operational system to search for extraterrestrial intelligence.

Building the NASA Program: Research, Development and Inauguration, 1983–1992

With funding at the level of about 1.5 million per year, NASA's Ames and JPL centers embarked on an intensive program, known initially as the Microwave Observing Project (MOP) and later, beginning in October 1992, as the High Resolution Microwave Survey (HRMS), to build the instrumentation necessary for their respective approaches to search for intelligent life. Building on the studies of the past decade, the goal of the Ames Targeted Search Element of the NASA SETI program was to search for artificial signals from eight hundred to a thousand solar-type stars within about one hundred light-years. Beginning with Arecibo, it would use the largest radio telescopes possible, observe each star for three hundred to a thousand seconds, and focus on the two billion channels in the one to three gigahertz region of the microwave spectrum. Because of practical limitations, it would process twenty megahertz of bandwidth at one time, necessitating that each star be observed one hundred times to cover the entire two gigahertz. The six simultaneous channel resolutions would range from one to twenty-eight hertz. The system would have the ability to detect either continuous wave or pulsed signals.

JPL's Sky Survey Element, on the other hand, made no assumptions about specific preferred targets in the sky but was designed to observe the entire sky at 1 to 10 GHz with smaller, thirty-four-meter class radio telescopes beginning with those of the Deep Space Network. Because it had a broader spectrum to cover (9 GHz rather than 2 GHz for the Targeted Search), the fully operational system was designed to process 320 MHz of bandwidth at the same time, with 20 Hz channels. The prototype system inaugurated in 1992 was capable of processing 20 MHz for each polarization. The Sky Survey observational strategy was to examine each spot in a tessellated "racetrack" pattern for only a few seconds at most, resulting in a sensitivity one hundred times less than the targeted search and losing the ability to detect any pulsed transmissions over time periods longer than its observation at a single spot. Each mosaic built up a "sky frame," and approximately twenty-five thousand sky frames would be required to cover all directions and frequencies, each taking about two hours to complete, for a total of about seven years for the complete survey. The targeted and sky survey strategies were in many ways complementary; only the observations would demonstrate which assumptions were best and which technique was most effective in terms of a successful detection.[23]

As envisioned in 1979, the components of both the Targeted and Sky Survey systems consisted of three chief elements: a wideband dual polarization receiver and low noise amplifier; a digital spectrum analyzer to break the signal down into many channels; and a signal processor to search for the intelligent signals. The heart of the system and the key to its success was the digital spectrum analyzer. In 1979 it was envisioned that the spectrum analyzer would be constructed of modules that could be configured for each of the two search strategies. In fact, as events developed, Ames and JPL developed separate spectrum analyzers, the Multi-Channel Spectrum Analyzer at Ames and the Wide Band Spectrum Analyzer (WBSA) at JPL, each suited to the particular needs of its observing program.

With this general description one can begin to see the daunting problems that faced the designers who actually had to produce the hardware and software that would make SETI work. Radio astronomy had never before attempted multi-channel spectrometers at the scale needed for the SETI search. Standard spectrometers had been developed for a wide range of requirements, from 200 Hz resolution over a band of 40 KHz (for studies of the OH hydroxyl radical emission), or 20 KHz resolution over a band of 3 MHz (for extragalactic twenty-one centimeter studies), but nothing approaching the resolution and millions of channels needed for SETI. The key to the new spectrometer was the advance of digital technology, and the specific application to SETI was worked out beginning with Alvin Despain of the University of California–Berkeley and Allen Peterson at Stanford. By 1976 Despain, who had done postgraduate work under Peterson at Stanford, had begun to collaborate with Peterson when they realized that work already under way in digital filter design for other purposes was applicable to the SETI problem. Work on the design of a 74,000-channel prototype MCSA with one-half hertz resolution had been begun already in 1977 at the Engineering College Laboratories at Stanford University headed by Peterson and was built under the immediate supervision of Ivan Linscott. This prototype, later known as MCSA 1.0, used wire-wrap technology together with commercial integrated circuits and was contained in a standard equipment rack the size of a refrigerator. Field tests of the MCSA prototype detector were conducted from 1985 to 1987, using the twenty-six-meter telescope at Goldstone's DSS 13. The detection of *Pioneer 10's* one-watt transmitter at a distance of 4.5 billion miles demonstrated the capabilities of the digital architecture. Beginning in early 1988, the prototype was further tested at Arecibo Observatory in Puerto Rico and also used for experiments in radio astronomy.[24]

Faced with the need to scale up this spectrum analyzer by more than a hundredfold to produce more than fourteen million one-hertz channels, MCSA 2.0 replaced the wire-wrap technology by a customized, very large integrated circuit chip. Initially designed by students from Stanford, this digital signal-processing chip was built under contract to NASA Ames by the Silicon Engines Company.[25] Its basic task was to perform Fourier transforms extremely fast, providing six simultaneous frequency resolutions ranging from one to thirty-two hertz. It was the upgraded version of MCSA 2.0, with a redesigned, more accurate signal processing chip on large format, multilayer boards which became operational at Arecibo on 12 October 1992.

Another crucial component to the SETI system was the method for extracting an extraterrestrial signal coming through the spectrum analyzer. While detection of signals from noisy data is a standard problem in communications, SETI presented a particular challenge because nothing is known with certainty about the nature of an artificial extraterrestrial signal. The signal detection team at Ames, headed by D. Kent Cullers, assumed that the signal would consist of narrowband carriers, single pulses, or pulse trains and designed its signal detection algorithms accordingly. Aside from detecting a continuous wave, the software algorithms searched for pulses over the range of 45 milliseconds to 1.5 seconds. Because the system had to reject any terrestrial radio frequency interference, this problem was studied extensively by both the Ames and JPL elements of the SETI project. Finally, because millions of channels were to be analyzed in real time, great demands were placed on the data acquisition system, which was specially designed for the project.[26]

As these events unfolded at Ames, parallel events had taken place at JPL. There Michael Klein (who had taken over from Edelson as head of the JPL SETI project in 1981) forged a collaboration with the Telecommunications and Data Acquisition Technology Development Office to use part of the Deep Space Network and to design and build an engineering development model of their system, including the Wide Band Spectrum Analyzer, the equivalent of Ames's MCSA. SETI drove the design of the spectrum analyzer, but the multi-mission users of the Deep Space Network would share in its use. The purpose of the JPL spectrum analyzer was in general the same as that of the MCSA, but its architecture was tailored to the needs of the Sky Survey. The prototype system used on 12 October consisted of a pipelined Fast Fourier Transform architecture that transformed 40 MHz of bandwidth into 20 Hz channels, for a total of two million channels. It could also be configured to analyze one million channels on each of two polarizations. As with the Targeted Search element, the Sky Survey had its own signal-processing and data acquisition problems to address.[27]

In 1985 Ames and JPL entered into a memorandum of understanding delineating the responsibilities of each group. The project underwent definition reviews in 1986 and 1987, and the formal Program Plan was adopted in March

1987.[28] In 1988 the Project Initiation Agreement was signed by NASA headquarters. Finally, with funding for FY 1989, SETI took on the status of an approved NASA project beyond the "Research and Development" phase and began "Final Development and Operations," to be completed by the year 2000 at a total cost of $108 million. Administratively, SETI had gone from a few people within a division at Ames in 1976 to two project offices in two NASA centers with a combined staff and subcontractors of about sixty-five in 1992. Fiscally, its annual budget had risen from a few hundreds of thousands of dollars in the early 1970s to over ten million in the 1990s. Conceptually, its strategy had been honed and reduced to politically realistic proportions since the visionary Cyclops days.

At NASA headquarters the SETI program had spent most of its lifetime (since 1978) in the Life Sciences Division. But in 1992 the Senate Appropriations Subcommittee directed NASA to rename the project the "High Resolution Microwave Survey" (HRMS) and move it to Space Science at headquarters, where it became the first element in the Solar System Exploration Division's "Toward Other Planetary Systems" (TOPS) program designed to detect other planetary systems (see chap. 7). The move was not popular among the TOPS team; as one member later wrote, "This was somewhat like trying to protect the life of a star witness in a high-stakes criminal case through a quick change of identity and a move to another state."[29] Nor was it popular among SETI scientists, who were apprehensive that it could be misconstrued as evasive action, as indeed it eventually was.

As the HRMS program began on 12 October 1992, the chief of the SETI office at Ames (since SETI's inception NASA's lead center for the project) was John Billingham, with Barney Oliver as his deputy chief. Jill Tarter (also located at Ames) was the overall project scientist. Tarter had come to Ames in 1975 on a postdoctoral fellowship from the National Research Council, having received her Ph.D. degree under Joseph Silk at Berkeley working on gas in large galaxy clusters and doing some of the earliest work on "brown dwarfs," substellar objects intermediate in mass between a star and planet. Her interest in SETI began while she was still a graduate student, when Stu Bowyer introduced her to the Cyclops report and invited her to join Berkeley's shoestring SETI program, known as SERENDIP. She arrived at Ames in time to become involved in the last two of the Morrison SETI workshops, and, when her NRC postdoc expired, John Billingham hired her to help with the budding NASA SETI program.

By choice, however, Tarter was not a civil servant and bridled at bureaucratic restrictions. She preferred to work out of Berkeley and brought in her own support money for SETI. This allowed her to travel extensively on various observing projects. As she recalled: "Early on I knew the best thing that I could do for the project was to do a lot of observing in a lot of different ways and try to understand the physical universe and what it looked like at high resolution, because that's where we were trying to build instruments to search. We really didn't know, when you got real granular on the astrophysical sources, what they looked like. If you started looking at masers with finer and finer resolution, do you see interesting things or, in fact, is there some lower limit to the width of a natural feature? Indeed, it looks like about 300 Hz. So we went for designing systems that could detect signals that are more narrow band than that, and thinking that if we found it, we'd either find a new [extraterrestrial] technology or we'd find a whole new branch of astrophysics." During the 1980s she became increasingly involved in the NASA SETI program, playing key roles in both the science and politics.[30]

Another crucial event during the 1980s was the beginning of the nonprofit SETI Institute, founded in 1984 with Frank Drake as its president and Tom Pierson as its executive officer. The SETI Institute was born out of the need to stretch funds for SETI. As SETI funding remained steady in the early 1980s, employees became more expensive, and the amount of R&D which could be done actually decreased. Many SETI employees were adjunct faculty at nearby universities, and almost half of NASA's 1.5 million SETI funding went to overhead charges at the universities. Enter Tom Pierson, who worked for the San Francisco State University's Research Foundation, managing research grants and contracts. Pierson had been handling the SETI contract for astronomer Charles Seeger (brother of the singer Pete Seeger). Given the problems SETI was having in stretching money, Seeger set up a meeting with Billingham and Oliver in June 1984 to discuss how to remedy the situation.

By September Oliver hired Pierson to study how SETI's fixed funds could be stretched. The conclusion of Pierson's study was to recommend forming a nonprofit institute that took adjunct faculty contracted from universities with high overhead rates and provided a professional home at a lower overhead rate, leaving more money for research. Unlike the Space Telescope Science Institute and the Lunar and Planetary Institute, NASA played no role in founding the SETI Institute, which was formed as a nonprofit corporation. On 20 December 1984 Pierson, Drake, Andrew Fraknoi, Jack Welch, and Roger Heyns held the founding board meeting for the SETI Institute. Among those who joined the institute immediately was Jill Tarter, who remained half-time with Berkeley. By 1992, when the NASA SETI program began observations, the institute had attracted some twenty members with about seven million dollars of grants from NASA and NSF, among others. Not only did it prove an efficient way to use funding, but members of the institute (unlike civil servants) were unencumbered in lobbying Congress for money, an important consideration. Over the years the SETI Institute provided essential support in logistics, funding, and education about SETI and exobiology in general. As we shall see, it soon proved crucial to the continuation of SETI.[31]

In June 1990 SETI advocates were taken by surprise when Congressmen Ronald Machtley (D-R.I.) and Silvio Conte (R-Mass.) introduced a motion on the floor of the House of Representatives to remove all funding for the NASA SETI program for FY 1991. Machtley declared, "we cannot spend money on curiosity today when we have a deficit." We have survived for fifteen billion years without knowing whether extraterrestrials exist, he said, and we can survive a few billion years more without knowing. Machtley suggested that, if Congress approved SETI, it might adopt a (Search for Congressional Intelligence (SCOTI) program. Conte concurred that "at a time when the good people of America can't find affordable housing, we shouldn't be spending precious dollars to look for little green men with misshapen heads." If one wanted to find out about aliens, he suggested, one could spend "75 cents to buy a tabloid at the local supermarket." Conte concluded by introducing into the *Congressional Record* several tabloid articles on UFOs and extraterrestrials.[32] Neither Machtley nor Conte had been briefed on the subject, but the members of the Senate Appropriations Subcommittee on Veterans Affairs, NASA, and Independent Agencies had been. With the support of the Senate Subcommittee chair, Barbara Mikulski (D-Mass.) and Senator Jake Garn (R-Utah) the full amount of 12.1 million dollars was appropriated. "In recommending the full budget request of $12,100,000 for the SETI program," the Senate report stated,

> the Committee reaffirms its support of the basic scientific merit of this experiment to monitor portions of the radio spectrum as an efficient means of exploring the possibility of the existence of intelligent extraterrestrial life. While this speculative venture stimulates widespread interest and imagination, the Committee's recommendation is based on its assessment of the technical and engineering advances associated with the development of the monitoring devices needed for the project and on the broad educational component of the program. The fundamental character of the SETI program provides unique opportunities to explain principles of such scientific disciplines as biology, astronomy, physics, and chemistry, in addition to exposing students to the development and application of microelectronic technology.[33]

In May 1991 Senator Richard Bryan (D-Nev.) assaulted SETI during Senate Authorization Committee deliberations. Although the funding made it through for FY 1992, it was an ominous warning of things to come.

Meanwhile, Billingham was attending to another facet of SETI. From early on he realized that the societal implications of SETI could be profound. One of the two splinter workshops from the 1975–1976 Morrison meetings was "The Evolution of Intelligent Species and Technological Civilizations," chaired by Nobelist Joshua Lederberg and held at Stanford. Fifteen years later, on the eve of the first NASA SETI observations, Billingham organized and chaired a full-scale series of workshops, dubbed "CASETI" (Cultural Aspects of SETI). With his penchant for interdisciplinary interaction, in 1991–1992 Billingham gathered a diverse group of two dozen scholars to consider the question, no longer academic, "What would be the cultural, social, and political consequences if NASA's HRMS project were to succeed at detecting evidence of and extraterrestrial civilization?" The resulting publication was a pioneering study that demonstrated how the social and behavioral sciences could add crucial insight to

SETI while at the same time demonstrating the complexity of the problem and its richness for further study. Not least, it showed how SETI had the capacity to bridge many disciplines even outside the natural sciences.[34]

In the face of numerous political hurdles, on 12 October 1992, symbolically the quincentennial of Columbus's landfall in the New World, the NASA HRMS was inaugurated amid considerable fanfare. On that date the 305-meter radio telescope at Arecibo, Puerto Rico, began the Ames Targeted Search, while the 34-meter antenna at the Venus station of the Deep Space Communications Complex at Goldstone in the Mohave Desert began the JPL All-Sky Survey (fig. 6.4). After more than fifteen years of sometimes sporadic planning and sixty million dollars of research and development, SETI was finally on the air.

The New World Has Been Canceled: Congress and SETI

The observations begun at both Arecibo and Goldstone in 1992 were to mark the beginning of an extended enterprise. Over the lifetime of the project the systems used there would be replicated or moved among observing sites by a Mobile Research Facility, consisting of a truck with spectrum analyzers and associated equipment. The Targeted Search would use telescopes in the United States, Australia, and possibly France, and in 1995 the 140-foot telescope at Green Bank was planned to become dedicated to SETI. The Sky Survey would use the Deep Space Network telescope in Tidbinbilla near Canberra, Australia, as well as Goldstone and the California Institute of Technology's Owens Valley Radio Observatory in California.

Despite the elaborate plans and high hopes, it was not to be. Senator Richard Bryan, a freshman Democrat from Nevada, had during FY 1992 and 1993 unsuccessfully introduced amendments to terminate SETI. On 22 September 1993 he offered an amendment to the NASA appropriation bill for FY 1994 to eliminate all $12.3 million in funding for the SETI program. By a vote of seventy-seven to twenty-three the Senate concurred. In a press release issued the same day from his office, Bryan was quoted as saying: "The Great Martian Chase may finally come to an end. As of today, millions have been spent and we have yet to bag a single little green fellow. Not a single Martian has said 'take me to your leader,' and not a single flying saucer has applied for FAA approval. It may be funny to some, except the punch line includes a $12.3 million price tag to the taxpayer." The same press release noted that Bryan had successfully eliminated Senate funding for the program in 1992, when the Senate Commerce Committee voted eleven to six in favor of his amendment to cut funding, and the full Senate concurred. According to Bryan, "To avoid the cut, NASA simply renamed the program from the original Search for Extraterrestrial Intelligence (SETI) to 'High Resolution Microwave Survey.'" Bryan left no doubt of his pique at his perception of what had happened, having either forgotten or being unaware that the Senate Appropriations Subcommittee had directed the name change when SETI became part of the TOPS program in 1992: "This is a hor-

FIGURE 6.4. Inauguration of the targeted search portion of the NASA SETI program with the thousand-foot radio telescope at Arecibo, Puerto Rico, on Columbus Day, 12 October 1992. Project Manager Dave Brocker in the control room is coordinating the simultaneous beginning of observations with the Deep Space Network telescopes in California for the sky survey portion of the search. Outside project scientist Jill Tarter lectures the public in front of the telescope dish. (Courtesy Seth Shostak.)

rendous case of bureaucratic arrogance that somehow by simply renaming the program NASA can avoid the cut. . . . NASA wants to spend more than $100 million and they have got to get the message that this program doesn't make the final cut."[35]

While many have wondered at Bryan's motivation for leading the fight to terminate SETI, he clearly played to his voting constituents when he wrote: "Only in Washington, D.C., is $100 million considered small change. This is a lot of money, and, frankly, I think this money could better be left unspent, which means we don't have to borrow as much and add to the debt. It really is that simple." It is possible that Bryan's motivation, playing to the voters and saving money, really was that simple. In any case, on October 1 a House-Senate conference committee approved the Senate plan, which included one million dollars for program termination costs. Recalling the SETI program's inauguration only a year earlier, one writer in the *New York Times* remarked, "It was as though the Great Navigator, having barely sailed beyond the Canary Islands, was yanked home by Queen Isabella, who decided that, on second thought, she'd rather keep her jewels.[36]

The termination of the taxpayer-funded SETI program must be seen in the context of other congressional action at the time. There is no doubt that in a climate of rapidly rising federal deficits Congress was looking for budget cuts. In the same session Congress had failed to kill two other NASA programs, the much maligned Space Station, which received the full $2.1 billion funding the president requested, and the $3 billion Advanced Solid Rocket Motor program. In light of the failure to make these cuts, some SETI proponents saw the termination of the much smaller (and therefore politically less supportable) SETI as a sacrificial lamb. Drake noted that one space shuttle launch cost $1 billion—"a century worth of SETI research"—while others noted that Stanford had just received a federal grant of $240 million for research on antimatter. Some saw the difference as the "giggle factor," a subject open to ridicule no matter how important. John Pike, of the Federation of American Scientists, noted that aliens were a frequent subject of the notorious *National Enquirer* tabloid and offered another theory: "The political problems SETI has demonstrate the way in which a member of Congress, in an irresponsible grab for headlines, can do serious damage to a program." One thing is clear: unlike the Superconducting Super Collider canceled in the same session of Congress, SETI was not terminated for bad management or cost overruns. One cannot, however, discount spillover bad feeling from the Hubble Space Telescope, then returning unfocused photographs due to a problem with its mirror, an embarrassment that better management might have caught.[37]

It should also be kept in mind that NASA overall came out of the congressional session in relatively good shape: the budget bill for FY 1994 (which began on 1 October 1993) provided less than NASA requested but more than many researchers expected. Overall, NASA received $14.5 billion, $200 million more than 1993. Included in this amount was an increase of $207 million, to $1.784 billion for space science, out of which SETI would have been funded. Despite the elimination of SETI and the cuts to a few other programs, NASA management could not have been too unhappy with its overall budget. Seldom does a government agency obtain funding for all its programs.[38]

The effect at the SETI level, however, was immediate. On 12 October 1993 Wesley Huntress, associate administrator for space science, wrote to Dale Compton and Ed Stone (directors of Ames and JPL, respectively), "Consistent with congressional direction, you are instructed to terminate the High Resolution Microwave Survey (HRMS) immediately." The directors were ordered to issue termination notices to contractors immediately, to provide a plan within one week to terminate the program within two months, but to preserve the hardware for potential use by others. The NASA SETI program was dead. Congress allowed one million dollars for termination costs, and NASA provided an additional million from FY 1993 funds in recognition of the real termination costs.[39]

The provision to preserve the SETI hardware for future use offered a glim-

FIGURE 6.5. SETI pioneers shown when the program was still headquartered at NASA Ames, 1989. *Left to right:* Vera Buescher, Charles Seeger, Jill Tarter, Frank Drake, Bernard Oliver, John "J.B." Billingham. (Courtesy SETI Institute.)

mer of hope that many years of research and development could be salvaged if funding could be found elsewhere. Although JPL's Sky Survey ended because it made use of the telescopes of the government-funded Deep Space Network, the Targeted Search was under no such constraint. Suddenly, the SETI Institute, which until now had played a supporting role, was crucial to the very existence of SETI. The institute was located only a few miles from the Ames Research Center. Targeted search personnel, including Billingham and Oliver, moved to the SETI Institute (Tarter and others were already there) and began to consider the possibility of private funding, which had a long if sporadic history of support for astronomy. The SETI Institute, after all, was located in the heart of Silicon Valley, and Barney Oliver had a long association with its oldest and most respected company, Hewlett Packard. Billingham, Tarter, Oliver, and Drake became fund raisers (fig. 6.5), and by December 1993 the institute had commitments of $4.4 million to continue a reduced-scope project with private funds.[40] Among the contributors were David Packard, William Hewlett, Paul Allen (cofounder of Microsoft), Gordon Moore (cofounder of Intel), and Mitch Kapor (founder of Lotus Development Corporation). Thus was Project Phoenix born, rising from the ashes of the NASA project. Its first observations were carried out in February 1995 at the Parkes Radio Telescope in Australia, later with the NRAO 140-foot telescope at Greenbank (a few hundred feet from Frank Drake's original observations for project Ozma), and at Arecibo whenever it could obtain telescope time. Even as Project Phoenix continued, but not content with sporadic telescope time, at the turn of the millennium the SETI Institute was deeply involved in planning a dedicated "Allen Telescope Array," funded by Paul Allen and his former Microsoft colleague Nathan Myhrvold. And an international consortium was designing an even more ambitious "Square Kilometer Array."

Although NASA had given SETI a major boost with its ten-year research and development program and had operated the world's flagship SETI effort for one year in 1992–1993, SETI survived after the loss of its chief patron. Not only did Project Phoenix continue the NASA project; other projects more limited in frequency and targets were carried on around the world. Especially notable were the Planetary Society program at Harvard and in Argentina, and the University of California–Berkeley Project SERENDIP, which had first piqued Jill Tarter's interest in SETI. Millions of ordinary citizens signed up for the SETI@home project, crunching SERENDIP data on their home computers, and the SETI League coordinated thousands of others to use their own radio dishes to form an amateur SETI network. Both these projects testify to the continuing popularity of the search. Whether popular or scientific, SETI's proponents argued that the question was too important to be sidetracked by politics or limited funding. Although the U.S. Congress proved unwilling to invest in such a long shot as extraterrestrial intelligence, national interest and human fascination with the subject suggests that, if a signal were actually found requiring a long-term funding effort to understand, NASA and Congress would once again be interested. In this sense the history of NASA and SETI may once again become intertwined in the future.

CHAPTER 7

The Search for Planetary Systems

Although NASA was very quick to latch onto Mars as a target for exobiology, the search for planetary systems was another matter. Compared to the stars, Mars was our next-door neighbor, an attainable goal for spacecraft. The search for planetary systems, by contrast, required new or improved ground-based techniques before one could even contemplate a search by spacecraft. And, although NASA did fund some ground-based astronomy in support of its Mars missions—ironically, Lowell Observatory was one of its primary beneficiaries—the National Science Foundation (NSF) had long been considered the government patron for telescopes on the surface of the Earth. Nevertheless, NASA eventually took up the challenge—and sooner than one might have predicted.

The search for planetary systems at NASA arose in three successive but overlapping contexts: the Search for Extraterrestrial Intelligence (SETI) in the 1970s, the expansion of planetary science in the 1980s, and studies in the 1990s which coalesced into the program known as the "Astronomical Search for Origins." What began as workshops and ad hoc discussions among small groups of scientists in the early 1970s ended a quarter-century later in some of the most complex programs NASA had ever conceived, involving large government-university-industry teams that produced detailed designs for real space missions. Unlike Mars missions, these spacecraft could not travel to their distant destinations but were designed to search for planetary systems from the vicinity of Earth. Not by accident, their goal of looking for Earths and unveiling our origins generated tremendous public interest. Planetary systems were portrayed as an integral part of cosmic evolution and thus an essential step in the search for life—and our place in the universe.

Early Discussions: Planetary Systems and NASA SETI

NASA's earliest official interest in other planetary systems arose out of its program to Search for Extraterrestrial Intelligence. After all, if one were going to search for intelligence in outer space, it would almost certainly be on the surface of a planet, unless one posited exotic life such as portrayed in Fred Hoyle's novel *The Black Cloud*. The existence of extrasolar planets was one of the crucial elements of the Drake equation, an essential parameter on the way to life. The 1971 NASA Ames summer study of a system for detecting extraterrestrial intelligence, headed by John Billingham and Bernard Oliver, contained a small section on planetary systems, which concluded that theoretical considerations pointed to a large number of planetary systems but that the actual observation of such systems was at the very limits of detectability. For observational evidence the authors did seize on the American astronomer Peter van de Kamp's announcement in 1963 of a possible planet around Barnard's star and several other borderline cases, but the stronger argument was that the nebular hypothesis predicted planet formation as a normal part of stellar evolution. Similarly, the series of lectures which Billingham organized at Ames during the summer of 1970 in connection with the embryonic SETI program had included only a theoretical discussion by A. G. W. Cameron.[1]

It is therefore not surprising that, as NASA's interest in SETI grew by the mid–1970s, experts were called in to assess the methods for detection of other planetary systems. The results of these discussions were reported in the pioneering "Morrison Report," *The Search for Extraterrestrial Intelligence* (1977), and were backed up by more detailed NASA reports. Such discussions were only the first of many that over the next quarter of a century would place NASA at the forefront of planetary system research, even though the early discoveries of actual planets in the 1990s were not a direct result of NASA programs. The goal of the workshops, which notably concentrated on observational techniques rather than theories of planetary formation, was "to define how observations might shed some light on the frequency of low-mass companions to stars."[2]

As the *Viking* spacecraft were approaching Mars and as the United States was approaching its bicentennial, two Extrasolar Planetary Detection Workshops were held under the auspices of NASA as part of its SETI investigations. The first convened in March 1976 at the University of California–Santa Cruz and the second two months later at NASA Ames, where the SETI project was making slow progress under John Billingham. The chair of the workshops was Jesse Greenstein, an established professor of astrophysics at Caltech, known for his pioneering work on the interstellar medium and stellar evolution. Not only was Greenstein "a very dominant scientific figure, a person with grand vision, and very smart," he also had a personal interest in planetary systems stemming from his own research. The executive secretary was David Black of NASA Ames. Black was much younger; only a few years earlier he had completed his doctoral work on meteorites at the University of Minnesota under Robert Pepin, which led to his interest in the primitive solar nebula and solar system formation. As a postdoc in 1971, he had argued that Peter van de Kamp's data on Barnard's star fit best if it were surrounded by two or three planets not orbiting in the same plane.[3]

Already at these early meetings a remarkably full complement of planetary detection techniques was discussed. The participants realized the extreme difficulty of the direct detection of an extrasolar planet by the light it reflects from its parent star. The difference in absolute visual magnitudes of Jupiter and the Sun, they noted, was 21 magnitudes (from 5 for the Sun to 26 for Jupiter), corresponding to a difference in brightness of 250 million between the two. Any attempt to find even a large planet around another star would be "washed out" by the brightness of the star. Nevertheless, the workshop tackled many possible approaches. Bernard Oliver, of future SETI fame, discussed "apodized" optics on a space telescope, the use of masking to block out some of the star's light. The problem could be made more tractable by using infrared (IR) wavelengths where Jupiter was only 4 orders of magnitude dimmer than the Sun; the workshop therefore suggested that a space system for infrared interferometry should be studied. Infrared observations could also be used to detect protoplanetary systems, extended disks of gas and dust that have a much larger area than the planets subsequently formed. Several participants discussed IR techniques, including Ronald Bracewell, a Stanford electrical engineer who had written on extraterrestrial intelligence and was thus inspired to invent better methods for planet detection.[4]

Of more immediate promise were the "indirect" methods, which detected the motion of a star due to a planetary companion, either back and forth in our line of sight (radially) or across our field of view (tangentially). Among these methods George Gatewood (of the Allegheny Observatory) and Kaj Strand (of the Naval Observatory) represented the classical "astrometric" community, the van de Kamp school, which had already used long-focus refractors and claimed detection of tangential stellar motion due to one or more planets around Barnard's star. The problems with this method were daunting. The displacement of the Sun due to Jupiter, as viewed from five parsecs, was only one milliarcsecond (a thousandth of an arcsecond), and the effect of the Earth was a thousand times smaller than that (one microarcsecond). The technology at the time might give three milliarcsecond accuracy after a year's observation, the workshop noted, but the method would take at least ten years and was on the very edge of detectability, even for Jupiters orbiting the nearest stars less massive than the Sun. At a special meeting convened at the Naval Observatory between the two planet detection workshops, astrometrists concluded that improvements in accuracy could result from the new charge-coupled device (CCD) detectors on ground-based telescopes, that ground-based optical interferometry might give fifty microarcsecond accuracy, and that space-borne telescopes might yield microarcsecond accuracies. The problem was that such technologies, with the exception of CCDs, would take decades to develop.[5]

As an extension of the classical astrometric method, Frank Drake discussed photoelectric astrometric techniques, while others discussed new techniques using optical, radio, and infrared interferometry.

The other major indirect approach to planetary detection was the less-developed but ultimately more successful technique of "radial velocities." As with the astrometric methods for detecting tangential motion of a perturbed star, the radial velocity method had daunting challenges. Jupiter causes a reflex motion of the Sun of about 12 meters per second, with a period of twelve years, and the Earth causes the Sun to move only about 0.09 meters per second. By comparison, the radial velocity systems then in use—for example, by Roger Griffin at Cambridge University—yielded accuracies of only 1,000 meters per second (1 km/sec). Griffin argued, and the workshop agreed, that accuracies of 10 meters per second were achievable, though they worried about noise due to surface motions of the star. For the latter reason the workshop was very interested in the work of American astronomers Robert Dicke and Henry Hill observing the surface pulsations of our Sun.

Despite the challenges, the conclusions of the workshop, as expressed in the final SETI report, were upbeat. "The prospects of increasing our confidence concerning the frequency and distribution of other planetary systems are good, if we are willing to invest the effort," Greenstein and Black concluded. "As a consequence of the Workshops, several novel approaches to the problem have come to light, as have potential improvements to classical means of detecting planets."[6]

Among the promising new techniques that Greenstein and Black mentioned in their summary was interferometry, a method routinely used in the 1970s with radio telescopes. By measuring incoming radio waves at several separated telescopes and then combining the two signals, astronomers could resolve and measure objects as if a single large telescope were being used. The method required meticulous detail in combining the waves but was more easily used with radio telescopes because radio waves were much longer than optical waves. Unfortunately, in order to find planets or their effects, one needed to observe in the optical or infrared region. At the urging of Billingham, a few weeks after the Extrasolar Planet Detection Workshops associated with SETI and five years after Billingham and Oliver had conducted Project Cyclops as a Stanford / NASA Ames summer study, Black conducted his own summer study in the same series to design a ground-based optical interferometer. "Project Orion," which was meant to build on the ideas of the Planet Detection Workshops, sought to apply new technology to develop a telescope that would increase the accuracy of astrometric measurements some ten to fifty times. Among the twenty-three participants were Bracewell, the expert on interferometry; Gatewood, the expert on astrometry; and Krzysztof Serkowski, an expert on radial velocity techniques. "We not only reviewed the evidence for other planetary systems, which was essentially non-existent at the time," Black recalled, "we also went to potential ways in which you could go out searching for what were the limitations on the various techniques, star spots, photometric noise, things of that nature." Out of these discussions the technique that emerged for the most focused study was a long baseline interferometer that sought direct detection of the planet's light. While the Orion design study team realized that the resulting "Imaging Stellar

Interferometer" was perhaps ahead of its time, it nevertheless recommended that a program to search for planetary systems, with its own budget and funding, should be included in NASA activities.[7]

These recommendations received a further boost at a NASA-sponsored workshop on planetary systems conducted in late 1978 and early 1979, in which Black again played a prominent role and which was again designed to take another step forward in planet detection techniques. With support from William E. Brunk at NASA headquarters, Black ran a small program in the late 1970s which funded Gatewood, Serkowski, and a young new player, Mike Shao at MIT, to work further on planet detection. All three and Jesse Greenstein, among about twenty others, contributed to the 1978–1979 workshops whose goal was to "begin to put together the scientific underpinning of what might be called a program." The workshop singled out six conclusions: (1) a scientifically valuable program to search for other planetary systems can be conducted with ground-based instrumentation; (2) significant gains in the accuracy of existing ground-based techniques can be made with modest application of current or near-term state-of-the-art technology; (3) existing telescopes are not currently a limiting factor for the accuracy of ground-based techniques; (4) none of the currently planned space-based systems is adequate for a comprehensive detection program, including NASA's Space Telescope and the European Hipparcos satellite; (5) a comprehensive program to detect planetary systems must use a multiplicity of techniques and instrumentation; and (6) a comprehensive effort to detect planetary systems will yield invaluable scientific results. In light of these findings, and with a view toward building a program, the workshops made four recommendations: (1) high-accuracy radial velocity studies of solar-type stars should be carried out with existing telescopes; (2) observational studies should be made of the Sun to study the effects of surface motions on radial velocity techniques; (3) speckle interferometry techniques should be used to search for planetary companions to binary stars; and (4) the development and testing of new instrumentation should be carried out as soon as possible.[8]

Workshops were one thing, but putting together a program supported by NASA was quite another. In doing so, the planet hunters had to confront practical political problems. They wanted to "sever the umbilical cord between SETI and planet detection" because SETI was at this time running into political problems with Senator Proxmire and the Golden Fleece Award. "It was at this point that we thought this was clearly a scientific endeavor," Black recalled, "not that SETI isn't, but [planet detection is] something you are measuring physical phenomena and you can tie to astrophysics." But then the problem was to find a home at NASA: "the only way you were ever going to get things like missions, which is of course the coin of the realm when it comes to NASA, was to get it fully embraced within a program. It slowly began."[9]

In trying to persuade NASA to pick up planet detection even as a fledgling program, Black and others ran into a common problem for new disciplines: the planetary scientists saw planet detection as astrophysics, and the astrophysicists viewed it as planetary science. Black made presentations to NASA headquarters and also to the National Academy of Sciences Committee on Planetary and Lunar Exploration (COMPLEX), arguing that "you are never going to understand the origin of this planetary system, which is a key part of what planetary is about, if you don't have this evidence [about other planetary systems]." Eventually, in a crucial meeting in 1980 with Ed Weiler, Brunk, and Angelo "Gus" Guastaferro, who headed planetary science at NASA headquarters, Guastaferro decided that planet detection would find its first home in planetary science. Weiler declined to commit funds, and "this went back and forth. Guastaferro basically almost slammed his fists on the table and said, enough of this, planetary will take it, and he got up and walked out. So that's how planetary detection got its planetary program." But it would not be the last time that planet detection had to seek a home in NASA. Black lobbied in other ways too: by writing a paper in *Space Science Reviews* and giving a review talk at the American Astronomical Society meeting the same year. "So gradually, I think there was more and more visibility and acceptance taking place in the science community that this was not only something worth doing but in fact not just a field full of loonies, but it was technically becoming possible to actually do this job."[10]

Another practical problem to confront was the level of funding. As George Field, director of Harvard's Center for Astrophysics, wrote in the foreword to the 1978–1979 workshops: "Few astronomers would be likely to take issue with the idea that some effort be expended in this direction. However, in view of the many competing claims on the research funds available, the questions of how much effort should be expended and when become critical ones. The answers depend on one's assessment of the chances of success, of the significance of the findings (whether positive or negative), and of the long-term prospects for more detailed observations of any planetary bodies that are detected."[11]

It was therefore in the context of SETI that all three NASA-sponsored discussions of planetary systems took place in the 1970s—the 1976 Greenstein workshops that fed into the Morrison SETI report, the 1976 Project Orion summer study, and the 1979 Black and Brunk workshop. It was at another SETI meeting—the NASA Ames conference on Life in the Universe, convened by John Billingham in the summer of 1979—that Black summarized the results of these three studies.[12] He concluded that improvements to both ground-based astrometric and radial velocity techniques, giving them the capability of detecting planetary systems, were possible and inevitable. In the case of astrometry it was not yet clear which technique would win out as the most efficient and accurate for a routine observational program, but interferometry with either one or two telescopes seemed promising.[13] Black found "little question" that radial velocity techniques would be improved to one meter per second necessary to detect planetary systems. As for space-based systems, Black made the prescient remark that the upcoming NASA Infrared Astronomical Satellite (IRAS) mission, while not searching for planetary systems, "might provide unexpected results," as indeed it did with the discovery of circumstellar material that might be interpreted as protoplanetary systems. Black was less optimistic about the capabilities of other space systems on the drawing board: the Space Telescope, while representing a vast improvement over Earth-based imaging, was not good enough to image planets, and the milliarcsecond astrometric capability of the Space Telescope and the European Space Agency satellite (later named Hipparcos) was not promising for detecting planetary systems. Both spacecraft were launched in the early 1990s, experienced early difficulties, but went on to perform flawlessly. But neither found any planetary systems.

There is thus no doubt that NASA's interest in the search for planetary systems was inspired by SETI in its early years. Precisely because of this association, it also had to battle the same political ridicule as did SETI and all endeavors associated with the search for extraterrestrial life. It is a telling sign of the times that at the beginning of the 1979 Ames meeting on Life in the Universe, NASA administrator Robert Frosch felt compelled to defend not only the search for life but also the general pursuit of knowledge for its own sake. The meeting, he remarked, "comes at a time in which we seem to have a faltering in global and national interest in knowledge for its own sake. We have become hyperpractical and are expected to explain the use of things we do not understand, before we understand them."[14] Intellectual risk taking, he argued, is an essential part of any groping for knowledge. The whole nature of science is "making errors, finding them, and disposing of them." In the search for planetary systems there would indeed be many errors and false starts, but, as the decade of the 1980s began, NASA had at least made a start.

Planetary Science Extends Its Realm

Although SETI had provided the context for the first discussions of planetary systems within NASA and although planetary systems would continue as a significant part of future SETI discussions, it was the better-established (and, in some opinions, more reputable) planetary sciences that would sustain the idea through the 1980s. As we have seen, it was in planetary science that planet detection found its first home at NASA. As SETI struggled with its own funding problems, during that time the planetary science community would carry the search for planetary systems "from the study phase to a level in which a program could be contemplated."[15] Both intellectual and practical reasons drove NASA's involvement. There was no doubt that the existence of planetary systems was a problem of the highest importance, the indispensable requirement for the existence of life beyond Earth. From the practical viewpoint NASA, like most government agencies, was always looking for new projects to push the frontiers of exploration (according to advocates) or to perpetuate itself (according to cynics). As spacecraft had been successfully dispatched one by one to the planets of our solar system during the 1970s and 1980s, NASA now sought more worlds to conquer. Both through its own committees and the advisory capacity of the National Academy of Sciences, it sought to extend the realm of the planetary sciences from our solar system to other planetary systems.

NASA sought this extension at a time when planetary exploration was in crisis. The golden era of solar system exploration, from *Mariner 2*'s first flyby of Venus in 1962 to *Voyager 2*'s final encounter with Saturn in 1981, was over. Already in the mid–1970s the resources for planetary exploration were in steep decline (fig. 7.1). Erratic funding and higher mission costs caused some to call into question the very survival of the planetary program at NASA. Under these circumstances, in 1980 Thomas A. Mutch, NASA's associate administrator for space science, recommended a fundamental review of NASA's planetary program. In the fall of that year administrator Robert Frosch obliged by establishing the Solar System Exploration Committee (SSEC) as a subcommittee of the NASA Advisory Council. Its report, published in May 1983 as *Planetary Exploration through the Year 2000,* focused tightly on space missions and barely mentioned the search for planetary systems. In doing so, it followed the lead of the National Academy of Sciences' Committee on Planetary and Lunar Exploration (COMPLEX), which had produced several reports that, while briefly placing solar systems studies in the context of planetary systems, made no recommendations to study them.[16]

Yet by 1986 an "augmented program" of planetary exploration also authored by the SSEC included an entire chapter on planetary systems, complete with recommendations. It was the knowledge of these recommendations before publication which triggered NASA's request for another COMPLEX study in 1985, specifically to include planetary systems. Planetary science managers at NASA knew that, if the process of extending the realm of planetary science to other solar systems were to succeed, the National Academy of Sciences, through the Space Science Board of its National Research Council, was an essential ally. From the beginnings of NASA the relationship with the Space Science Board had always been uneasy. Although NASA was not required to seek the advice of the council through its Space Science Board, for new programs and large projects the weight carried by an independent review of this National Academy body was often essential to success in arguing for funding.[17] Thus, the recommendation of COMPLEX regarding a program of research on other planetary systems was crucial.

The resulting COMPLEX report was everything NASA could have hoped for. Couching its report in terms of "a new opportunity for planetary sciences," the committee found that a coordinated program of astronomical observation, laboratory research, and theoretical development to study extrasolar planets and their stages of formation would be "a technologically feasible, scientifically exciting, and potentially richly rewarding extension of the study of bodies within the solar system." COMPLEX recommended to NASA's Office of Space Science and Applications that it initiate systematic observational planet searches using both astrometric and radial velocity (Doppler) techniques and, furthermore, that it study young stars for possible circumstellar material that could indicate

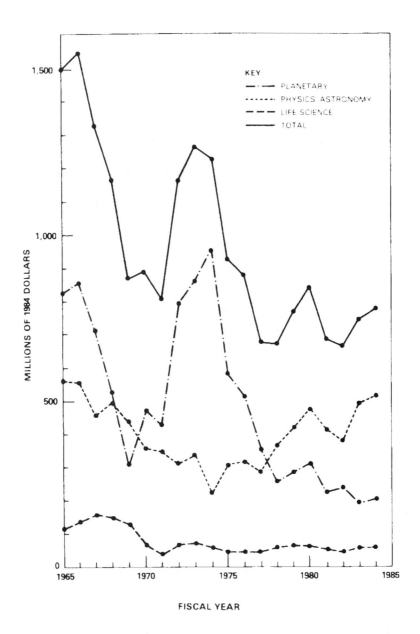

FIGURE 7.1. Space science funding by category. The decline in planetary exploration funding (*second plot, lower right*) in the mid–1970s is evident in this plot from *Planetary Exploration through the Year 2000: A Core Program* (May 1983). Life science funding (*bottom plot*) was holding steady, but physics and astronomy in general were on the upswing.

solar systems in various stages of formation.[18] One year later the National Academy's independent "decadal review" of astronomy (the "Bahcall Report") also gave major impetus to planetary systems science by identifying the field as a key area for scientific opportunity in the 1990s. Likening the problem of finding a planet to "trying to find from a distance of 100 miles a firefly glowing next to a brilliant searchlight," the reviewers concluded that optical or infrared ground- and space-based interferometers could survey hundreds of stars within five hundred light-years and detect Jupiter-mass planets. They also noted that such planets would produce velocity shifts in their parent stars "that should be detectable with sensitive instruments on the large ground-based telescopes to be built in the 1990s."[19] Thus, the mid–1980s were a turning point, as both committees of NASA and the National Academy took the study of planetary systems very seriously.

What had happened in the intervening few years to change the attitude toward planetary systems? One problem was that the search for planetary systems had simply been too expensive and too technically challenging. The 1986 NASA SSEC report (now chaired by David Morrison, a planetary scientist at the University of Hawaii and a student of Carl Sagan) described "missions of the highest scientific merit that lie outside the scope of the previously recommended Core Program because of their cost and technical challenge."[20] Three years did not make them less so, but, meanwhile, an astonishing discovery heightened awareness that real science could be done on the subject. The serendipitous discovery was made by NASA's Infrared Astronomical Satellite (IRAS), a joint project of the United States, England, and Holland. Launched in January 1983, the satellite's detector was still going through calibration tests when it found that Vega was shining ten to twenty times brighter than it should have at long infrared wavelengths, a phenomenon known as "infrared excess." Astronomers Hartmut Aumann of JPL and Fred Gillett of Kitt Peak National Observatory first feared there might be a problem with the detector, but further reflection and additional observations showed that the source of the infrared excess was a ring of dust surrounding Vega. In the fall of 1983 they announced their results in a landmark paper: the first direct evidence outside our solar system for "the growth of large particles from the residual of the prenatal cloud of gas and dust." The discovery was trumpeted on the front page of the *Washington Post* and newspapers around the world. Nor was this by any means a unique phenomenon; by mid–1984 some forty "circumstellar disks," or "protoplanetary systems," had been found, depending on the interpretation given to the infrared excess. The discoverers were careful to emphasize that planets had not been found; instead, "the presumption is that these rings will eventually condense into solar systems like our own; if so, that makes the Vega phenomenon the first semidirect evident that planets are indeed common in the universe."[21] By late 1984 one of the IRAS objects, Beta Pictoris, had been photographed by a ground-based optical telescope, producing one of the most famous images in astronomy which the new report did not fail to reproduce (fig. 7.2). Added to this excite-

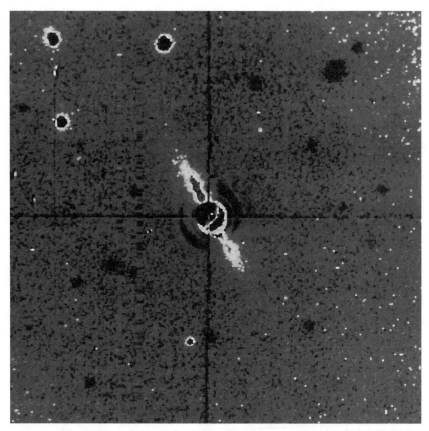

FIGURE 7.2. CCD image of a disk around Beta Pictoris (1984), early evidence for circumstellar material perhaps related to planet formation. The disk has been imaged many times in the last two decades, with indications of a warp that may be caused by planets or other objects. (Courtesy B. Smith, R. Terrile, and Jet Propulsion Laboratory.)

ment was the announcement of a "brown dwarf"—a substellar object intermediate in mass between a star and planet—around the star known as Van Biesbroeck 8. This implied that planet detection was only a little farther away and raised planet hunting to a fever pitch by the mid-1980s. Although the latter discovery turned out to be spurious, brown dwarf detections would not be much longer in coming.

Thus, it was not surprising to find in *Planetary Exploration through the Year 2000: An Augmented Program* an entire chapter on the search for new worlds beyond the solar system. "In the past few years it has become possible to make a rigorous search for planets around other stars, a search that will effectively open up a whole new area of science," the report stated. "The SSEC strongly recommends that such a search should go forward, augmenting limited ground-based methods by applying telescopes attached to the planned Space

Station." Theories of solar system formation were also advancing, the report noted, and predicted the existence of numerous planets. The Solar System Exploration Committee also argued that the search for planets was "a logical part of the NASA mandate, for it involves several major areas of current space science—the nature of the solar system, the mechanisms of star formation, and the possible existence of life elsewhere in the universe." In particular, the committee argued that such a search came under its purview because it addressed one of the division's fundamental goals: to understand the origin and evolution of our own solar system.[22]

The report concluded with seven recommendations, among them that the search for planetary systems was an activity properly coordinated by NASA's new Solar System Exploration Division. It recommended a ten- to twenty-year program to study about one hundred stars within ten parsecs of the Sun, capable of detecting Uranus-Neptune mass planets. It further recommended the capabilities of the Hubble Space Telescope (HST) and its infrared counterpart, the Space Infrared Telescope Facility (SIRTF), be used but that a space-based astrometric telescope be developed, possibly in conjunction with the Space Station, for which Black (now at headquarters) had become the chief scientist in 1985. Finally, it encouraged support of the study of a full range of techniques, including imaging, indirect detection by astrometry, photometry, and radial velocity searches as well as interferometry, whether from the Earth or space.

The 1986 report laid out an ambitious program, and its authors were particularly intrigued with the possibilities of an astrometric telescope in space: "It seems now that the most feasible and best-suited technical approach to planetary detection in the near future is a space-based astrometric telescope which can measure stellar positions to an accuracy of 10^{-5} seconds of arc," or 10 microarcseconds, they wrote. "This concept, which is now under study, should be examined in more detail in order to develop it as a possible experiment for the Initial Orbital Capability (IOC) phase of the Space Station." The idea for such an "Astrometric Telescope Facility" (ATF) originated at NASA Ames, where Black, Jeff Scargle, and Bill Borucki worked on it when it became clear in the wake of President Reagan's 1984 State of the Union Address that the Space Station would go forward. But when Ames management balked at taking on such a large space project, having recently had problems with its role in IRAS, JPL enthusiastically took over the project and used it as their entering wedge in the planet detection business. Although Charles Elachi and colleagues at JPL did Phase A studies for the ATF as a payload attached to the Space Station, both funding and technical problems prevented the project from proceeding. Among the technical problems was the realization that a manned Space Station might not be stable enough to make the extremely precise measurements for astrometry; the slightest human movement would set off vibrations that would spoil such delicate observations. There would be no lack of proposals for other astrometric space telescopes.[23]

As the writing of the Augmented Program was nearing completion, in December 1985, NASA's Solar System Exploration Division (SSED) established a Planetary Astronomy Committee to provide more specific advice on the future of planetary astronomy, including the search for other solar systems. Chaired by David Morrison at the University of Hawaii, the committee also included Black, among other planetary science experts from JPL, MIT, and a variety of other institutions. The committee urged the SSED to recognize a broad mandate for planetary astronomy, including "the search for other planetary systems and an improved understanding of the process of planet formation in other systems, as well as our own." Urging the detection and study of other planetary systems as a major new initiative for the division, the committee report pointed out that wider wavelength coverage, improved measurement precision, and the ability to probe circumstellar environments had created opportunities that would lead to a new field of "comparative planetary system studies." Ever mindful of the division's original scope, the report emphasized that this new field would be of great importance for understanding Earth and our own solar system.[24]

In carrying out its recommendation, the report recommended two strategies: first, that, for the sake of cost-effectiveness, the existing programs of NASA's Astrophysics Division (especially the Great Observatories, including the Space Telescope and SIRTF) were central to achieving its goals; and, second, that a variety of planet detection techniques be pursued, given that the best approach was not yet known. Among these techniques were the radial velocity and astrometric methods as well as space systems with direct-imaging telescopes and interferometers. An Astrometric Telescope Facility was envisioned for indirect planet detection by the motion of the parent star with respect to background stars and a Circumstellar Imaging Telescope for direct detection of circumstellar material and (less likely) planets themselves. The strength and weaknesses of each of these methods were weighed. In this report, for the first time, planetary systems was envisioned as fully integrated into planetary science. The search for planetary systems was "perhaps the most significant new initiative for planetary astronomy in the 1990s."[25]

Although the NASA and National Academy reports were not published until 1989 and 1990, respectively, by 1988 NASA had seen enough of their conclusions to act on the Planetary Astronomy Committee's recommendation to establish a Science Working Group (SWG) for planetary systems. Geoffrey Briggs, the head of the Solar System Exploration Division, established this committee, affectionately known as Planetary Systems Science Working Group (PSSWG), which temporarily transformed its name to Toward Other Planetary Systems Science Working Group (TOPSSWG) from late 1991 to late 1993 and would function until July 1995. The report of the group, chaired by MIT astronomer Bernard F. Burke, was issued in 1992, the same year in which planets were confirmed around a pulsar, a very un-Earth-like star. While pulsar planets could not harbor life, some enthusiasts argued that, if planets could form in the harsh environment of pulsars, they could form anywhere.[26]

The TOPS group (still known as PSSWG at the time) held its first meeting in April 1988 and decided that the scope of its work should include not only the detection of planetary systems but also studies of planetary formation and evolution as well as the study of circumstellar material in general. The first TOPS Workshop was held in January 1990 at the Lunar and Planetary Institute in Houston, whose new director was David Black. It resulted in a three-phase program, which was presented the following August to NASA Associate Administrator for Space Science, Lennard Fisk. The team recommended that TOPS-0, which focused on ground-based approaches, begin as soon as possible. They recommended that TOPS-1, proposing the development and launch of a space-based system, start by the end of the 1990s. The much more ambitious TOPS-2, construction of a major instrument to detect directly Earth-like planets and intensively study them, was so far in the future that no timeline was set.[27] The first two phases aimed to identify Jupiter-like planets around other stars and characterize their orbits, while the goal of phase 3 was to discover and study Earth-type planets. Phase 3 hoped to identify the nature of a planet's surface, temperature, and atmosphere. In retrospect the enunciation of these three phases of planetary searches was very important because within a few years they would be incorporated into a real program, known as "Origins."

When the Planetary Systems Science Working Group met in Houston in early 1990 plans for TOPS-0 drew largely on existing ground-based programs. The only ground-based astrometric search for planetary systems then in effect was known as the Multichannel Astrometric Photometer (MAP). The brainchild of Allegheny Observatory director George Gatewood, who had participated in the 1976 SETI planet detection workshops, MAP by this time had been used for five years on the Allegheny Thaw refractor. It had the capability of detecting Jupiter-sized planets around nearby stars but so far had found none. The other method involved radial velocities, also prominently discussed in 1976 but achievable then only at the level of one thousand meters per second. By 1990 Canadian and American groups had observational programs under way with long-term accuracies of less than one hundred meters per second. Among them were two astronomers at San Francisco State University (SFSU), Geoffrey Marcy and Paul Butler, who had been running a radial velocity program with Lick Observatory's three-meter telescope since May 1990. Groups from Harvard and Texas, using an instrument dubbed CORAVEL (Correlation Radial Velocities), which was a more classical radial velocity technique, had been obtaining measurements in the one hundred meters per second range, with hopes of soon reaching twenty-five to fifty meters per second. They had succeeded in detecting a small object that seemed to be not quite a star and not quite a planet. Thus, radial velocity technology was edging toward the level of about five meters per second, which most astronomers felt was needed to detect Jupiter-sized planets.[28]

By the time of the presentation to Fisk at NASA headquarters, however, the first phase of TOPS was centered on the W. M. Keck Observatory on Mauna Kea, Hawaii. The Keck Observatory housed the world's largest telescope, a ten-meter aperture consisting of thirty-six segmented mirrors, twice the size of the famous five-meter (two hundred–inch) telescope at Mt. Palomar in California, which reigned for more than forty years as the largest, until overtaken by Keck in 1993. By early 1990 a second ten-meter telescope was being considered for construction next to the first one, opening up another possibility: using the two in tandem for optical interferometry. But, in order to build the second Keck telescope, the University of California / CalTech consortium that operated it needed thirty-five million dollars, one-third of the cost of the telescope. The TOPS group recommended that NASA fund part of the Keck Observatory telescope as part of TOPS-0, and the Solar System Exploration Division and Fisk agreed. That was not, however, the same as getting the funding from Congress; in the end NASA had to come up with funding internally. NASA officially joined the partnership in October 1996, when the second Keck telescope became operational. Although the NSF had traditionally funded ground-based astronomy, there was precedent to do so at NASA because of the Infrared Telescope Facility already on Mauna Kea. Thus, construction of the largest pair of telescopes in the world was funded in part by the desire to find planetary systems. Eventually, the Keck telescopes would study protoplanetary systems and discover planets with the radial velocity equipment of Marcy and Butler. They even offered hope for the direct detection of massive substellar objects around stars.[29]

TOPS-1, the second phase of the program, considered three proposed space telescopes, each pushed by separate teams (fig. 7.3). Michael Shao, of JPL, pushed the Orbiting Stellar Interferometer design, at twenty meters in length the largest of the three instruments proposed. Robert Reasenberg, of the Harvard Smithsonian Center for Astrophysics, proposed the Precision Optical Interferometer in Space (POINTS). And Black and others proposed the Astrometric Imaging Telescope, a free-flying space telescope that was a slightly morphed version of their ATF. As interferometers, the first two were designed for indirect detection of the motion of a star caused by the gravitational pull of a planet; the latter (a two-meter-class telescope) could make either direct or indirect detections. The Hubble Space Telescope, launched in April 1990, had been touted as being possibly able to detect planets, but, almost simultaneously with its launch, Robert Brown and C. J. Burrows showed that the telescope was not capable of detecting planets, even after its spherical aberration problem was repaired. Hubble would return much wonderful data, but it would not confirm the existence of extrasolar planets.[30]

The competition for TOPS-1 heated up in 1991 with news that another NASA advisory committee was pushing for its own design for a space telescope, known as the Astrometric Interferometry Mission, which had already been favorably reviewed in the National Research Council's decadal survey, the Bahcall Report. The goal stated by the Bahcall Report was a thousand-fold increase in astrometric accuracy to about thirty microseconds for stars at twentieth magnitude. NASA's Astrophysics Division pushed this proposal, while the SSED pushed one of the three others proposed. The decision was supposed to have been made at the Woods Hole "shootout" in the summer 1991, where TOPS-0

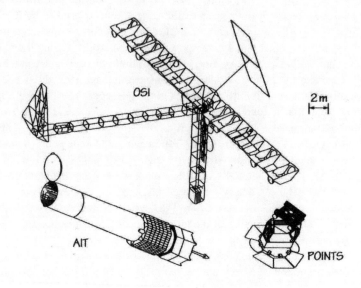

FIGURE 7.3. Three space telescopes proposed for detecting extrasolar planets: The Astrometric Imaging Telescope, the Orbiting Stellar Interferometer, and the Precision Optical Interferometer in Space. *(From* TOPS: Toward Other Planetary Systems [Washington, D.C.: NASA, Solar System Exploration Division, 1992*], 49.)*

was blessed, but no proposal for an astrometric telescope for TOPS-1 was approved.[31]

TOPS-2 envisioned the use of space- or lunar-based instruments to detect Earth-like planets directly. One possibility envisioned was a sixteen-meter infrared space telescope, in very high Earth orbit or on the Moon, with cooled optics. Another option was an interferometric array, perhaps on the Moon. Considering the normal horizon of NASA thinking, these were very imaginative proposals indeed.[32]

The obvious place to start was with TOPS-0 and the ground-based efforts already under way. Although some of the astrometric and ground-based teams received minimal funding, ironically it was SETI that became the first major funded element of TOPS-0, when Lennard Fisk tried to shield that program from congressional budget cuts in October 1992. When Congress terminated SETI one year later, the planet hunters changed TOPSSWG back to PSSWG, fearing that the entire TOPS program would be canceled. As attention focused again on TOPS-0 and the Keck Observatory, a battle took shape in 1993 over who would obtain funding for testing the Keck interferometry concept. JPL's Mike Shao proposed a facility on Mt. Palomar in California, but other universities had their own proposals and feared the worst from JPL, which depended on outside money for its funding. "The university-based scientists could see the TOPS program disappearing whole down the voracious mouth of JPL," wrote PSSWG member Alan Boss. Indeed, by giving JPL the programmatic responsibility of TOPS, NASA headquarters effectively gave Shao the go-ahead for his "Palomar Test Bed."[33] TOPS-1, the plans for an orbiting planet-search telescope, was delayed to the extent that no single design had yet been chosen from those proposed; such a selection was considered premature under the budgetary circumstances. And, with NASA's perpetual budget problems, TOPS-2 was off the radar screen for the foreseeable future.

Still, it is significant that such a far-reaching program as foreseen by TOPS had been proposed at all. Undoubtedly with an eye toward public relations and NASA funding but also from deep-seated personal feelings, the TOPS group was unusually forthright about the motivations for its proposed program. Humans, they emphasized, had a deep need to understand their relationship with the universe. The questions of the origins and frequency of planets which TOPS addressed had been asked for millennia by religion and philosophy but could now be tackled by science. And they were laying the groundwork for an even greater challenge, "the ultimate question engendered by the Copernican revolution: Does life exist on planets around other stars?" The group therefore had an impressive awareness that its recommendations were not only highly significant to science but were also of wider significance to humanity. Whether planetary systems are found to be common or rare, they concluded, "the results of TOPS investigations cannot fail to inform the human spirit and self-concept in a deep and fundamental way."[34]

While hopeful for the future of planetary systems science, as the TOPS group went out of business in the summer of 1995, it could not have known that the first detections of extrasolar planets around stars similar to our Sun were just around the corner. In retrospect it is interesting to assess the importance of two decades of NASA studies to the real landmark discoveries that began to be made in 1995. The judgment of history must be that NASA played a very minimal role in the early discoveries, which were made by the Swiss team of Michel Mayor and Didier Queloz, followed shortly by many more discoveries from Marcy and Butler. Marcy and Butler had begun their project in September 1986, aware of the pioneering work in Canada of Bruce Campbell and Gordon Walker using a hydrogen fluoride absorption cell to provide a stable wavelength metric against which to measure stellar radial velocities. As part of his 1987 master's thesis, Butler concluded that iodine provided a preferable absorption cell, and in May of that year he designed and built the cell with San Francisco State University glassblower Mylan Healy. This was the prototype for all subsequent iodine cells. Over the next four years, as the TOPS group was undertaking its studies (in which Marcy and Butler played no role), the SFSU team was unable to achieve long-term precision better than one hundred meters per second. After hundreds of blind alleys and innumerable dead ends, by early 1992 their long-term precision was down to twenty meters per second.[35]

Up to this point Marcy and Butler's work had been supported entirely by the NSF, and, when Marcy received his first three-year NASA grant beginning in 1992, it was not from the planetary science program but from an "Innovative Research program" designed to support risky but potentially high-yield projects. Even then, the NASA referees were skeptical of the prospects for success; the minimal grant paid Butler's first postdoc salary. With crucial improvements to the Lick-Hamilton spectrograph carried out by Steve Vogt in November 1994 and incremental improvements to the software, Marcy and Butler were able to reach three meters per second by May 1995. It was October when the Swiss team made its first announcement of a planet around 51 Pegasi, confirmed by Marcy and Butler about two weeks later. During the following years of continuous discoveries, the NSF continued to provide the bulk of the team's funding, with some support from NASA, most notably in continued access to the Keck telescopes. Looking back at fifteen years of work of the Marcy-Butler team, Butler was lavish in his praise of NSF funding and critical of NASA's conservative attitude. It was an interesting contrast to the biological component of exobiology, in which just the opposite had been true from the early 1960s.[36]

With many studies behind it, and despite its failure to back the team that actually cracked the problem in 1995, NASA would now embrace the search for planetary systems beyond the wildest dreams of the TOPS team. Dan Goldin's entry onto the stage as NASA's administrator on 1 April 1992 would prove crucial to this new direction for the space agency.

Planetary Systems and the Search for Origins

As the twentieth century neared its end, attention to the problem of planetary systems reached new heights. Researchers realized that technology was ripe to open a new field. Studies in increasingly greater detail were undertaken demonstrating how planets could be observed from Earth and from space, using a variety of technologies, including "normal" (filled aperture) space telescopes and space interferometry. Genuine results were also being announced. The discovery by the Swiss team of Michel Major and Didier Queloz in October 1995 of a planet around a Sun-like star, followed by a raft of similar discoveries by Marcy, Butler, and others, fed the new field and gave it intense excitement.[37] Observations of circumstellar disks, possible protoplanetary systems, were increasing again, after the initial discoveries of the Infrared Astronomical Satellite in the early 1980s. NASA continued to contribute to the field by funding researchers and with the Hubble Space Telescope's observations in 1994 of possible protoplanetary disks around 56 of 110 young stars in the Orion Nebula.[38] Beginning in the 1970s, NASA had also funded an important series of "Protostars and Planets" meetings that brought together researchers in the field; originally largely theoretical, these meetings increasingly reported observational results. Perhaps most important of all from a programmatic and funding viewpoint, the search for planetary systems became an important part of the bold new overarching program at NASA known as Origins. Under its banner planetary systems science was assured of continued attention and funding.

Three studies provided the backbone for the Origins program, although no one knew when the studies began that they would coalesce into a connected program. Even as the Solar System Exploration Division's TOPS group was meeting, the Astrophysics Division of NASA's Office of Space Science had created a Space Interferometry Science Working Group (SISWG) to follow up on the 1991 National Research Council Bahcall Report, which had recommended the start of an Astrometric Interferometry Mission, with the search for planetary systems being a major justification. This group was charged with deciding whether the JPL/Shao Orbiting Stellar Interferometer or Reasenberg's POINTS should be selected for development, a process at NASA known euphemistically as "downselecting." The committee met over the next four years and, after many twists and turns, received a revised charge in 1995 to decide on an instrument that could act as a technology precursor for interferometers being proposed by other committees for planet searches in the long term. The committee certified in the fall of 1995 that JPL's Orbiting Stellar Interferometer (OSI) satisfied the requirements and submitted its final report in the spring of 1996. The Astrometric Interferometry Mission of the Bahcall Report would take the form of JPL's OSI and was rechristened the Space Interferometry Mission (SIM). Planetary systems were a major part of the mission, scheduled for launch around 2010.[39]

Meanwhile, two other groups had been convened which would impact heavily on the planetary systems theme and eventually the Origins program; their results fed into the deliberations of the interferometry working group. The first was the "HST and Beyond" Committee, whose charge was to undertake a broad study of possible missions for ultraviolet, optical, and infrared astronomy in space for the first decades of the twenty-first century and to "initiate a process that will produce a new consensus vision of the long term goals of this scientific enterprise." This group was chartered in September 1993 by the Association of Universities for Research in Astronomy (AURA), through the Space Telescope Institute Council, with support from NASA. The eighteen members of the committee, chaired by Alan Dressler of the Carnegie Observatories, had broad experience with observations from space. The committee assumed that planned programs such as SIRTF and the Stratospheric Observatory for Infrared Astronomy (SOFIA) would be implemented; they were to look beyond that horizon, with full knowledge of the work of the Bahcall Report, the TOPS group, and discussions about a next generation of space telescope.

The group met three times, twice in 1994 and for the last time in May 1995, producing its report in May 1996, just a month after the SISWG group's report.[40] Taking the story of cosmic evolution as its broad background, the committee noted two crucial missing chapters: the detailed study of the birth and evolution of normal galaxies such as the Milky Way; and the detection of Earth-like planets around other stars and the search for evidence of life on them. To solve these problems the committee recommended a three-pronged approach for the decades beyond 2005. First, the HST observations, with its capabilities in the optical and ultraviolet, should be extended beyond 2005. Second, a new Space Telescope, optimized for infrared observations, should be built to follow in the footsteps of the HST. With a proposed four-meter aperture (compared to ninety-two inches for HST), it would be the first "facility class" instrument since the Advanced X-ray Astrophysics Facility (the x-ray satellite later christened "Chandra") and SIRTF, and would allow detailed studies of distant galaxies. This so-called Next Generation Space Telescope (NGST), which had already been studied since 1989, would end up on the drawing boards as an eight-meter telescope, thanks to the influence of the ubiquitous NASA administrator Dan Goldin, and eventually would settle on a six-meter mirror. Third, NASA should develop the capability for space interferometry, both in the optical and infrared regions. In the view of the committee infrared space interferometry, in particular, would be essential to the detection and study of extrasolar planets. These recommendations would increase support for the NGST, SIM, and a second-generation space interferometer even beyond the capabilities of SIM.

As the HST and Beyond group was in the midst of its work, another group was focusing much more specifically on planetary systems; in many ways its goal was to update the TOPS report of three years earlier. In March 1995 NASA chartered a group of scientists and engineers to lay out a roadmap for the Exploration of Neighboring Planetary Systems (ExNPS). In an activity coordinated by Charles Elachi, head of the Space and Earth Science Directorate at JPL, three independent teams developed roadmaps, which were completed in September 1995 and then synthesized into a single plan by an Integration Team. A blue-ribbon panel headed by Nobelist Charles Townes reviewed the roadmap on 4–5 October, the results were submitted to Dan Goldin on 7 November 1995, and the plan was published in August 1996.[41]

One measure of burgeoning interest in the subject is that some 135 scientists from 53 institutions participated in the ExNPS deliberations. They concluded that within twenty years a space-based observatory could detect Earth-like planets around the closest one thousand stars and characterize the atmospheres of the brightest ones. The ExNPS report laid out an entire program and timeline, ranging from the indirect detection of planets to "family portraits" of planetary systems and even detailed images of planets (fig. 7.4). Key to these goals, in addition to ground-based instruments and space missions already planned, were a space optical interferometer to detect wobbles in stars due to planets and a space infrared interferometer to detect and characterize Earth-like planets to thirteen parsecs. The optical interferometer would be SIM, Shao's proposal, which had just been selected by the SISWG. The more long-term infrared interferometer was envisioned as four or more 1.5-meter telescopes linked together on a 50- to 100-meter baseline and placed in a deep space orbit some 3 to 5 astronomical units (AU) from the Sun. It was based on studies by Roger Angel and Shao in 1990, using a "nulling" principle originating with Ronald Bracewell in 1978. That such an instrument could directly image and characterize Earth-like planets was the

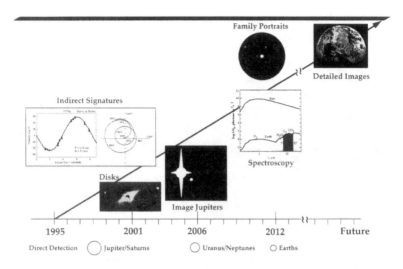

FIGURE 7.4. Program and timeline for exploring neighboring planetary systems *(From* A Roadmap for the Exploration of Neighboring Planetary Systems *[Washington, D.C.: NASA, 1996], 1–2.)*

"fundamental finding" of the ExNPS roadmap. Equivalent to the infrared space interferometer proposed in the TOPS report of 1992, it would soon be given the name Terrestrial Planet Finder (TPF).[42]

Following in the steps of the TOPS team four years earlier, the ExNPS team concluded that some of humanity's oldest questions were within scientific grasp, including the uniqueness of the Earth and life. "Our firm conclusion is that NASA can answer these questions within the next 10 to 20 years."[43] Although many reports gathered dust in NASA, the discovery of a planet around 51 Pegasi, announced in October 1995 between the Townes review and the presentation to Goldin, gave credence to the hope that planets actually existed and put ExNPS on a fast track. By the time the report was published in the summer of 1996, it included data for five possible planets around Sun-like stars and an HST image of a brown dwarf complete with a spectrum taken by the Keck telescope showing the presence of methane—an unambiguous indicator that this was no normal star (fig. 7.5). In addition, the HST had discovered protoplanetary systems.

Thus, in the period of a few months in 1996 three independent reports by the SISWG, HST and Beyond, and ExNPS teams were published. The conclusions of these groups were known well before publication, and Goldin lost little time capitalizing on them and the excitement of the discoveries of new extrasolar planets. In January 1996 he presented these results to more than a thousand astronomers at the winter meeting of the American Astronomical Society in San Antonio, Texas, where Marcy and Butler announced the discovery of two more

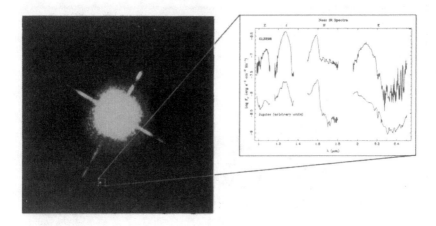

FIGURE 7.5. Hubble Space Telescope image of brown dwarf GL 229B. The large object is the star Gliese 229, and the brown dwarf is the tiny image, *lower right,* separated by 7.7 arcseconds. At *right* the spectrum of the brown dwarf indicates the presence of methane, similar to the gas giant planets of our solar system. *(From* A Roadmap for the Exploration of Neighboring Planetary Systems *[Washington, D.C.: NASA, 1996], 3–4.)*

planets. At this meeting Goldin wrapped together all of these studies as a connected program: NGST as an instrument for studying solar systems in formation, the Space Interferometry Mission for detecting planets, and the Terrestrial Planet Finder for studying the planetary characteristics.[44] The program was called "Origins."

During the course of 1996 the Origins theme was formalized during an administrative restructuring of the agency's Office of Space Science, when the former ultraviolet, visible, and infrared disciplines were combined into a single activity.[45] Wes Huntress, the associate administrator for space science, played an essential role in this reorganization, which made the "Astronomical Search for Origins" one of four themes in NASA's Office of Space Science, along with the Sun-Earth connection, solar system exploration, and the structure and evolution of the universe. The three independent reports published in 1996 on space interferometry, HST and Beyond, and ExNPS provided its essential foundation. Indeed, Dressler's HST and Beyond report contained a section on "The Scientific Case for the 'Origins' Program," with the word *Origins* still in quotation marks because the name had not yet been officially adopted for the program.

By summer 1997 a detailed "Origins Roadmap" was published by the Origins Subcommittee of NASA's Space Science Advisory Committee. The subcommittee was chaired by none other than David Black, the omnipresent figure in the field from the mid–1970s planetary systems workshops associated with SETI. The roadmap described three ambitious scientific goals for the Origins theme, dealing with galaxies, planets, and life, all keyed to the question "Where did we come from?" These goals—the epitome of cosmic evolution—were to understand how galaxies formed in the early universe and their role in the appearance of planetary systems and life; how stars and planetary systems form and whether life-sustaining planets exist around other stars; and how life originated on Earth and whether it exists elsewhere. SIM and NGST were set forward as the two mission candidates in the 1997 Origins roadmap.[46] Terrestrial Planet Finder was mentioned as a "long-term mission" that would not yet be ready for the 2000–2004 time frame. By the time the roadmap was updated three years later a fourth goal was added, distilled from the previous three: whether habitable or life-bearing planets exist around other stars in the solar neighborhood. Moreover, detailed studies had been done on TPF, and an upgraded version featuring four 3.5-meter free-flying telescopes stretched out along a kilometer baseline was incorporated into the 2000 roadmap.[47]

The 2000 Origins roadmap went even beyond the TPF. It envisioned a Life Finder (LF) to make detailed studies of any planets found by TPF. A Filled-Aperture Infrared (FAIR) telescope would anticipate the LF by developing technologies needed for the twenty-five-meter telescopes of LF. Finally, beyond the NGST, a Space Ultraviolet/Optical telescope would be developed. By combining all these missions into one program, each could build on the previous technologies.

All of these Origins programs represented missions that would be launched long term; SIM and NGST would not fly until about 2009 and TPF and LF after that. Meanwhile, more immediate missions emerged from other NASA programs. In late 2001 NASA chose the Kepler mission for launch in 2006. Although it was not formally part of the Origins program, Kepler was very much in the Origins tradition: in place of astrometry or the radial velocity method, it would use a photometric method to search for Earth-size planets as they "transited" in front of a star, dimming the starlight by extremely small amounts. The principle investigator for the mission was William Borucki, who had worked on the astrometric telescope project in the early 1980s at Ames. Still at Ames (where, as we shall see in chap. 9, an astrobiology program was in full swing), Borucki had been pushing such a mission for more than a decade. Now Kepler would be able to monitor one hundred thousand Sun-like stars for four years, looking for light variations that might indicate other Earths. In the planet-hunting tradition persistence paid off.

The progress in observational planetary systems science over twenty-five years was impressive. While the general search techniques were known even at the beginning of that period, by its end they had not only been greatly fleshed out, but planets and protoplanetary systems had actually been discovered. Just as early in its history the question of life on Mars drove much of NASA's space science effort, so now the question of planetary systems and life drove NASA's goals as never before. With HST returning spectacular pictures, SIRTF (the last of the Great Observatories) about to be launched, and Kepler, NGST, SIM, and TPF on the drawing boards, no one could accuse NASA of lacking vision. At least this was true in the space sciences, by contrast to human spaceflight, in which the space shuttle and space station were stuck in Earth's orbit. Curiously, the vision of space scientists—in part because of the lure of planets and life— was outmaneuvering the more expensive manned space flight, the latest episode in a long-running debate about the relative merits of the two approaches. For the planet search, the challenge was turning the vision into reality, a process that was a matter of NASA's internal priorities, public interest, and congressional funding.

CHAPTER 8

The Mars Rock

"Clues in Meteorite Seem to Show Signs of Life on Mars Long Ago: Startling Find of Organic Molecules from Space." The headline jumped out from the front page of the *New York Times*. It was Wednesday, 7 August 1996.[1] A few days later the top headline of the "Science" section of the *Times* declared, "After Mars Rock, a Revived Hunt for Otherworldly Organisms." Feature articles described the breaking news about Mars meteorite ALH84001[2] and also (with a high-resolution photo of Europa taken by the *Galileo* spacecraft) the possibility that "Jupiter's Moon Europa Could Be Habitat for Life."[3] The 7 August headlines were prompted by NASA calling a very sudden press conference at its Washington, D.C., headquarters, announcing findings from a Martian meteorite which suggested that microbial life may have existed on Mars over 3.5 billion years in the past; the two lead researchers were career NASA scientists. In close coordination with the NASA announcement, the White House issued further remarks. President Bill Clinton himself called this potentially one of the most important scientific discoveries in history; he called for a space summit in November to discuss future exploration of Mars. Vice President Al Gore began organizing a private conference for December to discuss the larger social implications if the discovery turned out to be true. In November and December NASA planned to launch the *Mars Global Surveyor* (an orbiter) and *Mars Pathfinder* (a lander, with a mini–surface rover called Sojourner) spacecraft, to arrive at Mars in the summer of 1997. In September planning was already well under way at JPL for a mission to return a Martian sample to Earth by 2005.[4] Not in the twenty years since *Viking* had Mars or NASA exobiology work generated this level of excitement. To most of the public it all seemed to come out of nowhere. As it turned out, even members of the research team working on the Mars meteorite had not originally planned to have their press conference until 15 August, the day before their published article would appear, and they were scrambling, in a rather unorthodox way for science, to break the story nine days ahead of publication (fig 8.1).[5] This surely ranks as one of the most dramatic moments in the history of NASA Exobiology, and it was the single most important impetus that led to the creation of astrobiology. No episode, not even the *Viking* search for life on Mars, demonstrates so

FIGURE 8.1. Three lead members of the team that authored the 1996 *Science* article arguing that biochemical and microscopic evidence from Mars meteorite ALH84001 suggested possible fossil life from ancient Mars. *Left to right:* Everett Gibson, Kathie Thomas-Keprta, and David McKay, posing with a globe of Mars in February 2000. In the background is a highly magnified image of the "nanostructure" that came to be dubbed the "worm." (Courtesy NASA.)

dramatically how integral public interest (and spending) has become to the science of exobiology; but how did it all come about? To find the roots of the story, we must go back almost all the way to *Viking* days.

Its Preposterous Heritage

In 1982 Donald Bogard and Pratt Johnson, two scientists at NASA's Johnson Space Center (JSC) in Houston, announced that they had liberated a sample of trapped gas from within glass inclusions in a meteorite picked up in Antarctica in 1979. The meteorite was named Elephant Moraine 79001 (from its location and the fact that it was the first one processed by scientists in 1979), or EETA79001. Upon analyzing the gas, they discovered that it matched almost perfectly the gas mixture of the atmosphere of Mars as measured by *Viking* in 1976.[6] When they published the detailed results,[7] the most likely explanation was an eye-opener: this rock had somehow been blasted off Mars some two hundred million years ago by an impact large enough to accelerate it to escape velocity (5 km/second). Then after a long time in space its orbit intersected Earth's, it landed, and there it lay, a stranger in a strange land, waiting only to be picked up once scientists became aware, beginning in 1969, of how many meteorites lay undamaged on the ice of Antarctic glaciers. Over forty-seven hundred of them had been collected by the end of 1980.[8]

Researchers had been thinking for some time that a group of rare meteorites called "Shergottite-Nakhlite-Chassignites" (SNCs),[9] though clearly extraterrestrial, were similar geochemically to terrestrial basalt and thus were from a parent body that had experienced complex melting and crystallization through vulcanism similar to Earth's. But the SNCs were all thought to have crystallized only 1.3 billion years ago, long after the asteroids and the Moon had cooled enough for volcanic activity to end. "Thus Mars and its relatively young lava flows seemed to be the most likely source. As Benton Clark of Martin Marietta Denver Aerospace showed . . . the chemical composition of Shergotty, the first of the four Shergottites to be found, provides the best match to the composition of Martian soil as determined by the Viking landers."[10] Still, the match by itself did not seem scientifically compelling. But Bogard and Johnson's 1982 analysis of noble gases within meteorite EETA79001, also a Shergottite, "brought sudden respectability, if not credibility, to the suggestion of a Martian origin."[11] The shock of the impact that blasted the rock off Mars formed the glass within, trapping gas from the Martian atmosphere in the glass.

At a conference on 17 March 1983 at the JSC, the idea received a further boost, albeit a psychological one. Even more convincing evidence, from direct geochemical comparison with *Apollo* lunar samples, showed another Antarctic meteorite to be undeniably from the Moon. The conceptual barrier to accepting the idea of intact escape of a rock from a planetary-sized body had been broken.[12] Afterward researchers refined their calculations and eventually concluded that the SNC meteorites were probably Martian, even if they could not prove right away how it was physically possible to get the original approximately ten-meter boulder (from which the meteorite must have come) off of Mars and up high enough to escape velocity without it being vaporized or pulverized. Interplanetary travel from Mars to Earth had occurred on several occasions, it seemed. (Earth's gravity is so much greater that it is a great deal less likely that an Earth meteorite could survive ejection to escape velocity and ever reach Mars.) It is currently believed that several Martian meteorites arrive on Earth every year, along with several from the Moon. The totals from each are about the same: even though the Moon is a much closer source, Mars is so much larger a target that it is struck more often by impacts large enough to eject rocks at escape velocity.[13]

By 1987 even University of Arizona geochemist Michael Drake, who was at first very skeptical, said of the Martian origin of the SNCs: "It's probable, but not proven; it's not likely to be incorrect. But short of going to Mars, no one will be absolutely convinced."[14] Evidently, the psychological barrier was not removed all at once at the 1983 JSC meeting; Richard Kerr noted in 1987: "perhaps more than anything, the passage of time has made a Martian origin an acceptable hypothesis. . . . Naturally enough, those working with the impressive geochemical data are most inclined to accept the idea, but support has broadened considerably."[15] Among those familiar with the data were geologist David McKay of JSC, Houston, and geochemist Harry McSween of the University of Tennessee. Subsequently, as of late 1999, a total of seventeen meteorites were known to have come from Mars; Bogard and Garrison showed that seven of them contained trapped Martian gases. (By April 2004 the number was thirty.)[16] In addition, University of Chicago isotope geochemist Robert Clayton recognized that all the SNCs have a unique nonterrestrial composition of oxygen isotopes in their silicate minerals which "shows they were from a unique oxygen reservoir within our solar system."[17]

By 1989 researchers at Britain's Open University thought they had discovered native organic matter in EETA79001.[18] This would have been extraordinary, since the *Viking* GCMS had shown no organic matter on the Martian surface, down to a few parts per billion. When other groups tried to replicate these results and failed, however, it was concluded that the organics in the meteorite must be Earthly contamination that seeped into it along with Antarctic meltwater during the thousands of years it lay exposed on the ice sheet there. Although this controversy attracted relatively little attention in the press, the result was that the scientific community still believed by the mid–1990s that Mars had no native organic matter, and, therefore, neither did Martian meteorites.[19] Even so, and notwithstanding the continued public disagreement of the majority of scientists with Gilbert Levin over the *Viking* LR results, the convening of an International Symposium on the Biological Evolution of Mars at Florida State University on 26–28 October 1990 showed that, whatever the public perception in the years after *Viking,* in the exobiology science community a hard core of interest in life on Mars remained very much alive and active. The conference was convened by Imre Friedmann and his ACME research group; other prominent participants included a wide sampling from the origin of life / exobiology field, including many senior researchers and administrators. Among them were Harvard Precambrian paleofossil expert Andrew Knoll; chemist Benton Clark, Leiden University (Netherlands) comet expert J. Mayo Greenberg; biological and prebiotic membrane specialist David Deamer; NASA Exobiology chief John Rummel; former NASA Exobiology chief Richard Young; Harold Klein of NASA Ames; National University of Mexico biologist Antonio Lazcano; chemoautolithotroph specialist and director of the Soviet Institute for Microbiology Mikhail Ivanov; biochemist Klaus Dose of Johann Gutenberg University in Mainz, Germany; planetary scientist Chris McKay; NASA Ames organic chemist and veteran of Moon rock analysis Sherwood Chang; and many others. It was a veritable who's who of the exobiology community in many countries and through at least two generations.

Little surprise, then, that analysis continued on the Mars meteorites, not only from a purely geochemical or planetary science point of view but, for some workers, with at least an occasional thought for exobiology. With continued study of meteorites and collection of new ones, more were recognized to be of the SNC class, and their Martian origin was more and more widely and certainly accepted. In 1993 David Mittlefehldt of NASA's JSC in Houston recognized for the first time that a 1.9 kilogram, potato-sized rock, the first meteorite collected in 1984 in the Allan Hills, near the Antarctic Dry Valleys (hence designated ALH84001), belonged to the Martian group. He sent a small chip of the meteorite to Robert Clayton's lab at the University of Chicago, where it was confirmed that ALH84001 had the unique Martian oxygen "isotopic fingerprint."[20] It was later found that the meteorite had been ejected from Mars sixteen million years ago and had landed in Antarctica thirteen thousand years ago.[21]

At the same time, in a lab across the hall from Mittlefehldt at JSC, NRC postdoc Chris Romanek, working in geochemist Everett Gibson's lab, was using a tightly focused laser beam on carbonaceous chondrites (including the Murchison meteorite) to measure the carbon isotope ratio at precise spots within the sample where they contained carbonate minerals. Romanek was a specialist in the formation of such minerals and wanted "to gain insights into whether those carbonate minerals were formed perhaps by biological processes and at what temperatures they formed."[22] Mittlefehldt was going over some images of small (1–250 μm diameter) globules of carbonate within ALH84001; knowing these were Romanek's special interest, he came across the hall and asked, "Hey Chris, do you want to see some really neat pictures of a meteorite that I'm working on?" He added that this was the latest addition to the family of Martian meteorites. Romanek was fascinated and immediately asked for a piece of the sample to include in his study on carbon isotope ratios, which Mittlefehldt supplied. This was the only one of the SNC meteorites known to have anything more than traces of carbonate minerals.

Romanek worked from 1993 to 1996 with geochemist Everett Gibson from JSC; they soon found that the carbon isotope ratios of the carbonate globules in ALH84001 were unlike any sample ever seen on Earth. They contacted the Open University group in Britain, Colin Pillinger, Ian Wright, and Monica Grady, knowing they were working on the same meteorite but using a different method, and asked what ratio they had measured. Both groups had independently arrived at a value (for $_{13}$C relative to $_{12}$C) of plus-forty per mil, using different methods; they agreed in 1994 to publish the finding together in *Nature*.[23] In this paper they also concluded that the stable oxygen isotope data supported a low-temperature (between 0 and 80°C) formation of the carbonate globules. This could indicate that they had resulted from biological activity; however, "petrographic and electron microprobe results indicated that the carbonates formed at relatively high temperatures (~700°C)."[24] These latter measurements were made by Case Western Reserve University geochemist Ralph Harvey and Harry McSween of the University of Tennessee. Clearly, this ambiguity had to be resolved before anything could be safely said about the origin of the globules. But, argued the JSC and Open University group, the unusual carbon isotope signature in the globules did suggest they had formed on Mars rather than Earth.

In trying to gain further insight into the temperature issue, Romanek decided to try an acid etching technique he had heard about in a talk by University of Texas geologist Robert Folk at the Geological Society of America. Folk had acid-etched carbonates that came from hot springs (on Earth), then used scanning electron microscopy (SEM) to image their surface features. So, Romanek tried the procedure on some of the ALH carbonate globules, using the SEM in the Solar System Exploration Division at JSC in Houston. In the original work Folk had seen "tiny features that he later characterized as nanobacteria; the fossilized remains of dwarf or miniature-sized bacteria that were trapped or entombed in these hot spring deposits."[25] Now in May 1994, when Romanek looked at the carbonate globules from the Mars rock, he saw features that looked strikingly similar to Folk's.

"In my estimation this is where the whole project began," he said. "I took those pictures down to Everett Gibson's office, and I showed him the pictures I got . . . and the pictures in Bob Folk's publication. I said you can see the difference between what you see in the meteorite and what we see for published nanobacteria in terrestrial rocks. . . . He immediately lit up, . . . and he said 'Chris, we need to go down and talk to Dave McKay.'"[26] McKay ran the SEM and transmission electron microscopy (TEM) lab at the Johnson Space Center; having been in on analysis of lunar soils from the very beginning, he was an expert on planetary regoliths. Once Gibson and Romanek showed him the photos and filled him in on the story, McKay became very interested but realized that he could not devote enough time to the project, so he asked if Gibson and Romanek would agree to bring in electron microscopy expert Kathie Thomas-Keprta, a contractor at JSC employed by Lockheed Martin Corporation nearby in Houston; they agreed. Both Gibson and McKay had been NASA Exobiology grantees before, though most of McKay's funding had come from the NASA Planetary Materials Program. McKay had also previously worked with Mittlefehldt's group, doing SEM petrography on thin sections of ALH84001 to see whether any Martian regolith was mixed into the less-dense, jumbled-up texture zones in the rock. Now Gibson and McKay applied for a new grant, specifically to look for signs of life in Martian meteorites. Their initial proposal was rejected, but another, submitted the next year (before the announcement of their work on ALH84001), was granted in the late summer of 1996.[27] They knew Chris Romanek's postdoc at JSC would soon end, so they made themselves, both career civil servants at JSC, the principal investigators on the grant applications.

When Kathie Thomas-Keprta was first approached, she was resistant to becoming involved in the Mars meteorite project; she already had a large workload in a project examining interplanetary dust particles (IDPs). When McKay explained what he wanted from her, she was highly skeptical, a "doubting Thomas" as she later described herself at the August 1996 press conference. But Romanek continued urging her, getting on the SEM with samples and showing them to her, and she slowly warmed to the project. Then, recalled Romanek, when she saw very tiny grains of the mineral magnetite in thin sections, located in the dark rims of the carbonate globules, "she became excited because she had . . . [seen] magnetites in other meteorites and in interplanetary dust particles . . . and knew that these magnetites in this meteorite were very different. . . . She started digging in the literature and realized—she's the one that came to the conclusion—these magnetites look exactly like magnetites that form from bacteria on Earth. And I think at that point it crystallized in her mind the significance of what she was working on and how much more work needed to be done."[28]

Early in 1995 Gibson invited J. William Schopf, the UCLA specialist in microfossils, to come to Houston and look at their images of putative nanobacteria. Schopf came in January; "he thought the morphological evidence was very interesting, but it was far from conclusive. . . . His main point . . . was that you will never convince anyone that these things are biologic unless you can find organic matter associated with them. And . . . that was kind of a big letdown for us, because we knew that there was no organic matter on Mars."[29]

McKay and Thomas-Keprta had previously worked with a team at Stanford University under Richard Zare to quantify carbon compounds in IDPs. Zare's team used a machine called a microprobe two-step laser mass spectrometer ($\mu L_2 MS$). Now Thomas-Keprta suggested their technique might be capable of finding organics in the carbonate globules, since it was capable of being focused down to a forty micron–diameter spot in a sample. She contacted Simon Clemett of the Stanford team and, without saying anything about the source of her samples, asked if Zare and Clemett's group could analyze them and tell her whether there was any carbon associated with them. The specialized Stanford mass spectrometer in March 1995 was tuned to look for a type of organic molecule called polycyclic aromatic hydrocarbons (PAHs), so, in order to avoid altering the settings, that is what they first looked for. PAHs are commonly found in interstellar matter, on meteorites, and in many other places, including on Earth. They can be formed by a variety of processes, both biological (in petroleum formation, in coals) and totally abiotic (in flame chemistry, auto exhaust, and interstellar gas), but one place they had never yet been detected was on Mars or on Mars meteorites. On each of three separate ALH84001 samples, PAHs were found to be quite common.

This was a major discovery in itself, since no organic molecules of any kind had been found on Mars. (The intellectual bias that would have resulted from that knowledge justified keeping the identity of the samples from the Stanford team until after it had made its measurements, according to Romanek.) The JSC team carried out numerous control experiments to demonstrate conclusively that the PAHs did not get into the sample in the Houston lab, the Stanford lab, or in transport between the two. Simon Clemett even showed that the concentration of the molecules increased from outside the meteorite to the inside, strongly presumptive evidence that the PAHs were native to the inside of the Mars rock.[30]

Because of their ubiquitous distribution in the universe from abiotic as well as biological chemistry, the molecules were not ideal as markers of biogenic organic matter, what the JSC team was initially seeking. But the very finding of organics in a Martian sample where no one believed there would be any was a big boost to the team's hopes that the morphological findings in the rock might have biological significance. Romanek had gotten a job when his postdoc ended and moved in March 1995 to the University of Georgia's Savannah River Ecology Lab. Therefore, an additional team member was recruited at JSC to do more intensive TEM work, Hojatollah Vali, a McGill University Ph.D. graduate in electron microscopy who was at JSC on an NRC fellowship. Thomas-Keprta and Vali worked hard to get the clearest, most unambiguous electron micrographs possible of the "nanostructures."

At the March 1995 annual Lunar and Planetary Science Conference at JSC, Thomas-Keprta gave a paper on interim thinking on the project, barely hinting at the idea that the evidence to date might be of biogenic origin. The title was "Organics Indigenous to Mars or Terrestrial Contamination?" as the controls had not yet been done; the press showed little interest, as a result. Most researchers outside the team assumed that, because *Viking* had shown no organics on Mars, the PAHs must be Earthly contamination. One exception was a reporter from the *Houston Chronicle,* Carlos Byars, who seemed to catch a whiff of where the finding of organic matter might be headed. After starting his new job at Savannah River, Romanek stayed in constant telephone contact with the Houston group and returned to work intensively on the project for two weeks in June, three weeks in December 1995, and then as a visiting faculty member for the summer of 1996. McKay obtained a lot of very high-resolution SEM images of the nanostructures on the carbonate globules using a field emission gun (FEG SEM) at the NASA Houston facility.

By late 1995 the members of the team began to think that they might be close to having enough data after almost three years of work to submit a paper to *Science* or *Nature,* arguing for a possible biological explanation for the data. Because the igneous rock had crystallized on Mars 4.5 billion years ago (much older than any of the other SNC meteorites) and the carbonate globules seemed to have formed within the rock between 1.3 billion and 3.6 billion years ago, their argument would amount to hypothesizing that microscopic life had existed on Mars sometime between 1.3 and 3.6 billion years ago (probably at the earliest end of that period, since Mars began to dry up and lose its atmosphere by 3 billion years ago). As the carbonate globules formed, Romanek thought that, possibly under the influence of some biogenic process in an aqueous environment at a temperature below 80°C, some microbes (at least the extremely tiny ones, only 100 to 380 nm long—i.e., only 0.100 to 0.380 μm) became trapped in the globules and later fossilized there.

From the beginning of work on the paper, the team members realized that none of their lines of evidence was conclusive by itself; all had ambiguities that allowed for an abiotic explanation as readily as a biogenic one. Thus, they began constructing their argument according to an unusual line of reasoning:

whereas each of several different lines of evidence was not in itself conclusive proof of biogenic activity, "when they are considered collectively, particularly in light of their spatial association, we conclude that they are evidence for primitive life on early Mars."[31] This reasoning (perhaps used only out of lack of choice) was representative of the historical process of the investigation, rather than the much more common rationalist reconstruction used in scientific papers to make it look as though the entire investigation unfolded in a logical sequence according to rational hypotheses and their tidy, sequential testing. According to Chris Romanek, the published version of the paper was actually substantially more cautious and qualified in its claims than what was first submitted, which he considered an excellent outcome—the scientific process working just the way it should.[32] As we shall see, however, in the minds of a great many scientists, the kind of reasoning in the *Science* paper weakened the case and made it suspect from the outset.[33]

Its Sudden Fame

McKay, Gibson, and their colleagues submitted their paper to *Science* on 5 April 1996, later revised it, and had it accepted on 16 July 1996; on 7 August of that year they announced its findings in a NASA press conference, and the paper was finally published nine days later. It opened with two major qualifying statements: "Our task is difficult because we only have a small piece of rock from Mars and we are searching for Martian biomarkers on the basis of what we know about life on Earth. Therefore, if there is a Martian biomarker, we may not be able to recognize it, unless it is similar to an earthly biomarker. Additionally, no information is available on the geologic context of this rock on Mars."[34] The first point was a constant occupational hazard that had dogged exobiology from its beginning. The last point, about the rock being studied in complete absence of its geological context, has recently been shown to be a problem well worth mentioning up front. We will return to this at the end of this chapter.

The authors then laid out four main lines of evidence to indicate possible biogenic activity, which they later summed up as: "1) the presence of carbonate globules which had been formed at temperatures favorable for life, 2) the presence of biominerals (magnetites and sulfides) with characteristics nearly identical to those formed by certain bacteria, 3) the presence of indigenous reduced carbon within Martian materials, and 4) the presence in the carbonate globules of features similar in morphology to biological structures."[35] These lines of evidence were not simply to be considered in an additive fashion, they argued; because so much of the independently suggestive molecules all existed in the carbonate globules or their immediate vicinity, the presumption of all having been caused by biogenic activity in that locale was strengthened in a synergistic way. This "spatial association" argument was important: a large number of observers were willing to dismiss the case out of hand based on each of the lines considered separately because in not one of those cases had the team shown the biogenic explanation to be significantly more persuasive than one or more abiotic explanations. Many skeptics who said they still kept an open mind on the question said it was the spatial association argument that gave them pause.

Because some of the carbonate globules were "shock-faulted," which must have occurred on Mars or in space, the authors argued, this ruled out an Earthly origin for the globules. On Earth such fine-grained carbonates usually form under water and most often by biologically mediated processes; in addition, Thomas-Keprta found minerals in their rims that were often associated with microbial activity (magnetite, pyrrhotite, and other iron sulfides such as greigite). The *Science* paper argued that the redox and pH conditions usually required for the inorganic deposition of fine-grained carbonates, magnetite, pyrrhotite, and greigite were largely incompatible with one another; it would require a strained and extremely unlikely combination of circumstances to explain the formation of all these minerals in the same place by purely abiotic means.[36]

The paper carefully ran through the control experiments that had been carried out to rule out contamination at JSC, in transit, or at Stanford as the source of the PAHs. The authors had cultured chips of the meteorite in standard microbial media, both aerobically and anaerobically, and had found the chips to be sterile.[37] Regarding the possibility that the molecules represented terrestrial contamination from before the meteorite was ever collected in the Antarctic, they argued that the outside crust was almost totally devoid of the PAHs. Furthermore, their concentration rose going in toward the center; it was highest in the immediate vicinity of the carbonate globules. The authors took this to be suggestive of a common (biogenic) process of origin for the globules and the PAHs.

In the published paper (unlike at the press conference, where some more recent and more dramatic SEM images were also shown) the least was made of the putative "nanobacteria." They were described for the most part using the neutral description "ovoid and elongated forms." Only a single paragraph compares them to Folk's nanobacteria and states that they "resemble some forms of fossilized filamentous bacteria in the terrestrial fossil record," noting, however, that those microfossils are "more than an order of magnitude larger than the forms seen in the ALH84001 carbonates."[38] Predictably enough, the press and the public watching on television responded much more strongly to visual images that looked like familiar bacterial shapes than to arcane arguments about isotope chemistry or little-heard-of molecules such as PAHs. To one not familiar with microbial biochemistry there was no obvious reason why a lower limit on bacterial size, if it existed, would fall above these structures, whose shape was so compellingly lifelike.

Above and beyond the scientific evidence or logic, another factor that may potentially have predisposed some observers to be skeptical was the JSC team's unusually secretive behavior during the time the work was being done and even after the paper had been submitted and was under review for publication in *Science*. Everett Gibson has stated that the team considered the Clinton Administration and the bureaucracy at NASA headquarters in Washington to be a "sieve," systematically subject to press leaks of any important story. Thus, after the summer or fall of 1995 the team deliberately did not keep NASA managers in the usual chain of command informed of their work; they simply considered the story so potentially big that, without secrecy, leak(s) would be inevitable. They informed their immediate supervisor, Doug Blanchard, as well as Carol Huntoon, in the director's office at JSC, but no other "higher ups."[39]

David McKay has also said that members of the group wanted to gather as much evidence as they could before publicizing their argument, to be sure they were right before going out on a limb with such an extraordinary claim. Schopf's January 1995 comments had certainly sensitized them to this possibility, in addition to their own scientific training about what makes compelling evidence. Furthermore, according to McKay, "we knew a hundred other groups had this meteorite and we didn't want to be scooped by one of them, and we knew if we started talking about this openly at meetings and so forth, everybody would turn to it and start looking at it, and so we wanted to be first really."[40] These circumstances come with the territory of exploring a truly exciting new discovery; how to handle them is not spelled out in any simple set of rules in a handbook, so scientists attempt to negotiate these treacherous waters on a case-by-case basis when they discover themselves in such situations. Concern for priority, if not ubiquitous, is at least very common; given the grant-based, peer-review-driven process of modern science it could hardly be otherwise.[41]

In the event, the concerns of the JSC team turned out to be justified in a more bizarre way than any of its members foresaw. When *Science* officially accepted the paper on 16 July, top NASA administrator Dan Goldin finally got wind of what the JSC team had been working on and of the news that it was to appear in print in the most prestigious science journal in the country in one month. He immediately contacted associate administrator Wes Huntress and told him to get Gibson and McKay to Washington, D.C., and into his office as quickly as possible. Within days the two had been ordered to do a command performance before their most senior of bosses. In Goldin's office at NASA headquarters in late July, Huntress watched as Goldin grilled the two scientists mercilessly, probing the strengths and weaknesses of their soon-to-be-published argument. Goldin recognized that the entire prestige of NASA, not merely of these scientists, was riding on the publication of such a spectacular claim. The October 1993 cancellation of all SETI funds by Congress, after Nevada Senator Richard Bryan convinced his colleagues that it was a frivolous "great Martian chase," was a wound that still smarted. And a major congressional vote on renewed NASA funding was coming up in September.

After two hours or more Gibson and McKay had satisfied Goldin that the ALH84001 paper made its claims with proper scientific caution and had secure and provocative evidence for how far it pressed the case for past life on Mars. He congratulated the two men and told them henceforth to communicate any news directly with him or his deputy, skipping over intermediate officials in the hierarchy.[42] Then he eagerly went to work, first to notify the president and vice president of what could potentially be the most important scientific story of all time. He instigated planning of a major news conference for 15 August, just prior to publication, to announce the results to the press and the world and to explain the evidence and its limitations carefully. So concerned was Goldin to avoid the impression that NASA was being grandiose and unscientific that he arranged for J. William Schopf of UCLA to give a formal presentation at the press conference of the case for why he (Schopf) and many other scientists were skeptical and felt the evidence did *not* justify the conclusion of past life on Mars.[43]

President Bill Clinton took great interest in the findings; Vice President Gore even more so. Among others who were briefed by Clinton was his closest political advisor, Dick Morris. The reader may recall that in mid-August of 1996 a scandal arose in the White House when it came out that Morris had an ongoing relationship with a girlfriend who was a prostitute. In late July, just prior to those revelations, one of the last pieces of inside information Morris's girlfriend became privy to was the Mars meteorite findings. She immediately set about calling up newspapers, including a British tabloid, trying to sell the story. According to Gibson, he had given a copy of the galley proofs of the *Science* manuscript, which had his initials on it, to Goldin. Goldin had sent it to the White House, "and it went from Al Gore, Bill Clinton to Richard Morris to the hooker who tried to sell it, and it ended up in a colleague's hands in England who called me [before any public announcement] and said I know your initials."[44]

NASA headquarters began receiving calls from the news media around 1 August, inquiring if there was any substance to the story. "When the story got out, there were press people who had galley proofs!" observed NASA Exobiology chief Michael Meyer.[45] Goldin realized that an even worse public relations debacle was in the making than he had feared initially; he quickly attempted emergency damage control by pushing up the press conference eight days, to 7 August, the soonest it seemed possible to assemble at least the key players at NASA headquarters. (Romanek was en route from Houston back to Savannah River when CNN broke the news on television on the night of 6 August and said that the press conference in Washington was now scheduled for the next day at 12:30 or 1 P.M. He happened to be watching the news report and thus learned of the change in barely enough time to rework his plans and get a plane to Washington in the middle of the night. By the morning of the seventh the story had appeared on the front page of the *New York Times* and the *Washington Post*. Romanek was in a cab from the airport, trying to get to NASA headquarters—never having been there, he was at the mercy of a cab driver's knowledge.)[46]

The scientific community looks with profound unease upon efforts that seem to be "headline grabbing." It is considered acceptable behavior to publicize one's work to the press only *after* (or simultaneously with) the publication of the findings and after they have undergone a formal peer review process. The shunning of Pons and Fleischmann by the scientific community after they chose to announce their "cold fusion" discovery by way of a press conference well before any paper had completed the prepublication process reveals just how strong a behavioral norm this practice has become. Thus, Goldin was taking a calculated risk in making an early announcement on a topic with the long, publicity-charged history of life on Mars, even only nine days early.

NASA officials had feared that, because the original 15 August date came during the 1996 Republican presidential convention, "there's a worry that this is going to backfire, this is going to look like orchestration at the highest level."[47] It would be directly competing for headlines with Bob Dole's announcement of his running mate. But moving the date up was also not good etiquette in science rather than politics; NASA did indeed take heat in the press for this choice. Speculation was rife that Goldin was trying to influence the congressional budget vote for NASA in September; there was no obvious reason otherwise to broach a sacred behavioral norm of science, and without details Goldin's vague assertions about an imminent news leak did not sound convincing enough to justify the impropriety. At the very least some said NASA was still, as in the *Viking* days, unable to resist the temptation for "grandstanding."[48] In retrospect, now knowing the source of the potential leak, Goldin's calculation seems perfectly reasonable, even wise.

Its Disputed Meaning

Independent of its slightly unorthodox debut (and its near-miss with an even more scandalous career), the scientific case for "possible relic biogenic activity in ALH84001" received a great deal of attention from the scientific community, most of it in the nature of real scientific examination and critical review. The Mars meteorite soon became "the most intensively studied two kilograms of rock in history," with $2.3 million in NASA and NSF funding allocated for its analysis by November 1998.[49] NASA Exobiology chief Michael Meyer felt the paper was a positive contribution to science. Of its authors he said: "They're honest scientists, and they didn't jump the gun. They did good research, and looking at all the lines of evidence they had, that's what their conclusion was. It's a bold conclusion, and most people would be more conservative. But it's their honest conclusion. . . . It's generated a lot of interest already. We're going to learn more about what we know and don't know, and my suspicion is we'll end up two years from now saying, 'well, the odds are . . . , but we don't know.' So we have to go to Mars."[50] Many were fascinated by the findings; a great many felt much the same as Meyer about the process of science in action, even among those who were extremely doubtful of the biogenic explanation of the findings. There was no shortage of such critics, nor were they silent about their views.

J. William Schopf had the earliest opportunity (after those who reviewed the paper for *Science*) to respond. He was among the harshest critics, for whom the "spatial association" argument held no persuasive value at all. He describes the entire body of evidence as "circumstantial," saying that in science it simply would not constitute proof. In his colloquial terminology nothing less than a "smoking gun" was an adequate standard of proof.[51]

When Dan Goldin first invited Schopf to be part of the NASA press conference announcing the findings, Schopf had replied with a trademark line of Carl Sagan's which he often used when criticizing less-than-convincing paleofossils claims: "extraordinary claims require extraordinary evidence." But he tried to turn down the invitation politely; he thought in this case that the evidence "was not even close." Schopf opines that, because Goldin was "a Sagan fan (and was said to have been pleased by the quote)," this might account for why Goldin "had personally pegged me for the job" and prevailed upon Schopf until he agreed to participate. It seems likely that Schopf's involvement in the story since January 1995 as critical outside referee also played a part in Goldin's choice. But in any case, applying the "Sagan standard," Schopf believed in the case of a claim as extraordinary as life on Mars (even "possible relic life"), the evidence must be more extraordinary than even for paleofossils on Earth. The gun must not only be smoking, but there must also be a ballistics match.[52] Six years later Schopf would discover that the Sagan standard could be used in ways less to his liking, as we shall see.

In his presentation at the press conference Schopf objected to each one of the lines of evidence. The carbonate globules did not appear to him, a longtime specialist in microfossils and paleofossils, to have any characteristics that compelled him to think they were likely to have been made by living processes. The morphology and micrographs of the nanostructures were indeed striking, he said, but they were so tiny that they could not possibly contain even the minimum requirements to be alive. The most striking micrographs shown at the press conference, showing among other things a structure that came to be called the "Worm," had not been peer reviewed, as had the paper, Schopf pointed out. Finally, PAHs were so ubiquitous, even on meteorites, that Schopf said they did not have any biotic implications at all. The members of the Mars meteorite team had made clear in the paper that they knew about the ubiquitous distribution of PAHs; nonetheless, a great many more critics very quickly attacked their case on this point. John Oró was one of them; to him it seemed that the team's members simply did not understand what this meant. If they did, they would share the opinion of himself, Schopf, and many others that the meteorite PAHs were consistent with abiotic processes.[53] Their emphasis was on the opposite side of the coin that there was "nothing inconsistent with biogenic origin." To Schopf and Oró that was precisely the extraordinary claim that the scientific method prohibited without extraordinary evidence. Because the greatest danger in science was, as Norman Horowitz had emphasized during the planning of *Viking* and physicist Dick Feynman famously warned: "You must not fool yourself, and [when it comes to things you want very much to believe] you are the easiest person (for you) to fool."[54] Romanek, by May 1997, was willing to say, "I agree with people that say that PAHs are probably one of the worst things to look at as a type of biomarker compound."[55]

Early rounds of critical reaction began to appear in print very quickly.[56] Many cited the work by Ralph Harvey and Harry McSween from July 1996 which implied that the carbonate globules formed by a high-temperature process, in excess of 650°C, ruling out life. Romanek had mentioned this in the initial paper but stated that his measurements by an alternate method suggested a low-temperature origin; therefore, this dispute to some extent amounted to trusting one lab or method over another. At the annual Lunar and Planetary Science Conference (LPSC) in March and April 1997 at the JSC evidence was presented from many labs, but the results were evenly divided in favor of a low-temperature and high-temperature origin.[57] This issue still remains unresolved, but there is sufficient evidence to make a low-temperature origin a viable possibility.[58]

At the 1997 LPSC two further criticisms had been fielded: first, Harvey and McSween said they did observe the kind of magnetite crystals which the McKay team had described. They said, however, that in addition to those shapes (sometimes associated with biogenic activity) they saw a "whole zoo" of different shapes of magnetite crystals.[59] Furthermore, many of the crystals, including the supposedly biogenic type, contained defects of a kind that should not be present if they were crystallized in the stable environment inside a cell. CalTech specialist in paleomagnetism Joseph Kirschvink, who had studied the magnetites made by terrestrial bacteria in great detail, objected that sometimes biogenic crystals were produced outside the cells, resulting in a fairly wide range of shapes. This would support the McKay team's interpretation. But Harvey "highlighted a particular defect called a 'screw dislocation' . . . that has never been linked to biogenic magnetite."[60] Defects that serious were a difficult problem for the McKay team. They could maintain that not all the magnetite crystals originally targeted as biogenic had to be biogenic, but the more strained the argument became in this way, the less convincing it was, even to those who had not initially been deeply skeptical.

In addition to the temperature and the magnetite, John Bradley of Georgia Tech, Harvey, and McSween advanced a detailed argument explaining how the visual nanostructures in the electron micrographs of ALH84001 could be entirely explained, they claimed, as side and angled views of finely layered crystal structures and protruding ledges along fracture planes in pyroxene and "magnetite whisker" minerals. These appearances were further stilted in a deceptive direction by the gold/palladium coating used for electron microscopy, which can produce segmented-looking coatings like that of the compelling image that had been dubbed the Worm (see fig 8.1, image in background). This critique was published a few months later, in December 1997.[61] Four of the Mars meteorite authors responded in the same venue; they showed that the suggested artifactual explanation was by no means conclusive, though any but a technical expert in microscopy and/or mineralogy might be left wondering which argument was more persuasive.[62] Apparently at the scale of observation in question, phenomena are quite complex and ambiguities in interpreting the data common. A conference was held at JSC on 2–4 November 1998 on the state of the evidence,

"Martian Meteorites: Where Do We Stand and Where Are We Going?" By that time the McKay team did seem to accept that a certain number of its original putative nanobacteria images, especially those in which multiple cells appeared to be oriented in parallel, probably *were* examples of that kind of artifact.[63]

From the beginning much of the criticism was directed at the entire concept of nanobacteria. Although varying in the degree to which they thought it impermissible to speculate, most scientists echoed the original criticism Schopf had brought forward at the August 1996 press conference: namely, something as small as a rod 20 nanometers wide and 100 nanometers long is simply so small that it has no space for even the minimal required biochemical molecules to be alive.[64] "Such an 'organism' would be two orders of magnitude smaller than the smallest known one-celled organisms on Earth, mycoplasma," said Harold Morowitz.[65] Robert Folk and several others in the geology community had reported such tiny structures, but at least some reports from the biomedical community also supported the claim that nanobacteria might exist.[66] New reports began to come in and to receive much more attention because of the controversy generated by the Mars meteorite claims.[67] Kuopio University, Finland, microbiologist Olavi Kajander said that it had been difficult even to get such observations published before; peer reviewers simply rejected them out of hand rather than allowing them into print, where they could be judged in the court of public science.[68] In October 1998 the National Academy of Sciences, at the request of NASA, convened an expert panel to review existing evidence and come to some conclusions about what the minimum size range credible for life really is.[69]

The NAS panel included eighteen experts on microbial life, among them Norman Pace and John Baross. After a month of deliberations they embraced a lower cutoff size for life equivalent to the volume of a sphere 200 nanometers in diameter. And at the NASA Martian meteorites meeting of early November 1998 it sounded as though the ALH84001 team had moved a considerable way in that direction. At that meeting David McKay said, of anything smaller than a 100-nanometer sphere, "We simply don't believe [it] is indicative of bacteria." *Science* commentator Richard Kerr noted, "That criterion eliminates the objects in the [1996] *Science* paper as well as 'The Worm,' which is 250 nm long but too slender to make the cut."[70] McKay, however, did not completely abandon the claim of possible nanobacteria. "We think there are large objects that are still candidates," he said, though he demurred on providing any specific evidence of examples at that time. He also opined that the original "ovoids" and rods might be parts of Martian bacteria.[71] If this sounds like top-of-the-head improvising by one stuck in a tight corner, we must also note that, by the time the NAS panel's report on nanobacteria appeared at the end of 1999, their own 200 nanometer published figure was also being finessed to leave some "wiggle room," particularly on account of Philippa Uwins's reported nanobes (in 1998) from Australian rocks. They held that "known terrestrial bacteria in the range of 200 nm probably marked the lower size limit for current life, but held out the possibility that primitive unknown microbes might have been as small as 50 nm, about the size of the Australian nanobes."[72] John Baross, interviewed by the *New York Times*, repeatedly emphasized a 100 nanometer bottom line, exactly where McKay had left his claim a year previously.[73] And, again unintentionally echoing McKay, Baross speculated: "'We have to think about them [nanobes] in a different way, and one is that they are components' that function as a living organism only in totality, the whole being greater than the sum of the parts."

In the report a colleague on the NAS panel, Pittsburgh University biologist Jeffrey Lawrence, "laid out a detailed analysis of such hypothetical community life made up of extraordinarily tiny components, calling the aggregate a meta-cell."[74] In a similar vein, after an April 1997 JSC meeting on the Early History of Mars, one thought about the nanostructures in ALH84001 was "whether the 20 nm structures could represent not fully functioning microbes but important nonliving prebiotic structures, such as membrane-defined structures, on the road to life."[75] In many ways NASA Exobiology–funded work prepared the way for this kind of novel reconceptualizing about life. Consider Margulis's work on understanding eukaryotic cells as endosymbiotic communities in an analogous way as well work on microbial mats as holistic ecological communities and on biofilms. But suffice it to say: the jury is still out on nanobacteria.

Frances Westall, a JSC colleague who worked on electron microscopy of very small potential microfossils, became interested in the ALH84001 results and began collaborating with the McKay team on trying to study in detail the processes by which microfossils form (e.g., silicification of bacterial cells)[76] in order to develop a set of criteria for recognizing extraterrestrial microfossils.[77] Similarly, a persistent and constructive skeptic of the ALH84001 claims,[78] cosmochemist and meteorite specialist Peter Buseck of the Geology and Chemistry Departments at Arizona State University in 2002 launched a project under NASA Astrobiology funds to study "nanoscale minerals as biomarkers."[79] Under another concurrent grant from NASA Cosmochemistry, Buseck is investigating "the reactions and distribution of polycyclic aromatic hydrocarbons and fullerenes in extraterrestrial material."[80] Whatever the outcome on nanobacteria per se, the Mars meteorite claim does seem to be driving crucial parts of the science of exobiology forward. This sentiment was expressed in a prominent editorial in the journal *Meteoritics and Planetary Science* by editors Derek Sears and William Hartmann: "The Antarctic meteorite Allan Hills 84001 may be at the center of a revolution in our thinking about the origin of life on Earth, Mars and perhaps elsewhere. This is not because of the attention given by non-scientists to last summer's paper on this meteorite, but because it has forced a reexamination of the importance of microbes in the ecosystem, the nature of the smallest possible life forms, the nature of organic materials and structures that led to the origins of life and the temperature regime at which life originated."[81]

To return to this very fruitful criticism: the McKay team was frequently criticized for not citing in their paper the 1989 "false alarm" on PAHs in EETA79001. Jeffrey Bada of the exobiology NSCORT in San Diego, the chief critic of that earlier claim, who convinced most scientists that those PAHs were contaminants that had seeped into the earlier Mars rock with Antarctic meltwater, now attacked the ALH84001 evidence on the same grounds. Because ALH84001 contained a limited assortment of PAHs quite similar to the ones reported in the earlier meteorite claim and because Bada's team showed that suite of PAH molecules to be present also in samples of Antarctic ice, Bada's group suggested terrestrial contamination was just as likely this time to be the source. Regarding the fact that the concentration of PAHs was greatest in association with the carbonate globules and practically nil on the outermost layer of the meteorite, Bada suggested that a chemical explanation was more likely than shared biogenic origin: PAH molecules preferentially adsorb to carbonates by a purely physico-chemical affinity.[82] Romanek replied:

> Well, that's true, but PAHs are hydrophobic molecules; they don't like water. They want to be adsorbed to anything that is non-aqueous. And so what needs to be done now is . . . to look at other components of the meteorite—the fusion crust, the orthopyroxene ground mass—and perform these same experiments and see if PAH is preferentially adsorbed to those materials. I . . . strongly suspect that they will, because of this hydrophobic nature. . . . And so that kind of casts doubt . . . into whether this process of transporting PAHs into the meteorite from the Antarctic ice is the actual process that generated these concentrations that we measured in the carbonates. At this point in time, I'm not convinced of that at all. If these experiments do come out and show [what I predict] . . . , I've got to go with the idea that they're indigenous to Mars.[83]

As the individual lines of evidence began to fray and seemed increasingly strained, the "spatial arrangement" argument also lost favor. *Science* reporter Kerr, apparently himself fairly skeptical, noted at the November 1998 NASA Mars meteorite conference at JSC, "even two years ago, many researchers were unimpressed with that holistic argument. 'I never bought the reasoning that the compounding of inconclusive arguments is conclusive,' says petrologist Edward Stolper of [CalTech]. And it was clear at the workshop that now, as pieces of the argument weaken, it is losing its grip over the rest of the community."[84]

Despite the skepticism of the Bada group and others, there can be no doubt that, in the best tradition of science, the ALH84001 results provoked them to do a lot of new work, searching for indigenous and/or contaminant organics in Mars meteorites. And, indeed, they found what appeared to be almost entirely contaminant (overwhelmingly the L-isomer) amino acids in both ALH84001 and Nakhla.[85] This represented substantial progress, however, in understanding Mars meteorites. More than that, the Bada team observed that "the rapid amino acid contamination of Martian meteorites after direct exposure to the terrestrial environment has important implications for Mars sample-return missions and the curation of the samples from the time of their delivery to Earth."[86] They suggested that any strategy for seeking organics on Mars must focus only "on compounds that are readily synthesized under plausible prebiotic conditions, are abundant in carbonaceous meteorites, and play an essential role in biochemistry."[87]

Similarly critical, longtime meteorite researcher and NASA Exobiology grantee John Kerridge of UCSD concluded from the ALH84001 debate that Martian sedimentary rocks precipitated from solution were by far the most likely to be fossiliferous rocks worth sampling. Thus, Kerridge urged, finding sites from orbit that are clearly dried up sea- or lake-beds should precede any attempt at sample collection.[88]

Furthermore, he noted, the remarkable popular interest generated by the 1996 announcement was an important contribution in itself. Even the enthusiasm for life on Mars which convinced the taxpayers to spend a billion dollars on *Viking* was not nearly as great as the outpouring of interest since August 1996, opined Kerridge. And in a science in which public funding was crucial, this was no side issue. The 1993 congressional cancellation of NASA SETI funding was a constant reminder of the flip side of this same coin. Six months after the initial press conference he thought about McKay's group that "they demonstrated beyond a shadow of a doubt that the public wants us to do this. And that is going to make it much easier for us to get money out of Capitol Hill than we've ever done before."[89]

Less deeply skeptical about the science of the JSC team, former Exobiology chief Donald DeVincenzi came to almost the same conclusions in May 1997. The ALH84001 paper produced debate of the healthiest kind, he thought: "It's . . . absolutely amazing. It has stimulated so much research, . . . a whole new field of research. It's demonstrated that we're going to have our hands full when we get a protected Mars sample back on Earth. Here we've got the thing [i.e., the meteorite] in our hands with all the power on this planet, and we still don't know if [the 1996 claim is] right or wrong yet, we really don't. And to me that's a tremendously important non-finding, that nine months later we still don't know the answer. And here we are saying, jeez, we really want to get some Mars sample back here in 2005, and we know what to do with it. Yeah, right. I would think we don't yet, but we will by then. I think this is a good case in point."[90]

The lessons from ALH84001 will surely vastly improve preparedness for obtaining informative Mars samples, no matter who turns out to be correct about different aspects of the original 1996 claim. Even after only a few years the debate has already had a large salutary effect in this direction.

DeVincenzi also compared the ALH84001 findings to the first results of the *Viking* lander biology experiments, noting many striking parallels. The first appearance of the evidence was strikingly biological in both cases. "And then three years later they were still arguing about that [the LR results], but now after three years of intensive research, there was a new theory. And it's a chemistry explanation. But it's not simple, it's complicated, and you need three different oxidants in order to explain all the results. Three. Not one. . . . I think maybe that's what's going to happen here, that it really is going to take a lot of different lines of evidence, and if it does come up negative it's going to be like the Viking thing; there'll be more or less an extraordinary negative explanation for these extraordinary results. . . . It's not going to be just a simple explanation, I don't think."[91]

It should be noted, however, that the issues involved in the controversy have turned out to be much more complex than either side initially envisioned. Even given three or four separate lines of evidence in dispute, opinions that the debate would be resolved within a year or two have turned out to be excessively optimistic. By the November 1998 NASA meeting McKay thought that sorting out the ALH84001 results might be work for the next five, maybe ten, years. A majority within the exobiology research community probably currently considers that the ALH84001 evidence leans strongly against biogenic activity as the most likely explanation. But this consensus appeared even more strongly negative in late 1998 than a mere four years later.[92] And in February 2001 an independent research team under Imre Friedmann produced new evidence about the magnetite crystals, which gave new vigor (if not complete resuscitation) to the possibility that the Mars rock actually might contain microfossils.[93]

A team led by Kathie Thomas-Keprta also published the results of new, much more detailed studies on the magnetite grains in the Mars rock, arguing that they "were likely produced as a biogenic process." As such, they argued, the crystals represented "Martian magnetofossils and constitute evidence of the oldest life yet found."[94] Friedmann's group found one of the things critics of the biogenic magnetite had been demanding: in samples in which magnetotactic bacteria produced the granules, they were found in the dead cells, just as in life, lined up in chains. Thomas-Keprta's group said that some 75 percent of the magnetite crystals in the carbonate globule rims were, as critics alleged, of inorganic origin. They still held that 25 percent of the crystals were so identical in shape and structure to those from known magnetotactic bacteria that they were overwhelmingly likely to be of biogenic origin. Some life was breathed back into the Mars rock, it seemed, at least initially.[95]

Yet many remained cautious about the Martian "pearl chains."[96] For those who had watched the original four lines of evidence weakened one by one, as biochemist and meteorite organics expert John Cronin saw it, "as to the magnetite chains, it seems that the life of ALH84001 now hangs by these slim chains, a miniscule component of the meteorite, even of the meteorite total magnetite. At best, I doubt that they will ever fully meet the Sagan requirement of extraordinary evidence for an extraordinary claim. ALH 84001 was born with a bang but seems destined to die with a whimper."[97] Cronin's opinion was largely shared by Peter Buseck, who studied these magnetites in some detail and was launching into a new, and it was hoped, definitive study in early 2002.[98] By contrast,

Joseph Kirschvink, the magnetite expert at CalTech, was now a supporter of the biogenic view.

The complexity of the issues, pushing the limits of available technology, has only been one dimension of the Mars rock story. Science historian and philosopher Iris Fry has observed, "At the same time, the persistence of McKay's team in its original contention despite the harsh criticism addressed against it clearly transcends the empirical issues involved and demonstrates the sociology of science at work. A great deal is at stake here in addition to the major question being addressed. . . . money, ambition and politics are all involved in this project."[99]

One might add that the degree of invective among their opponents also illustrates commitments above and beyond the evidence. It has taken two to tango in jacking up the level of personal sensitivity in the debate. And most of the opponents, as well as the McKay team, are NASA grantees; neither side has lost work from NASA by taking one side or the other. Given the level of public interest in the topic, that situation seems likely to continue.

As if to emphasize that controversy is the norm in science, one of J. William Schopf's most renowned discoveries, the 3.45 billion–year–old Apex Chert microfossils (discussed in chap. 5), was called into question even as the Mars rock outcome remained unresolved. Much to the surprise of the exobiology community, a paleofossil research group led by Martin Brasier of Oxford University announced in March 2002 that the fossils listed as the world's oldest in the *Guinness Book of World Records* might not be fossils at all but mere inorganic deposits of graphite or of organic matter produced abiotically by a Miller-Urey-type synthesis in hydrothermal vent waters.[100] Examining the original type specimens Schopf had deposited at the Natural History Museum in London as well as the rocks in their original geological setting, Brasier's group claimed that Schopf had incorrectly believed the rocks to be from a shallow sea bottom and the putative microfossils to be cyanobacteria. They also found many of the supposed bacterial filaments to be irregularly branched and/or folded in ways not seen in those organisms; the "fossils," they thought, were much more likely deposits of organic material around the edges of crystals which gave the appearance of living cells in much the same way that Bradley, Harvey, and McSween had posited for the Mars rock "nanofossils." The Schopf group at UCLA and another group at the University of Alabama–Birmingham were informed about the Brasier results, submitted to *Nature* on 14 February 2001. They had begun studying the Apex chert fossils with Laser-Raman spectroscopy to determine the nature of the organic material of the fossils in situ and differentiate it from that of the surrounding rock matrix. They submitted a manuscript to *Nature* which effectively addressed the Brasier claims, and the papers were published side by side in the same issue.[101] Jill Pasteris, a Washington University scientist with twenty years of experience in Laser-Raman spectroscopy, has expressed skepticism about Schopf's interpretation of its results. Thus, the controversy continued.[102]

More than one commentator noted the irony that for Schopf, who had built his reputation on debunking mistaken microfossil claims and establishing the criteria to determine fossil from artifact, the "extraordinary claims" shoe now seemed to be on the other foot.[103] Some argued that, because Schopf's fossils were from Earth, not Mars, his claim was not "extraordinary" in the same way as the McKay team's and thus should not require the same extraordinary standard of proof. But for one who had so freely wielded the argument in his 1999 book as well as against those whose terrestrial paleofossil claims he disagreed with, this did not appear quite symmetrical to many observers. Some claimed that at the very least Schopf's implication that the Apex chert organisms were photosynthetic was no longer valid; if the formation was a deep-sea hydrothermal vent, there would have been insufficient light for photosynthesis.[104]

In a second episode with some parallel features geologists Chris Fedo and Martin Whitehouse took a much closer look at another recent spectacular claim about the most ancient evidence for life on Earth. In 1996 a team at the NASA NSCORT led by Gustaf Arrhenius's student Steve Mojzsis claimed to have found carbon isotope evidence for biotic organic carbon in the 3.85 billion–year–old rocks of Akilia Island, Greenland, pushing the date for presumptive life on Earth back farther than Schopf's fossils by another 400 million years, to the time immediately after the heavy bombardment of Earth by meteorites ceased.[105] Mojzsis accepted previous identifications of the rock layer as a sedimentary banded iron formation (BIF), generally thought credible at that time. He and his team argued that the apatite crystals in which the carbon was found would be resistant to meta-morphism.

When Fedo and Whitehouse closely examined the rocks in question in their geological context, however, they found persuasive evidence that the rocks were highly metamorphosed and not sedimentary in origin. No fossils could possibly have been preserved in that rock, they claimed; any carbon left would be so altered from metamorphism over almost four billion years that it would be unsafe to draw any conclusions about its origin. Their paper in the 24 May 2002 *Science* cautioned that any rock needs to be studied in the field in its full context, rather than just in the laboratory. Although the controversy is still unresolved, it seems clear at this point that the interpretation of the rocks and any carbon they contain is more ambiguous and open to multiple readings than was first thought.[106]

This episode strikingly echoes the qualifier with which the Mars meteorite group opened its 1996 paper: that the researchers knew nothing about the geological context on Mars from which the rock originally came. Science writer Richard Kerr of *Science* found Fedo and Whitehouse's criticisms credible and drew several parallels between all three cases: Schopf's Apex chert claims, ALH84001, and the Mojzsis claim.[107] Still, on the greater lesson for exobiology and for science in general, all parties are in striking agreement. George Cody of the Carnegie Institute, Washington, D.C., says: "I don't believe any of the evidence from the Martian meteorite, . . . but it's been the biggest boon for space science. It got us thinking."[108] The McKay team sees the same big picture. "Whether we are right or wrong," says Everett Gibson, "the scientific community will be better prepared for that day when samples from Mars will be returned to Earth for study. In addition, new ways are being developed which permit the scientific community to seek the signatures for life. We feel a bite of personal pride inside because of what we have accomplished."[109]

CHAPTER 9

Renaissance

FROM EXOBIOLOGY TO ASTROBIOLOGY

The year 1995 looms large in the history of exobiology. In that year, seven months before the announcement of the first planet around a Sun-like star and more than a year before the infamous Mars rock episode, the young discipline began to reinvent itself based primarily on the threat of a deep administrative upheaval at NASA. Out of a NASA-wide reevaluation of the agency known as the "zero-base review," and the resulting tumultuous experience for NASA Ames Research Center in California, emerged a new word in the exobiology lexicon, *astrobiology,* which redefined the boundaries and the concept of exobiology. By 1996 a workshop had made a first attempt to define *astrobiology,* by spring 1998 a virtual Astrobiology Institute embraced a geographically diverse number of institutions and individuals, and by late 1998 scientists from a variety of fields had constructed a general roadmap for the discipline. The buildup of astrobiology was remarkably swift, fed by the intense excitement surrounding the discovery of planetary systems, the controversy over the Mars rock, the possibility of an ocean on Europa, and research on life in extreme environments among other developments, including the biotech revolution spawned by the Human Genome Project. While the ultimate outcome of this activity was still in doubt at the turn of the millennium, it is clear that in the aftermath of these events exobiology would never again be the same. These unexpected events not only mark the latest chapter in the four-decade history of exobiology; they also provide a further revealing window on scientific discipline building and hint at a "great age of discovery" which aims to place life in a cosmic context.

Crisis at Ames

In the mid–1990s NASA was facing massive budget cuts from Congress. Administrator Daniel Goldin had submitted a budget for fiscal 1994 which reduced NASA's budget by fifteen billion dollars over five years—a significant cut for a budget then running at about fourteen billion dollars annually. Two years later he reduced NASA's budget again by ordering the redesign of the International Space Station and canceling programs. But Congress kept the pressure on NASA's budget, and Goldin decided to streamline NASA's structure through a zero-base review, one that started from ground zero rather than from the previous year's budget.[1]

It was in this context that, on 2 February 1995, a NASA "Red Team" white paper was produced that immediately spread fear across the agency. Entitled "A Budget Reduction Strategy" and drafted by NASA deputy chief of spaceflight Richard Wisniesk, the purpose of the paper was "to provide a starting point for discussions on a proposed realignment of center roles and missions." The self-described driving force for the paper was the constrained budget environment, and the paper was meant to communicate "NASA's commitment for revolutionary change" across the agency. Among the overarching principles of the plan were that NASA would maintain its in-house capabilities to perform research and development and that operations would be accomplished through the commercial sector. But the report stated pointedly that "the luxury, and perhaps the wisdom, of overlapping roles at the Field Centers is no longer an option." As part of the streamlining of functions, Ames was to remain the lead center for aerodynamics and aviation human factors. But Ames was to drop its programs in Mission to Planet Earth and in life and planetary sciences. Equally large changes were to take place at other field centers. NASA teams already in place, the paper ominously promised, would fully review and evaluate the proposals for feasibility.[2]

At Ames, center director Ken Munechika assigned Bill Berry, acting director of the Space Directorate, the task of taking action under the "ZBR" guidelines. Taking those guidelines seriously, he had little choice but to develop what amounted to a going-out-of-business plan for his directorate, which included life, space, and Earth sciences. Because Goddard had a big Earth science contingent, JPL a big planetary science / space science contingent, and Johnson Space Center a very large life science group, the plan was to parse each of these functions out to other centers, consistent with the aims of the zero-base review team. But, when Berry circulated a draft of the plan to his division chiefs in mid-March, they balked. Lynn Harper, then acting chief of the Advanced Life Support Division at Ames, resisted the drastic implications and urged a new strategy: to argue that the manifold activities at Ames were not a weakness but a strength, that interdisciplinary research was more important, indeed more productive, than fencing research within traditional disciplinary boxes, provided that Ames use this strength to focus on a single topic—life in the universe.

Such a strategy was not new; Harper recalled that it was part of the philosophy enunciated by John Billingham in connection with the NASA SETI program he had headed at Ames beginning in the 1970s: "Billingham was always convinced, and convinced me, that if you attempt to understand life in the universe then you have to have all of the pieces—life on the cosmic scale, the planetary scale, the organism scale, and the volition or the purpose or the intelligence piece of it that manages evolution if it wants to do so. Those pieces were so powerful and important, both as a scientific discipline and for what it offers to humanity, offers to the future of my kids, that it would be wrong to break up that unique capability." In support of this philosophy Billingham had organized numerous workshops, including the influential ECHO report on the Evolution of Complex and Higher Organisms which foreshadowed some of astrobiology's themes. In this sense Billingham may be considered the father, or one of several parents, of astrobiology.[3] The tools to carry out such a research program were now much advanced over the 1970s, and the opportunity was at hand if only it were seized.

Ames management, faced with convincing Dan Goldin and other high-level administrators in NASA that Ames's expertise in life, Earth, and space sciences was unique within the agency, seized on a redefined exobiology to play a crucial integrating role. This strategy was risky at best, both personally for the individuals involved and for Ames as an institution. As we have seen in chapter 2, from the early 1960s Ames had always been NASA's focus in exobiology, a focus that admittedly had become fuzzy and weakened in the disappointing aftermath of *Viking*. As one NASA insider put it, space science, with its flashy results, was the glittering jewel of NASA, while life science was somewhere down in the pond scum. Yet exobiology remained the very definition of an interdisciplinary endeavor, and, if that activity could be revamped, strengthened, and put in the context of real space missions, it could be the savior of the Ames Research Center. It was in recognition of the capability for mission-oriented multidisciplinary research across all three lines, Ames management argued, that NASA should not only keep Ames open but should assign to it a newly strengthened endeavor termed *life in the universe*. Luckily, their emphasis on biology was attuned to Dan Goldin's thinking, and as administrator his opinion counted for a great deal.[4]

Such an argument was entirely counter to the guidelines of the zero-base review. But it was exactly the argument Ames managers made at an extraordinary weekend meeting at Ames on 26–27 March 1995, when they briefed NASA chief scientist France Cordova, the associate administrators for Space Science (Wes Huntress), Life and Microgravity Science (Harry Holloway), and Earth Science (Bill Townsend), and others who had gathered to decide how Ames was going to dispose of the pieces of its program. This fateful meeting, at which Berry made the key presentation (written primarily by Lynn Harper, who integrated discipline-specific input from the Ames Science Advisory Council), was a turning point and the origin of Ames's mission lead for astrobiology. Instead of presenting a going-out-of-business plan, Berry presented a "Life in the Universe" plan, backed up by the Ames Science Advisory Council. The council, chaired by Muriel Ross, gave in-depth technical presentations based on their study of what science could be done if disciplines were merged at Ames with no barriers to drawing on talent and resources. The arguments found favor with Huntress, Cordova, and eventually Goldin. It was at this meeting that Huntress remarked that he disliked the term *life in the universe* and suggested that *astrobiology* be used instead. In April the zero-base review team at NASA headquarters in Washington, D.C., recommended that Ames be given the lead in astrobiology, and on 19 May Goldin made the formal announcement. At the same time, Ames was also given the lead in information sciences, on which the new biology of the biotech revolution was heavily dependent.[5]

A Dear Colleague letter dated 30 May from Associate Administrator for Space Science Wes Huntress, entitled "Space Science and the Zero Base Review," introduced another new concept while making the first official use at NASA of the word *astrobiology*. The Space Science program at Ames, it held, would be privatized by forming an institute through a consortium of Bay Area universities and local industry. The virtual institute concept was initiated because it was unlikely that Ames would ever get the hiring authority needed to do the job. Harper and Kathleen Connell did the feasibility assessments in April 1995, including the legal precedents that would allow the creation of the institute. In November 1995 David Morrison, Scott Hubbard, Joan Vernikos, and Estelle Condon were among the Ames personnel who served on formal committees to create the institute. Although the nature of the organization would later be redefined, this was the beginning of the idea of an Astrobiology Institute. The letter further defined the scope of the field, stating that the new entity would "have prime responsibility for the 'Origin and Distribution of Life in the Universe' theme, and will be the lead NASA Center for astrobiology and astrochemistry, areas in which ARC has developed unique, world-class expertise. Specialty areas include cosmochemistry, chemical evolution, the origin and evolution of life, planetary biology and chemistry, formation of stars and planets (space science), and expansion of terrestrial life into space."[6]

Defining Astrobiology and Building a Program

In a four-month period from February to May 1995 Ames had escaped disaster. Instead of drastically reducing the scope of its work, the center now set about building the new program in astrobiology. Essential to that process was defining *astrobiology*. Already in the 1996 *NASA Strategic Plan,* in which the word *astrobiology* was used for the first time in a published agency document after Huntress's unpublished letter to colleagues, the focus was on the key questions, recognizing that too broad a program was no program at all when it came to limitations of funding. Astrobiology was the "study of the living universe" to be sure, but in particular it was seen as providing the scientific foundation for the study of the origin and distribution of life in the universe, the role of gravity in living systems, and the study of the Earth's atmosphere and ecosystems. These three programs were already in existence, but astrobiology was to go beyond them, asking questions that require the sharing of knowledge, resources, and talents of existing programs and striking out in new directions as well.[7]

Even the focus on key questions left a broad scope and much room for interpretation. In mid–1997 Don DeVincenzi, head of the Space Sciences Division at Ames, admitted: "I have a fairly good view of what astrobiology is. But I don't know that anybody else particularly subscribes to my definition. Everybody's got their own definition, you know. Some people look at it as an umbrella for everything; from the big bang to today, and I don't take that view, I don't think that's what Goldin meant, and I don't think that's what is appropriate." In DeVincenzi's view the Origins program was the broad umbrella, while astrobiology was intended to be a more limited program to focus on biology and the origin, evolution, and distribution of life. It was to be broader than the old exobiology but more confined than the whole of Origins. Exobiology as funded from headquarters had not paid much attention to the origin of planets but had been following the history of carbon. Exobiology funding from NASA had traditionally ended with the earliest ecologies on the planet, about 3.5 billion years ago. By contrast astrobiology wished to place the origin of life in the context of the environment in which it happened. In this sense planetary origins and evolution became an essential component of astrobiology, at least as they related to the conditions of habitability. Furthermore, astrobiology aspired to address questions beyond early ecologies to the origin and evolution of higher life forms. In other words, exobiology was the core of astrobiology but would now be placed in the context of evolving planetary environments. One could ask how gravity and radiation shape the origin and evolution of life on Earth and elsewhere, address the origin and evolution of ecosystems and global biospheres, and even hope in the future to look for spectroscopic signatures of life in the atmospheres of extrasolar planets.[8]

One thing is certain: in distinguishing exobiology from astrobiology, the difference between a concept and a funded program was essential. As Lynn Harper at Ames put it: "the sea change between exobiology and astrobiology was the inclusion of Earth sciences and life sciences as part of the portfolio. Conceptually, exobiology had always recognized them, but practically it didn't develop them within that program umbrella. Astrobiology pulled them in hard and made some conceptual advances based on the synergies between Earth sciences and space sciences or Earth sciences and life sciences that had never occurred before."[9] The definition and scope of astrobiology were not entirely academic questions, for they played heavily into how NASA would build its program. Indeed, some consensus on what astrobiology should become was necessary to proceed at all.

The astrobiology plan was therefore much broader than exobiology as previously conceived in NASA. The exobiology program managed out of NASA headquarters still thrived, under the management of Michael Meyer, at the level of $8.4 million in 1997. This money funded about one hundred principal investigator proposals per year, and about one-third of the funding came to the exobiology effort at Ames, which had to compete for the money in the same peer-review process as everyone else. A shift in emphasis had occurred in 1995, when the exobiology NASA Research Announcement (NRA) indicated that the program was seeking fewer proposals on the evolution of the biogenic elements, because so much research had been done on the subject that the origin and evolution of those elements was fairly well understood. "We wanted more constraint to the program than that," Meyer recalled, "because we were getting too many proposals. And most of them, although very good studies, wouldn't help very much to answer 'How do you get life started in a planetary system?'" Exobiology was recentered more on the origin of life—how polymers get put together, how to get cell membranes, and the minimal living organism—as well as on trying to understand Earth's early evolution.[10]

Defining *astrobiology* would be an ongoing process. Meanwhile, with the 1996 NASA Strategic Plan as the enabling document giving Ames the astrobiology mission, NASA went about building the discipline in several ways: by developing internal consensus and funding, by involving the outside professional community, and by engaging the public. None of these were easy or entirely separable activities, but all were essential for success in the broadened discipline.

Inside NASA an essential element for the rapid rise of astrobiology was the strong support of NASA administrator Dan Goldin. Goldin believed biology was the science for the twenty-first century, advocated astrobiology enthusiastically in his speeches, and provided moral support. David Morrison, director of space at Ames and one of the architects of astrobiology, remarked in 1997 that "the major commitment that Administrator Dan Goldin has made to biology within NASA, to the Origins Program, to understanding the origin of life on Earth, to exploiting the space station and its biological research capabilities, to searching for habitable planets around other stars, as well as Mars exploration, has all served to greatly invigorate exobiology and astrobiology in the last year or two." "Goldin was pivotal," Lynn Harper recalled a few years later. "He prevented us from being crushed or pulled apart by the organization. . . . He basically said this is something he wants to see work . . . and then he spoke about it well in places that needed to hear it and really helped make astrobiology happen. He never came through with money, but he helped." In late 1997 Goldin was still lamenting that "the biological revolution has passed the space program by." He wanted to change that, telling his Advisory Council he would like funding for the Astrobiology Institute to reach one hundred million dollars eventually. "You just wait for the screaming from the physical scientists [when that happens]," he said.[11]

From all appearances Goldin was truly interested, but the problem of funding, left to astrobiology's managers at a lower level, called for creative thinking. It was one thing to declare that astrobiology should join Earth, space, and life sciences in a common endeavor; it was quite another to secure funding commitments from those three distinct organizational elements at NASA headquarters.

Life and Microgravity Sciences, now under Arnauld Nicogossian at headquarters, was initially opposed to astrobiology. Nicogossian had his own programs to fund and saw astrobiology as a competing program. The early reaction from Earth Science was similar. Astrobiology found its first allies in Space Science under Wesley Huntress, who, after all, had coined the word *astrobiology* and given the go-ahead for it to proceed at Ames. There the Advanced Concepts and Technology Division, under Peter Ulrich and Rick Howard, provided early funding for astrobiology at the level of about $100,000, parallel to the way in which early SETI funding had come from the Office of Aeronautics and Space Technology (OAST) at NASA headquarters. The traditional exobiology program, also under Space Science, was a logical source of funding, but its funds were committed for traditional areas of research, and in these early days its head, Michael Meyer, may well have felt that what was happening at Ames in astrobiology was beyond his control. Thus, for several years funding for astrobiology was kluged together from a variety of sources whose managers believed in astrobiology's promise and acted as its advocates. Astrobiology was able to succeed because a number of people each committed relatively small but important amounts of funds to make specific activities succeed. Personalities and professional connections played a considerable role in this process. Mel Averner, who had managed the biosphere program at NASA and arrived at Ames as program manager of fundamental biology in the midst of astrobiology's development, acted as a kind of link to life sciences back at headquarters. He was also essential in providing funds from his program for astrobiology, especially those needed to fund an essential series of workshops.[12]

At NASA Ames the action in senior management fell to Henry McDonald, Scott Hubbard, David Morrison, and Donald DeVincenzi. McDonald, who replaced Munechika in spring 1996 as Ames director, was an active advocate for astrobiology—an essential advocacy if the discipline was to get off the ground at Ames. Morrison, DeVincenzi, and Hubbard would each play essential roles in their own way. Lynn Harper led the Astrobiology Advanced Missions and Technology (AAMAT) group until September 1999, when Greg Schmidt took over as head of what would be called the Astrobiology Integration Office. It was the early AAMAT effort that commissioned the workshops, paid for initial feasibility studies, and in general acted as the engine for moving astrobiology more rapidly forward. The AAMAT group encouraged its members to recruit science talent beyond the traditional NASA boundaries. With this encouragement Emily Holton recruited two Nobel Prize winners, Baruch Blumberg and Walter Gilbert, to chair one of the sessions at the Astrobiology Roadmap Workshop. It would be a historic meeting. Holton would again recruit Blumberg and another Nobelist, Richard Roberts, to cochair a follow-on workshop to the Roadmap, called Genomics on the International Space Station. This was commissioned by Harper, cofunded by AAMAT and Averner, and paved the way for Blumberg's eventual decision to head the Astrobiology Institute. A host of managers and scientists helped guide astrobiology through its early birth, whether in organizing workshops, providing money, doing research, or using their professional contacts to advance the new discipline. If early astrobiology seems a jumble of names with a variety of backgrounds and motivations and no central brain, this is an accurate reflection of its origins; as Harper put it, astrobiology was about constellations, not superstars.

Cooperation was necessary to make astrobiology work as an interdisciplinary endeavor. David Morrison, an early student of Carl Sagan and a pioneer in planetary science, was pivotal in this regard as one of the conceptual leaders of astrobiology. As Harper recalled, Morrison "embraced the broad view right from the beginning, and could see how all the pieces contributing together provided some discovery opportunities scientifically that separating them really didn't. These opportunities were exciting and they were new and they were important. . . . Morrison was able to articulate them in a very compelling way, and helped in the communication of astrobiology to everybody, regardless of their backgrounds." Moreover, "he was evenhanded with all of the [internal] organizations. Astrobiology was such a fragile thing when it started. If Morrison had supported space science at the expense of life science astrobiology would have cratered, but he didn't . . . he was the glue that held all of the pieces together. Morrison really was the lead in important ways of the integration of the effort."[13]

An important exercise in consensus building occurred in September 1996, when Ames hosted the first Astrobiology Workshop. DeVincenzi, who had a long history in exobiology management and planetary contamination issues, played a leading role in organizing this workshop. NASA's first attempt to court the Earth, space, and life sciences in one gathering brought about one hundred invited attendees, including twenty-three physicists and astronomers, thirty-seven Earth and planetary scientists, and thirty-eight life scientists. The meeting was organized around five major questions: (1) How does life originate? (2) Where and how are other habitable worlds formed? (3) How have the Earth and its biosphere influenced each other over time? (4) Can terrestrial life be sustained beyond our planet? and (5) How can we expand the human presence to Mars?[14] It is notable that at this stage of discipline building the sole stated goal was to stimulate cross-disciplinary thinking and new ideas for research. The organizers made no attempt to reach consensus on research priorities, recommendations, or funding requirements.

As another step in consensus building, Wes Huntress at headquarters dispatched Gerald Soffen, the former *Viking* science leader and now director of University Programs at NASA's Goddard Spaceflight Center, around the country to build consensus on what astrobiology should be. Soffen consulted hundreds of researchers and program managers, inside and outside NASA and by mid–1997 had drafted a program plan. Noting that "we are entering a great age of discovery in biology," the internal report viewed NASA's exobiology program as being subsumed under the new field of astrobiology, noted that the time was ripe because of recent discoveries, and advocated an increasing role for

NASA because the agency's missions and technology would be needed to answer some of astrobiology's fundamental questions. In viewgraphs that distilled the program plan for headquarters discussions, Soffen enunciated six points for action: (1) develop the scientific questions; (2) form a virtual institute; (3) find the leaders; (4) develop young talent; (5) relate to NASA Mission where appropriate; and (6) relate to the rest of biology.[15]

Meanwhile, activities at Ames were defining roles inside NASA. Lynn Harper, a past SETI program manager at headquarters who had also worked in exobiology, Earth sciences, and life sciences and appreciated the value of multidisciplinary work, was one of the principal behind-the-scenes architects of the astrobiology program. It was she who first articulated many of the principles under which astrobiology operated as part of an "Astrobiology Development Plan" written during 1997. Incorporating input from many other scientists both inside and outside NASA, the document set forth the recommendations of Ames for the science and technical content of a national program in astrobiology and how it should be implemented. The program was to be built on NASA's four "Strategic Enterprises" as set forth in the 1996 NASA Strategic Plan: Earth Science, Space Science, Human Exploration and Development of Space, and Aerospace Technology. The development plan viewed astrobiology as an emerging "superdiscipline" that cut across many disciplinary boundaries. Its scope once again was defined as the origin, evolution, and destiny of life, where *destiny* was defined as "making the long term-occupation of space a reality and laying the foundation for understanding and managing changes in Earth's environment." The program implementation was to involve ground-based, airborne, and space flight research and technology, spread across the Earth, life, and space sciences, with education and public outreach as fully integrated elements of the program.[16]

Under the general scope of the origin, evolution, and destiny of life, the development plan set forth a breathtaking array of eleven "scientific challenges," ranging from understanding the formation of planetary systems to the evolution of Earth's biosphere for its first billion years, the evolution of life beyond Earth, and the ability to sustain life beyond Earth. In keeping with its space mission cutting across all NASA strategic enterprises, the plan emphasized how its goals could be accomplished with missions planned or already in development. In studying how to sustain life beyond Earth, the International Space Station was seen as "an essential evolutionary test-bed" for research on the effect of the space environment in biological evolution. The Mars Sample Return mission had the potential to provide an unambiguous answer about extant or extinct life on Mars. And the human exploration of Mars tapped into a long-held part of the American psyche fed from Lowell to Bradbury to *Viking*. The mission details, however, were yet to be developed. Scott Hubbard, who had been the originator of the *Mars Pathfinder* during its formative stages at Ames and had served as the mission manager for the equally successful *Lunar Prospector*, played a key role in this regard, providing expertise in relating astrobiology to real missions. Mission relatedness also provided astrobiology credibility within

NASA; any concept that could not utilize spaceflight was a hard sell within a space agency. Astrobiology's first mission, an airborne sortie to observe the Leonid meteors predicted to "storm" in November 1998, was a good example of the extended reach of the new discipline. "The central theme of this mission was astrobiology," said principal investigator Peter Jenneskins. "We were especially interested in learning the composition of [comet] Tempel-Tuttle's debris, the molecules that were created during the meteors' interaction with Earth's atmosphere, and the composition and chemistry of the atoms, molecules and particles detected in the meteors' path. We hope this will help us understand how extraterrestrial materials may have helped create the conditions on Earth necessary for the origin of life. The mission also sought clues about how biogenic compounds formed in stars are eventually incorporated into planets."[17]

From a content point of view Ames's Astrobiology Development Plan envisioned building on the traditional exobiology program, as well as a new initiative in evolutionary biology, while integrating Earth, life, and space sciences. It envisioned strong collaboration with the university community to develop undergraduate and graduate training for the next generation of multidisciplinary scientists. And, although Ames was to be NASA's lead center for astrobiology, JPL, the Johnson Space Center, and the Goddard Spaceflight Center would also be primary participants. If Edison invented the modern research laboratory and E. O. Lawrence the modern large-scale multipurpose national laboratory, the plan saw itself as creating a national "superlaboratory" that built on the advances of information technology to enable a truly multidisciplinary approach to astrobiology. The Astrobiology Institute would embody that new step in multidisciplinary cooperation.[18]

Important as input for the Astrobiology Development Plan and in the longer term for defining the scope and limits of the new discipline were a series of workshops held at Ames beginning in 1996. The earliest actually preceded the first astrobiology workshop by several months and was dubbed the "Pale Blue Dot" workshop, referring to planet Earth as described in Carl Sagan's 1994 book with the same title. (Although Sagan was not directly involved in the development of astrobiology at Ames, he was in many ways a guiding spirit, even after his early death in 1996.) The goal of the Pale Blue Dot workshop was to find and characterize habitable planets in other solar systems, other "pale blue dots," with whatever techniques could be mustered. Related to this goal was an "exozodiacal dust" workshop, held in 1997, which focused on the problem of dust interfering with the detection of planets.[19]

In 1998, as it became evident that serious funding for astrobiology might be forthcoming, the pace of workshops accelerated, and their scope widened. A flurry of workshops commissioned by Harper and the co-leader she recruited, Greg Schmidt, were led by Ames scientists and attended by government, university, and industry representatives. A "Piggyback Missions" workshop identified opportunities for near-term astrobiology payloads on missions already planned and evaluated the readiness of candidate payload technologies. A workshop on "Advanced Measurement Systems" characterized the state of technologies usable for astrobiology and brought in Defense Advanced Research Project Agency (DARPA) superstars, with their ultraminiaturized detection systems. Another meeting on "Evolution and Development" evaluated astrobiology opportunities related to the coevolution of life and the environment as well as rapid change and ecosystem evolution. At the same time, a "Beyond Planet of Origin" workshop evaluated mission opportunities to determine how life (including terrestrial life) would evolve beyond its home planet. Also in 1998 two workshops were held related to astrobiology and Mars and, in 1999, one on "Genomics and the International Space Station"; the latter brought in Baruch Blumberg again and led to his agreement to lead the institute. These workshops played a key role in bringing people together from a variety of backgrounds and crystallizing support for a broadly conceived astrobiology program. In some cases they led to important and long-range elements for the astrobiology program: the Advanced Measurement and Piggyback Missions workshops, chaired by John Hines and K. R. Sridhar, resulted in the programs known as Astrobiology Science and Technology for Exploring Planets (ASTEP) and Astrobiology Science and Technology Instrument Development (ASTID). These programs, which Schmidt, Michael Meyer, and David Lavery shepherded through Congress, provided astrobiology the critically needed resources for adapting the latest technology for mission use.

In addition to these workshops, other regularly scheduled meetings fed into the new field and were in turn affected by it. In November 1997 the Sixth Symposium on Chemical Evolution and the Origin and Evolution of Life met at Ames. Because this triennial meeting involved most of the principal investigators in NASA's exobiology program reporting on their recent research results, it provided a good opportunity for early discussion of astrobiology. Indeed, in opening remarks headquarters discipline scientist for exobiology, Michael Meyer, discussed "Astrobiology and Exobiology," and characterized the Exobiology program as "a key element of NASA's nascent Astrobiology Initiative."[20]

At the same time, ever mindful of funding issues, the tremendous public interest was not lost on NASA officials. "We're not going to find the cure for cancer by doing this," DeVincenzi remarked, "but the payback to the American public and the worldwide public is a continuing new perspective on ourselves, on our role, how our environment shapes us and we shape the environment. The impact is more of a philosophical impact than a practical impact. And it will affect our education, it will affect what stimulated new science and technology developments, and that's what basic research is all about." Key in involving the outside world was Kathleen Connell, who as Astrobiology's outreach manager at Ames made sure that astrobiology received a hearing in Washington political circles. This was done in a variety of ways, through the Internet, the Aerospace States Association, with its many contacts on Capitol Hill, and well-placed briefings. As with all NASA missions, the Astrobiology Institute carried out its own Education and Public Outreach program, mandated at 1–2 percent of the total mission funding. These activities cannot be underestimated in astrobiology's meteoric rise. As the interest among students and the benefits to education became increasingly apparent, the educational component of astrobiology was correspondingly strengthened.

The Astrobiology Institute

The idea of an Astrobiology Institute was the product of constrained budgets at NASA as well as Goldin's desire that NASA should leverage its contacts with the academic community for scientific research and do less in-house research and more collaborative efforts with academia. JPL, a NASA center with no civil servants, run by CalTech, was an example. NASA already had two other institutes, the Goddard Space Institute in New York and an institute that Marshall Spaceflight Center had formed with the University of Huntsville in Alabama. In an extreme form of the proposal for Ames civil servants would have been fired and transferred to an astrobiology institute, but this idea did not reach legislative action in Congress. Nevertheless for a year a team consisting of members from NASA headquarters and its centers studied the idea of an institute in some form, visiting institutes such as the National Center for Atmospheric Research (NCAR) as benchmarks. In the end emerged the Biomedical Institute at Johnson Space Center, the Microgravity Institute at Lewis Center in Cleveland, and the Astrobiology Institute at Ames.[21]

The initial development of the Astrobiology Institute concept fell mainly to Scott Hubbard, the deputy director of space at Ames, working with Michael Meyer at headquarters and Hubbard's colleagues David Morrison and Lynn Harper, among others at Ames. Gerald Soffen was also essential as an advocate at NASA headquarters for the institute, convincing—some might say strong-arming—life and Earth sciences to contribute substantial funding. In April 1997 Ames personnel wrote a first draft of the concept for an institute, in which more could be done with less. The draft was widely circulated to the scientific community, with comments and questions to be considered until 29 August. The Cooperative Agreement Notice (CAN) soliciting proposals for members of the NASA Astrobiology Institute was released in September 1997, for selection in early 1998. Among the innovative features of the institute was its "virtual" nature: its members were to be geographically dispersed and not individuals but organizations, ranging from industry, universities, and nonprofit groups to NASA centers and other government agencies. Organizations were encouraged to form cooperative partnerships. The virtual institute members would be tied together by the "Next Generation Internet" (NGI); by personnel exchanges; by series of workshops, seminars, and courses; and by sharing common research interests. The resulting research would complement work carried out by individual principal investigators in NASA's Exobiology and Evolutionary Biology grant programs.[22] The CAN also clarified the relation of astrobiology to the Origins program, emphasizing that it "has substantial overlap with the Origins program, and extends beyond it to encompass questions dealing with the adaptability of terrestrial biology to nonterrestrial environments and the development and evolution of ecologies and their interaction with their changing environments, especially when those changes are rapid."

In addition to multidisciplinary research, the institute was charged with developing new program directions and mission and technology requirements, developing a new generation of astrobiologists, and "capitalizing on the great public appeal of Astrobiology by building an education and outreach program to share the excitement of discovery with the people who pay for it." Its goal of using the Next Generation Internet as a tool for conducting research and fostering scientific exchange dovetailed nicely with Ames's designation as NASA's Center of Excellence in Information Technology, charged as the NASA lead in a multiagency effort to develop the NGI.

On 19 May 1998 NASA headquarters announced the selection of eleven academic and research institutions as the first members of the Astrobiology Institute and billed it as "launching a major component of NASA's Origins Program." The competition had been intense; fifty-three "uniformly first-class proposals" had been submitted. The eleven winners, expanded to fifteen in 2001(table 9.1), included five universities, three research institutions, and three NASA centers, including Ames, Johnson Space Center, and JPL. The inclusion of three NASA centers made sense: JPL was the lead center for the Origins program, Johnson was the center for the team that had announced the Mars rock, and Ames had its long history of exobiological research and was astrobiology's parent. In a memo sent to all staff the same day, Ames director Harry McDonald congratulated the team submitting Ames's proposal, remarking that it had been "earned by years of making significant contributions to the subject matter. . . . We are very proud of our astrobiologists!" The original eleven institutions divided some four million dollars for fiscal 1998, looked forward to nine million in 1999, and hoped eventually to grow to one hundred million per year.[23]

The establishment of the new institute generated an enormous amount of excitement, especially among the winners. Harvard paleontologist Andrew Knoll saw it as "providing for the first time a comfortable intellectual home for these kinds of investigations." But establishing a new institute of such scope again raised funding issues similar to those three years before, when astrobiology was first broached. The search for sustained funding caused considerable tensions within the NASA bureaucracy, as some players refused to participate by contributing money from their already established programs. Among other administrative issues was the question of choosing a director. The top choice to head the institute, departing NASA associate administrator for space science Wes Huntress, declined, and, for the better part of a year, first Gerald Soffen and then Scott Hubbard served as the interim directors of the Astrobiology Institute.[24]

Only in May 1999 did Goldin announce that Nobelist Baruch S. Blumberg would take over in September as head of the Astrobiology Institute, headquartered at Ames (fig. 9.1).[25] In appointing Blumberg at age seventy-three, Goldin

TABLE 9.1 *NASA Astrobiology Institute Members and International Partners*

Institution	Research Focus
Eleven institutions announced, 19 May 1998[a]	
Arizona State University, Tempe	Organic synthesis
Carnegie Institution of Washington	Life in hydrothermal systems
Harvard University, Cambridge	Geochemistry and paleontology
Pennsylvania State University	Coevolution of Earth's biota
Scripps Research Institute	Self-replicating systems
University of California, Los Angeles	Paleomicrobiology; early ecosystems
University of Colorado, Boulder	Origin/habitability of planets; RNA catalysis; philosophical aspects
Marine Biological Laboratory, Woods Hole	Microbial diversity; origins of proteins
Ames Research Center, Mountain View	Planet formation; Earth-biosphere interaction
Jet Propulsion Laboratory, Pasadena, Calif.	Biosignatures of life
Johnson Space Center, Houston, Tex.	Biomarkers in rocks
Four additional institutions announced, 19 March 2001	
Michigan State University, East Lansing	Earth analogs to life on Mars and Europa
University of Rhode Island, Kingston	Extremophiles in deep biosphere
University of Washington, Seattle	Earliest life on Earth; extrasolar planetary life
Jet Propulsion Laboratory, Pasadena, Calif.	Recognizing biospheres of extrasolar planets
International partners	
Centro de Astrobiologia, Torrejon de Ardoz, Spain	
United Kingdom Astrobiology Forum and Network, Cambridge, UK	
Australian Centre for Astrobiology, Sydney, Australia	
Grupement des Recherches en Exobiology, Paris	

[a]Agreements were for a period of five years. In 2003 twelve new teams were chosen, some at the same institutions but with different topics. Six institutions added at this time were Indiana University, the SETI Institute, NASA Goddard Space Flight Center, University of Arizona (Tucson), University of California (Berkeley), and University of Hawaii (Manoa). At that time Arizona State, Harvard, Scripps Research Institute, the first Jet Propulsion Lab team, and Johnson Space Center ended their tenure as members. By this time the European Exo/Astrobiology Network Association had also been added as an international partner.

secured a man with a sterling reputation in science but no background in exobiology. Blumberg was a biochemist who had received the 1976 Nobel Prize in Physiology and Medicine for his discovery of the hepatitis B virus and the development of a vaccine. But he had made contributions to a broad array of problems in human biology, biochemistry, and genomics. And genomics was envisioned as one of the core fields for astrobiology. It was Blumberg who had chaired the Ames workshop "Genomics on the International Space Station" five months before he was named to head the institute.[26] His participation in this

FIGURE 9.1. Daniel Goldin, Harry McDonald, and Baruch Blumberg at the 18 May 1999 press conference at which Goldin announced Blumberg's appointment as head of the Astrobiology Institute. McDonald was the director of NASA Ames, where the institute was headquartered. (Courtesy NASA Ames Research Center.)

workshop showed real insight into astrobiology and a genuine love of multidisciplinary research. He was excited by space flight and believed in the importance of the new astrobiology program. Blumberg was also a field biologist who understood space missions intuitively because he had made scientific discoveries in deep Africa using only the equipment he could carry on his back. He related immediately to the Antarctic exobiology researchers. He also was a believer in the value of research in extreme environments, including the extremely low gravity environment of space. There is no doubt that the appointment of such a luminary, who also assembled a luminous board of advisors, was an important landmark for astrobiology.

Blumberg's appointment also provided an opportunity for Goldin to give astrobiology a rhetorical, if not a monetary, boost. Astrobiology, NASA's administrator remarked in ceremonies at Ames, was "the cornerstone to NASA's mission in the new millennium." Comparing the understanding of the origin and evolution of life to the generational effort of cathedral building a thousand years earlier and hoping to bring a new level of knowledge to biology as had been done in physics over the previous fifty years, Goldin remarked that "quite possibly the rewards from this pursuit of Astrobiology may eclipse the societal and economic benefits of all prior NASA activity." One of the reasons for locating the institute at Ames was to enable the synergy between information technology and astrobiology, not only at Ames but with the surrounding Silicon Valley as well. Astrobiology, Goldin noted, "is a revolution that will require its own revolution . . . in communications, networking, information technology, computing and scientific thinking." Noting the collaboration of government, industry, and academia within the Astrobiology Institute, Goldin saw their goal as "trying to discover if there is a thread of life beyond Earth. It is a powerful concept. And it is a concept whose time has come." Blumberg agreed, foreseeing a "flowering of biology" in the next century. Not to be left out, chemists also showed interest in joining the institute.[27]

Thus, exactly four years after Ames was given the lead for astrobiology in May 1995, and one year after the institute's first eleven members were chosen in May 1998, the Astrobiology Institute was well on its way to becoming an important institutional home for the new field of astrobiology. Meanwhile, one other element had been put in place, a more detailed plan for astrobiology's future. By summer 1998 astrobiology management at Ames, feeling the program was ready to gel, convened an all-important roadmap meeting.

The Roadmap

Three years of hope, hype, and hard work culminated on 20–22 July 1998, when 150 scientists met at Ames to draft a roadmap for astrobiology for the next twenty years, with emphasis on the first five. The invitation letter from David Morrison (cochair of the meeting with Michael Meyer) billed the workshop as "a critical planning activity to delineate NASA's role in the new field of Astrobiology, spanning elements of space, life and earth science." The task was to proceed from astrobiology's basic questions to "a more detailed plan of how and when we will answer these questions." Starting with the fundamental questions developed in the first astrobiology workshop of September 1996 and subsequently refined in the Astrobiology Institute CAN, the workshop was to articulate "a visionary set of science goals to be achieved in the coming decades in this new field, as well as the intermediate science objectives that must be met to realize these goals." Furthermore, it was to derive requirements for laboratory and theoretical research, for missions, and for the technologies to accomplish these goals. This, in turn, would lead to a decision about where astrobiology's goals would fit in with, and where necessary modify, existing programs such as the Mars program, the *Discovery* program, and the International Space Station.[28]

The concept of a "roadmap" can be traced back only to 1995 at NASA, when three teams were assembled to put together the Exploration of Neighboring Planetary Systems (ExNPS) roadmap, an effort coordinated by JPL and published in 1996 (described in chap. 7). The idea of a roadmap was not to set down detailed milestones or even to map goals onto missions but to provide guidance for research and technology development over the long term. Within NASA veterans knew that astrobiology could not be a purely intellectual endeavor; it had to be tied to what NASA did best: space missions. Exactly how astrobiology would be integrated into NASA's space science, Earth science, and human space exploration enterprises would take years to work out, and in doing so astrobiology's goals had to be kept constantly in mind.

The roadmap workshop began with opening remarks from Ames director Henry McDonald, administrator Goldin (by videophone, since he was tied up with budget issues in Washington), Michael Meyer, David Morrison, and Scott Hubbard, who was then the interim manager for the Astrobiology Institute. After brief tutorials on various aspects of astrobiology, breakout sessions were held centering on astrobiology's driving questions and how they might be answered by existing or future NASA missions.

The final Astrobiology Roadmap, released on 6 January 1999, identified four principles, ten goals, and seventeen objectives for astrobiology. The operating principles were as follows:

1. Astrobiology is multidisciplinary, and achieving our goals will require the cooperation of different scientific disciplines and programs.
2. Astrobiology encourages planetary stewardship, through an emphasis on protection against biological contamination and recognition of the ethical issues surrounding the export of terrestrial life beyond Earth.
3. Astrobiology recognizes a broad societal interest in our subject, especially in areas such as the search for extraterrestrial life and the potential to engineer new life forms adapted to live on other worlds.
4. In view of the intrinsic excitement and wide public interest in our subject, astrobiology includes a strong element of education and public outreach.

Astrobiology's goals as perceived at this meeting were more specific (table 9.2). All, of course, were related to the three fundamental questions that had been enunciated early in the development of the concept of astrobiology: (1) How does life begin and evolve? (2) Does life exist elsewhere in the universe? (3) What is life's future on Earth and beyond? The roadmap further spelled out how each of the goals might be met through even more specific objectives (see app. D) and implementation examples.[29]

One of the unexpected events of the meeting was the development of a significant splinter discussion by a small but diverse group of participants: "How will astrobiology affect and interact with human societies and cultures?" Participants in this discussion group, inspired by Astrobiology's third operating principle, proposed that a multidisciplinary approach be used to understand the consequences of the search for life on Earth and beyond, the explanation of life beyond Earth, and the discovery of life beyond Earth. This question became the object of controversy, with some claiming that social science had no place in NASA, especially if it were going to divert funding. A few of the scientists, including planetary scientist Bruce Jakosky from the University of Colorado, were sympathetic; they argued that to a large extent philosophical questions were the intellectual drivers behind astrobiology and that it was incumbent on the scientific community to work through the issues of what the results of astrobiology

TABLE 9.2 *Astrobiology Goals*

1. Understand how life arose on Earth
2. Determine the general principles governing the organization of matter into systems
3. Explore how life evolves on the molecular, organism, and ecosystem level
4. Determine how the terrestrial biosphere has coevolved with the Earth
5. Establish limits for life in environments that provide analogues for conditions on other worlds
6. Determine what makes a planet habitable and how common these worlds are in the universe
7. Determine how to recognize the signature of life on other worlds
8. Determine whether there is (or once was) life elsewhere in our solar system, particularly on Mars and Europa
9. Determine how ecosystems respond to environmental change on time scales relevant to human life on Earth
10. Understand the response of terrestrial life to conditions in space or on other planets

Source: From Astrobiology Roadmap, released 6 January 1999.

meant to society. In the end the three goals the group proposed were not included in the final report. Nevertheless, astrobiology's third operating principle, recognizing "a broad societal interest in our subject," did sanction such discussions, and in 1999 Ames sponsored a workshop on cultural aspects of astrobiology. There was precedent for this activity—SETI pioneers beginning with Philip Morrison had discussed societal implications; John Billingham championed such discussion by organizing a series of workshops in 1991–1992, and exobiology meetings occasionally entertained, and even featured, the subject. The roadmap workshop itself encouraged such discussion when, as if the science were not mind-expanding enough, the organizers brought in futurist Alvin Toffler to engage in a dialogue about "The Fourth Wave and Astrobiology." Toffler became one of the participants in the cultural aspects discussion group, along with science fiction writer Ben Bova.

In the wake of the roadmap Ames redoubled its advocacy for the program for which it was now the lead. In October 1999 Kathleen Connell organized (via the Aerospace States Association) an astrobiology symposium with a difference: this one was held in the Dirksen Senate Office Building, featured several members of Congress as speakers in addition to Blumberg and other astrobiology luminaries, and had a largely political audience. In his remarks Blumberg struck a "Lewis and Clark" theme, emphasizing that astrobiology was about exploration, a defining feature of American culture. There were other indications of the up-and-coming status of astrobiology. Soffen was instrumental in establishing an "Astrobiology Academy" at Ames, an internship for a dozen students during the summer. Postdoctoral awards, sponsored by NASA and administered by the National Research Council, were given for the Astrobiology Institute beginning in 2000. The University of Washington developed the first graduate program in astrobiology, and several astrobiology textbooks were being written. And, with increasing interest and research overseas, astrobiology was becoming internationalized, with some institutions becoming associated with the Astrobiology Institute (see table 9.1).

As the end of the millennium approached, many of the elements were in place for a reinvigorated discipline of astrobiology: a definition, a roadmap, a virtual institute, enthusiasm, and minimal funding. How these elements, and the lofty principles, goals, and objectives of astrobiology translated into real science, and whether they would usher in Soffen's great age of discovery, remained for the future to determine. In the epilogue we can offer only a glimpse of the shape of things to come but no hint at all of the discipline's ultimate answer to the question of the past, present, and future of life in the universe.

Epilogue

Astrobiology Science

Into the Great Age of Discovery?

*I*n an emerging scientific discipline the political skills needed for fund raising, convening workshops, and providing the myriad details of administration are necessary precursors, not ends in themselves. The ultimate goal of all these activities is to foster world-class science. The scientific questions of astrobiology, long-standing mysteries with potentially great societal impact, were the primary motivator for expanding the horizons of exobiology. The Astrobiology Institute, although virtual in concept, was the collaborative engine that would drive the new discipline and, it was hoped, spark it onward toward the development of innovative techniques and into what Gerald Soffen optimistically called the "Great Age of Discovery." There were no guaranteed outcomes, either for discovering life beyond Earth or for finding the optimal administrative and technical methods to reach that goal. Although there would be many advances made along the way, in the end the emergent discipline of astrobiology as a means to reach the ultimate discovery itself remained a great experiment.

In this respect, although workshops, funding, and administrative challenges were nothing new, the contrast between exobiology as conceived in the 1960s and astrobiology at the turn of the century was quite striking. To be sure, exobiology and astrobiology shared the core concerns of origins of life research and the search for life beyond Earth. But astrobiology placed life in the context of its planetary history, encompassing the search for planetary systems, the study of biosignatures, and the past, present, and future of life. Astrobiology science added new techniques and concepts to exobiology's repertoire, raised multidisciplinary work to a new level, and was motivated by new and tantalizing evidence for life beyond Earth. In addition to comparing astrobiology to the Lewis and Clark exploration, Astrobiology Institute director Baruch Blumberg was fond of pointing out that astrobiology was different from most science in that, instead of becoming more and more specialized, it was increasingly generalized, making use of many specialties to tackle a very broad set of questions.

Exactly how astrobiology would develop was anyone's guess when it was invented, the roadmap notwithstanding. In its early stages perhaps the best gauge was the biennial astrobiology science conference, the first of which was held at Ames in April 2000. In the inaugural meeting, consisting of three days of oral and poster presentations, more than 350 participants demonstrated, as Ames director Henry McDonald remarked, that "astrobiology is already a real and exciting science."[1] For those worried about the scope of astrobiology David Morrison offered an operational and practical definition: "Astrobiology will be defined in time by what astrobiologists do." Baruch Blumberg agreed that astrobiology would incorporate new objectives as new interests and opportunities developed and emphasized that astrobiology was a generational endeavor, analogous to cathedral building, not only in terms of such activities as a mission to Europa but also in incrementally increasing knowledge of astrobiology's major questions. Failure to discover extraterrestrial life, he felt, would be a step back from the Copernican revolution. Conference organizer Lynn Rothschild exulted that astrobiology "liberates us from disciplinary boundaries." And Exobiology Discipline scientist Michael Meyer added what everyone wanted to hear—that the budget for astrobiology at headquarters was on an upward curve.

Notwithstanding Morrison's open-ended definition of *astrobiology,* limits were evident in this first meeting. No papers were presented on the Big Bang and the origin of the universe, none on galaxy formation and dynamics, not even any on the large-scale structure of our own galaxy. Rather, the discussion began (logically though purposely not in order of presentation) with solar system dynamics and planetary detection, proceeded to cosmic chemistry and the origin of life, continued through the evolution of the genome, metabolism, and microbial communities, and ended with the evolution of advanced "metazoan" life. In this discussion Mars played a large role, including its geology, climatology, and oxidants; the latest research on the Mars meteorite; and planned Mars missions. The single greatest interest was shown in laboratory and theoretical studies of prebiological chemistry, perhaps still an artifact of funding in the old exobiology program. But interest in new research on biomarkers and on life in extreme environments was also very strong. Aside from a paper given by Bruce Jakosky (the chair of the Scientific Organizing Committee), the roadmap's renegade question on the cultural impact of astrobiology was entirely absent, perhaps equal parts a reflection of the difficulty of getting social scientists involved and the lack of encouragement from natural scientists. And SETI was notably lacking, except for a handful of poster papers, one of which was dedicated to education. With respect to SETI, the meeting starkly demonstrated how government funding, or lack thereof, could shape an entire field. Altogether, however, some thirty categories of the emerging science were represented aside from SETI (app. C). And this was just the *first* astrobiology science meeting.

The second astrobiology science conference, held on 7–11 April 2002 at Ames, revealed an even more thriving discipline. The venue was the soaring 1930s "Hangar 1" dirigible building, a necessity in order to accommodate the seven hundred participants but also a symbol of astrobiology's lofty aspirations. (One would not want to carry the metaphor too far; the hangar became obsolete in the 1930s, when dirigibles began crashing, a reminder that astrobiology was always in danger of losing funding.) The unofficial theme, enunciated by Michael Meyer as the meeting opened, was that "astrobiology has arrived." Baruch Blumberg had sounded the same theme in a special issue of *Ad Astra,* the magazine of the National Space Society, circulated at the meeting. Assessing "Astrobiology at T + 5 Years," Blumberg wrote: "In five short years, Astrobiology has been transformed from a buzz word one had to explain into an overarching research and exploration paradigm that people from diverse backgrounds can intuitively and easily grasp. Its influence can clearly be seen in a variety of Earth-based and space-based research projects." He concluded: "Astrobiology has arrived. And we've only just started."[2]

Blumberg's statement was true in a variety of ways. The Astrobiology Institute budget had by now increased to some forty million dollars, 90 percent of it from Space Science at headquarters and the remainder from Earth and Life Science. Six "focus groups" had sprung up to coordinate and enhance research efforts: evolutionary genomics, astromaterials, mission to early Earth, mixed microbial ecogenomics, Mars, and Europa. In another sign of an emergent discipline, in addition to the relatively venerable *Origins of Life and Evolution of the Biosphere,* two journals now vied for prominence at the meeting: the *International Journal of Astrobiology,* published by Cambridge University Press, and the American journal *Astrobiology.* By the second meeting in 2002 the Astrobiology Institute had grown to fifteen members and four international associate or affiliate members (see table 9.1). Most important of all, the scientific basis for astrobiology was growing more solid, as evident in the number and quality of the papers and in new techniques. Although the categories at the second astrobiology science conference had been conflated from thirty-one to thirteen more general topics, the scope was still the same. In part (and only in part, given that international partners received no NASA funding, and not all American researchers in the field did either) the astrobiology science meetings represent a command performance, a chance for astrobiology researchers to show NASA funders what they are getting for their money. More than that, they were an important means of communication and socialization, especially critical for such a multidisciplinary endeavor. Along with published research the meetings demonstrate how astrobiology was developing, as reflected in the research undertaken at the member organizations of the Astrobiology Institute and in universities and laboratories around the world. Finally, these gatherings, taken together with published research, give an early sense of how, and in what relative proportions, astrobiology is addressing its three guiding questions: how does life begin and evolve? Does life exist elsewhere in the universe? And what is life's future on Earth and beyond?

As we have repeatedly stressed, and as astrobiologists themselves explicitly acknowledge as part of their motivation, these are fundamental questions that humanity has asked in increasingly subtle and refined forms over millennia. It is now fair to inquire, as critics often do, whether any progress has been made in addressing these questions, especially since NASA's involvement began, over four decades ago, with its infusion of government funding.

It must be said that forty years of research on the origin and evolution of life has resulted in great advances in understanding while leaving the ultimate questions unresolved.[3] The basic problem of whether organics originated on Earth, from space, or some combination of the two, is still very much open. Laboratory and theoretical studies of prebiotic chemistry—the bread and butter of the exobiology program from the beginning—remain a strong research program in astrobiology. Work on laboratory models for replicating systems, metabolism in primitive living systems, and microbial ecology are advancing in ways unforeseen. Of all the new techniques, especially genomics—use of the gene database for clues to evolution—has opened entirely new vistas of research. Still, consensus on ultimate origins remains elusive, even as the questions were refined.

Under these circumstances there was considerable scope and hope for a great age of discovery. The specific scientific tasks needed to answer the questions of origins were embedded in astrobiology's objectives as stated in the original roadmap (app. D). Work on these tasks, spread not only among the members of the Astrobiology Institute but also in laboratories around the world, is advancing unevenly and sometimes excruciatingly slowly. But taken together they represent a unified attack on one of the great problems of science, the first component of astrobiology's ambitious agenda.

Questions of the origin and evolution of life were addressed using several broad complementary methods: laboratory studies, real-life Earth history, and astronomical observations, often used in combination. All were well represented at the Astrobiology science conferences. The origin of complex organics is a case in point. At both conferences Louis Allamandola and his team at NASA Ames reported on their work combining laboratory simulations with infrared studies of interstellar molecules and ices to show how complex organics such as poylcyclic aromatic hydrocarbons (PAHs) were formed through the interaction of ultraviolet light and cosmic rays.[4] Such a "cold start" for complex organics in space was a stark contrast to the hot dilute soup assumed by early exobiologists. And both contrasted in technique and concept with the "hydrothermal vent" scenario for the origin of life, based on hot springs and undersea vents, the latter completely unknown when exobiology began its career. Carl Woese's work on phylogenetic relationships from gene sequencing fingered "Archaea" as in some ways the most primitive and perhaps earliest organisms, giving rise to the possibility of a "hot start" for life in the energy-rich environment of water and minerals recycling at mid-ocean ridges where seafloor spreading was taking place. Thus, one could choose whether the rudiments of life began in outer space, on or below the Earth's surface, or deep undersea in the realm of the extremophiles.

Because the evidence of "what actually happened" at the origin of life was forever lost, laboratory studies were especially crucial for the early stages of biogenesis. One major problem remained the emergence of self-replicating molecules, a crucial step on the way from inanimate chemical reactions to the chemistry of living systems. Here the "RNA world" model was pitted against the more traditional protein-based model. Several laboratories, including those at the University of Colorado and the Scripps Research Institute, in collaboration with the University of Florida and the University of California–Riverside, worked to test these models. Beyond self-replication the laboratory approach also shed light on a variety of other steps in biogenesis. At the University of California–Santa Cruz, for example, David Deamer has studied for decades the self-assembly of membranes from "amphiphilic" components that he showed in 1989 to be present in carbonaceous meteorites such as Murchison, an essential step in explaining the membrane-bounded cell perhaps essential for the origin of life. By 2001 Jason Dworkin, working with Deamer and Allamandola, among others, synthesized self-assembling amphiphilic molecules in simulated interstellar/precometary ices. The origin of prokaryotes in Earth's early biosphere, eukaryote origins, and evolution of cellular complexity all were part of the astrobiology effort at the venerable Marine Biological Laboratory in Massachusetts, among others. Laboratory studies of models of simple cellular systems were under way.

Beyond the laboratory researchers employed a variety of techniques to study Earth's earliest life. Led by J. William Schopf, part of UCLA's astrobiology effort concentrated on the geobiology and geochemistry of the oldest records of life on Earth, some 3.46 billion years old. (As an indication of the difficulty of such work, at the 2002 astrobiology science meeting Martin Brasier from Oxford University and his team questioned the validity of the morphological evidence on which Schopf's claim was based.) At Penn State a broad array of researchers focused on the coevolution of life and the environment 4.5 billion to 500 million years ago, especially the chemistry of the atmosphere. As an example of this work, James Kasting, a pioneer in the field of coevolution of life and the environment, reported at the 2002 meeting on the relationship of cyanobacteria to the rise of atmospheric oxygen around 2.3 billion years into Earth's history. Another research area centered around the new field of "biogeochemistry," in particular the study of modern cyanobacterial mats as analogues to ancient mats that left stromatolitic fossils on primitive Earth. Such microbial mats have a 3.46 billion–year fossil record and represent the oldest known ecosystems. David Demarais, one of the pioneers in this area, was also one of the founders of astrobiology at Ames and constantly emphasized the multifaceted relevance of biogeochemistry to astrobiology, including its role in generating biosignatures in Earth's atmosphere. At Harvard a broad range of studies were under way, concentrating on three transition periods in Earth history believed to be critical to the evolution of life: the Archean-Proterozoic period 2.5 to 2 billion years ago, when bacteria with aerobic metabolism and eukaryotes with mitochondria evolved; the Proterozoic-Cambrian period 800 to 509 million years ago, when large multicellular life emerged; and the Permian-Triassic period 251 million years ago, when a major mass extinction occurred. In search of life's extremes, astrobiology's researchers trekked to the Antarctic dry valleys, the Chilean Atacama desert, the Siberian permafrost, and surface hot springs; dived thousands of feet under the ocean's surface to hydrothermal vents at the mid-ocean ridges; and explored exotic cave ecosystems.

Whether in the lab or during empirical investigations, the use of 16s ribosomal RNA shed light on the relationship of the earliest organisms in ways undreamed of a few decades earlier. Woese's use of this method to discover the tripartite structure of the living world as composed of bacteria, archaea, and eukarya was now used to define the many branches of the universal tree of life. One such study at the 2002 astrobiology science conference demonstrated the genetic diversity and dynamics of microbial populations of cyanobacteria associated with stromatolites, believed to be analogues of early life on Earth. Using similar phylogenetic studies, at the same meeting University of Colorado researchers reported on endolithic microbial communities as a function of the type of rock (sandstone, limestone, and granite) in which the microbes are found. So promising was the new technique that the Astrobiology Institute formed a focus group on exactly this field of "evolutionary genomics"—the analysis of genomes with the goal of understanding how life originated and evolved on Earth. An understanding of how life on Earth changed with the Earth's environment might provide a basis for developing biomarkers on other habitable planets. This "evogenomic" group was complemented by the "ecogenomics" group, which studied the relationships among gene expression, microbial diversity, and biogeochemical processes. In particular, gene expression in microbial mat organisms was compared in various environments. Ames and the Marine Biological Laboratory led this effort, which was inconceivable at the beginning of the space age.

Such research represents only the tip of the iceberg of research in astrobiology's "origins" question. Such research was by no means confined to official members of the Astrobiology Institute. But the benefit of the institute was that it fostered interdisciplinary collaboration both within and among institutions, perhaps with mixed success because that goal was so challenging. Often members of a single institution studied not only many aspects of the origin of life but also other parts of the cosmic evolution puzzle as well. Astronomers and biologists were not accustomed to talking to one another, even at the same institution, a problem compounded between institutions. The laudable goal of the Astrobiology Institute was to create synergy not only from a unified research program and new techniques but also from increased interactions among researchers.

The theme kept constantly in mind in origins research was that what had happened on Earth might have happened on other planets, that the past could illuminate the present, shedding light on life on other worlds. The Earth was a great petri dish, and so too were other planets. Astrobiology's second great question was not only whether life *could* exist in the universe but whether it actually *does* exist.

The bottom line in the quest for life on other worlds—the oldest component of astrobiology—was that it had still not been found, claims for Martian microfossils notwithstanding. At the same time the question had been entirely transformed compared to exobiology forty years before. To be sure, there was the usual theme of life on Mars, now immeasurably enhanced by better spacecraft data of the planet. There was also the theoretical and empirical work on habitable planets, now transformed from embryonic planetary formation theories and Peter van de Kamp's single (and spurious) claim for a planet around Barnard's star to much more robust theories of planet formation and the discovery of almost one hundred extrasolar planets. But there were entirely new areas almost unheard of at the beginning of the space age: work on life in extreme environments, possible organic molecules and even habitats for life in the outer solar system (notably Europa), and biomarkers for detecting life on extrasolar planets. In addition, the idea of panspermia, the spread of life from planet to planet, was given great impetus by the controversy over the Mars rock.

The renewed search for life habitats in the solar system was a remarkable reversal of fortune in the wake of the disappointing *Viking* results. The infamous Mars rock and its still-disputed fossils certainly played an important role in this revival. But new data from missions to Mars, the first since *Viking*, played a crucial role as well. Although it carried no life detection experiments, *Mars Pathfinder* reinvigorated interest in Mars in 1997, with its daredevil bouncing landing, Sojourner rover, and raft of science data ranging from Martian geochemistry to meteorology. In the summer of 2001 *Mars Global Surveyor* revealed numerous gullies on Martian cliffs and crater walls and evidence of geologically recent liquid water (fig. Epi.1). Within months of beginning its mission in February 2002, *Mars Odyssey* gave strong evidence that large quantities of water were present within three feet of the surface of Mars at latitudes from the south pole to 60 degrees south. In 2004 the European Mars Express Orbiter returned data indicating the presence of methane in the Martian atmosphere, possibly of biogenic origin. Meanwhile Opportunity, one of the American Mars Express Rovers (MERs), examined an outcrop of salt-laden sediment and found thin intersecting layers interpreted as sand ripples, perhaps shaped by flowing water in a huge shallow sea.

Even more surprising than Mars was the astrobiological potential of the Jovian satellite Europa. The *Voyager 2* spacecraft in 1979 had originally discovered the fractured nature of Europa's surface. In 1996, the same month that the Mars rock fossils were claimed, the *Galileo* spacecraft gave added impetus to the theory that these fractures could be cracks in an ice-covered planet. Moreover, the *Galileo* spacecraft supported the claim that Europa might harbor a liquid ocean below the ice (fig. Epi.2). And where there was water, there could be life. Not even Arthur C. Clarke's science fiction had dreamed of this scenario when NASA's exobiology program began in the early 1960s, though Clarke did broach

EPILOGUE FIGURE 1. Gullies on Mars, believed to be less than a million years old, indicate that water may still exist just under the surface of the planet. (Mars Global Surveyor image courtesy NASA / JPL / Malin Space Systems.)

it in his novel *2010*. The controversy raged over how thick the ice was, whether life could originate in an ocean, and how to reach it. The National Research Council of the National Academy of Sciences drew up a science strategy for exploring Europa and for preventing its contamination, and NASA even contemplated a Europa mission. The Astrobiology Institute Europa focus group was only one of many that addressed these questions. With all the excitement and an increasing number of published papers, Europa had a small but steady presence at the astrobiology science meetings.[5] And beyond the moons of Jupiter loomed Saturn and its enigmatic moon Titan, whose secrets (including possible complex organic molecules) might be revealed in 2004 when the Huygens probe of the *Cassini* spacecraft entered its atmosphere.

Beyond the solar system an important focus of the astrobiology science

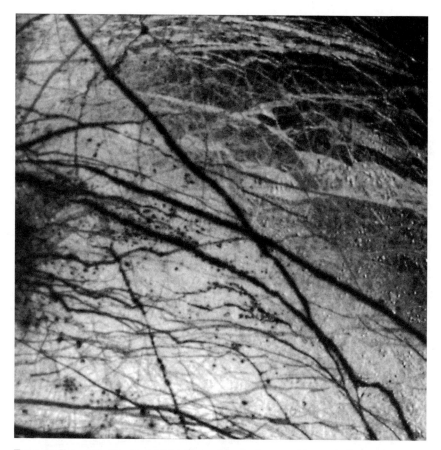

EPILOGUE FIGURE 2. The fractured surface of Jupiter's moon Europa indicates that water may exist below the ice. (Courtesy NASA/JPL.)

meetings and members of the Astrobiology Institute was the search for extrasolar planets, now rapidly advancing using a variety of techniques. Although NASA had been slow to support the ground-based observations that had netted about one hundred gas giant planets by the turn of the millennium, in 2002 it plunged fully into the planet search when it funded the Kepler mission, a method of searching for "transiting" planets by measuring the diminution of light when the planet passed in front of its parent star. Bill Borucki, an astronomer at Ames, had been the longtime champion of this method; he remained its principal investigator but now headed a team that would be responsible for launching the spacecraft in 2007 and analyzing the data thereafter. Meanwhile, along with the wider astronomical community, members of the Astrobiology Institute tackled other aspects of what some had dubbed the new "planetary systems science." At UCLA, the University of Colorado, and the Carnegie Institute of Washington, among other institutions, researchers studied the formation of stars and planets and planetary habitability. And, whereas a decade earlier no planets had been known at all, at JPL work was already being undertaken on the longer-term problem of recognizing the biospheres of extrasolar planets. By June 2002, when NASA and the Carnegie Institution sponsored a meeting in Washington, D.C., on "Scientific Frontiers in Research on Extrasolar Planets," it drew some 250 researchers on this subject alone, many of them just entering a field they well recognized as ripe for innovation and discovery.

In addition to missions to search for planets and life, new techniques played a crucial role in reinvigorating the exploration for life in possible solar system habitats—just as they did in the related field of origin of life studies. Extremophiles research at Ames and elsewhere probed the limits of life as it might exist on other planets. The Carnegie Institute of Washington and Arizona State University undertook laboratory investigations of organic chemical systems as a means of understanding hydrothermal systems. Such systems were potential analogues to solar system bodies and potential sites for the origin of life on Earth. The Johnson Space Center, center of the Mars rock controversy, studied the problem of biomarkers in astromaterials, including meteorites, interplanetary dust particles, and future sample returns. A field once almost abandoned in the post-*Viking* era was now more robust than ever.

The future of life on Earth and beyond—a question hardly enunciated in early exobiology—remained the most undeveloped of astrobiology's three questions. Many scientists were not accustomed to dealing with the future, and it is no surprise that this aspect of astrobiology was least represented at its science meetings. Nevertheless, precisely because of the lack of attention, the potential for new thinking and important discoveries was great. As the astrobiology roadmap had stated, NASA had much to contribute to global problems such as ecosystem response to rapid environmental change and Earth's future habitability in terms of interactions between the biosphere and the chemistry and radiation balance of the atmosphere. It was uniquely suited to understanding the human-directed processes by which life could evolve beyond Earth. And it was charged with initiating and refining planetary protection guidelines both for other planets that its spacecraft visited and for the Earth itself as sample return missions were contemplated. Problems such as terraforming Mars were indeed problems of the future but nonetheless important for that. NASA tackled such problems with a greater or lesser degree of enthusiasm, which depended to a great extent on individuals willing to lead the charge in these areas. Even in a fourteen billion–dollar agency with thousands of employees, much still rested on individual initiative.

Tied into this lack of enthusiasm for studies of the future of life was the lack of attention to the societal implications of astrobiology. Only one of the Astrobiology Institute's members, the University of Colorado, had "philosophical aspects" as part of its official charter, due largely to the personal interest of planetary scientist Bruce Jakosky. As the institute was gearing up, NASA did sponsor a workshop in 1999 on "Societal Implications of Astrobiology" which drew a small but diverse group of scholars.[6] This was an outgrowth of the interest in cultural implications expressed at the roadmap workshop. But progress in this endeavor remained difficult; papers were an occasional feature of the triennial bioastronomy meetings, and only one paper at the second astrobiology conference dealt with cultural evolution and its effect on the future of humanity.[7] In part for long-standing reasons of the "two cultures" divide, the melding of the social sciences with the natural sciences proved even more difficult than the joining of biological and physical sciences in exobiology's earlier history. There was hope, however, for using NASA resources to study these problems. Such studies were certainly in line with the statement of NASA's new administrator, Sean O'Keefe, that "in broad terms, our mandate is to pioneer the future, to push the envelope, to do what has never been done before. An amazing charter indeed. NASA is what Americans, and the people of the world, think of when the conversation turns to the future.... So in the end NASA is about creating the future."[8] Moreover, under the O'Keefe administration NASA's vision for the future was "to improve life here. To extend life to there. To find life beyond." The future of astrobiology seemed bright.

At the beginning of the new century astrobiology was thriving, with the old concerns of exobiology at its core and the Origins program of cosmic evolution as its ultimate context. Forty years after Harold Klein inaugurated Life Sciences at NASA's Ames Research Center, it remained a center for astrobiology in terms of numbers of researchers, laboratories, and sheer breadth. But now the Astrobiology Institute, reinvigorated by new institutional members (see table 9.1) each funded at one million dollars per year and with refined objectives, immensely multiplied and leveraged those factors. And beyond the institute a worldwide effort was under way to answer one of science's oldest questions. In four decades the effort had grown beyond the wildest expectations of exobiology's founders.

Would all these activities have been enough to silence critics such as evolutionist George Gaylord Simpson, who in the 1960s had declared exobiology a science without a subject? After forty years, was exobiology a scientific discipline or not? Implicitly or explicitly, astrobiologists took a practical operational stand. "Whenever anything comes up about exobiology you treat it as a discipline, just like we're going to treat astrobiology as a discipline," Ames's Donald DeVincenzi remarked in 1997. "Strictly speaking is it? I don't know. If we say it is and we treat it that way, then it is, for these purposes.... These names basically tell us how to manage things, we know how to manage disciplines. So if we're going to invent some brand new branch of science we're going to call it a discipline. So there will be an astrobiology discipline, I can guarantee you. And it will be run just like any other NASA discipline: geochemistry, geophysics, planetary atmospheres. But you're not going to find it in a textbook probably, or in a department chairmanship."[9] A few years later astrobiology textbooks and university courses and programs in the subject, if not departments, were a fact.

Despite the excitement, astrobiology's future remained unclear at the dawning of the twenty-first century. Like the dirigible hangar in which the second astrobiology conference was held, it was possible that the new science could become obsolete by failure—failure of funding, failure of imagination, or failure to answer its core questions of the origins and ubiquity of life. Some astrobiologists worried that the field was in danger of fragmenting and becoming too narrow.[10] It was hard to imagine, however, that it would fail through lack of interest, whether public or scientific. Although the fundamental questions of astrobiology remained unanswered, the desire to find answers was stronger than ever.

Appendix A: Unpublished Sources

Oral History Interviews

Interviewee	Date	Interviewer
Peter Backus	16 September 1992	SD
John Billingham	12 September 1990; 1 June 1992; 17 September 1992; 8 June 1993	SD
David Black	30 January 2001	SD
Martin Brasier	16 May 2002	JS
Thomas Brock	15 February 1999	JS
David Brocker	16 September 1992	SD
Melvin Calvin	2 June 1992	SD
Glenn Carle	13 May 1997	SD
Sherwood Chang	21 November 1997	SD
Erwin Chargaff	6 February 1999	JS
Kathleen Connell	13 June 1997; 6 April 2000	SD
Gary Coulter	27 July 1992; 30 September 1993	SD
John Cronin	28 January 1997; 6 December 2000	JS
William Day	17 August 1998	JS
David Deamer	25 June; 11 July 1997	JS
David DesMarais	12 May 1997	SD
Donald DeVincenzi	8, 21, and 28 January; 4 and 11 February 1997	JS
	12 May 1997	SD
Frank Drake	29–30 May 1992	SD
Jack Farmer	19 November 1997	SD
Sidney Fox	27 January; and 1 February 1993	JS
Imre Friedmann	18 November 1997	SD
Everett Gibson	16 and 23 January 2002	JS
Sam Gulkis	26 June 1992	SD
Lynn [Griffiths] Harper	13 May; and 21 November 1997	SD
Lawrence Hochstein	15 May 1997	SD
Norman Horowitz	15 January 1999	JS
John Jungck	10 November 1999	JS
Nicolai Kardashev	9 August 1988	SD
John Kerridge	15 February 1997	SD

Interviewee	Date	Interviewer
Bishun Khare	16 May 1997	SD
H. P. Klein	15 September 1992; 14 May 1997	SD
	28 November 2000	JS
Michael Klein	10 August 1988; 26 June 1992	SD
Keith Kvenvolden	4 January 2002	JS
Antonio Lazcano	27 February 1997	JS
Joshua Lederberg	12 November 1992	SD
	15 January 1999	JS
Gilbert Levin	21 February 2001	JS
James Lovelock	23 March 2000	JS
Lynn Margulis	23–24 June 1998	JS
Gene McDonald	6 December 1999	SD
Chris McKay	12 May 1997	SD
David McKay	19 November 1997	SD
Michael Meyer	4 February 1997; 27 December 2000	SD
Stanley Miller	18 February 1997; 23 February 1999	JS
Carleton Moore	9 January 2002	JS
Harold Morowitz	20 March 2003	JS
David Morrison	18 November 1997	SD
Barnard M. Oliver	1 June 1992	SD
Edward Olsen	8 January 1993	SD
Juan Oró	28 January; and 5 February 1997	JS
Bonnie Packer	2 September 2002	JS
Michael Papagiannis	5 August 1988	SD
Yvonne Pendelton	3 November 1997	SD
Katherine Pering	8 and 11 January 2002	JS
Tom Pierson	16 September 1992; 16 May 1997	SD
Cyril Ponnamperuma	24 May 1982	WH
Chris Romanek	12 May 1997	SD
John Rummel	16 November 1997	SD
	2 September 1998	JS
Carl Sagan	6 January 1993	SD
Greg Schmidt	6 April 2000	SD
Alan Schwartz	21–22 February 1999	JS
Charles Seeger	31 May 1992	SD
Adolph Smith	30 January 1997; 27 September 1998	JS
William Stillwell	4 September 1998	JS
Jill Tarter	15 September 1992	SD
Richard S. Young	25 May 1982	WH

Sources of Unpublished Materials

Krishna Bahadur papers, courtesy of Adolph Smith
Elso Barghoorn papers, courtesy of Lynn Margulis
A. Graham Cairns-Smith papers, courtesy of A. G. Cairns-Smith
Sidney Fox papers, courtesy of the late S. Fox
Norman Horowitz papers, California Institute of Technology Archives, Pasadena, Calif.
Harold P. Klein papers, courtesy of the late H. P. Klein
Sol Kramer papers, University of Florida Special Collections, Gainesville, Fla.
Joshua Lederberg papers, National Library of Medicine, Bethesda, Md.

Gilbert Levin papers, courtesy of G. Levin
James Lovelock papers, courtesy of J. Lovelock
Lynn Margulis papers, courtesy of L. Margulis
Harold Morowitz papers, George Mason University Archives, Fairfax, Va.
National Academy of Sciences Archives, papers on Space Sciences Board
NASA History Office, files on Exobiology Program
SETI Institute Archives
Carl Woese letter to R. Young, courtesy of C. Woese

APPENDIX B: NASA LEADERSHIP IN EXOBIOLOGY

NASA Administrators

T. Keith Glennan	19 August 1958–20 January 1961
James E. Webb	14 February 1961–7 October 1968
Thomas O. Paine	21 March 1969–15 September 1970
James C. Fletcher	27 April 1971–1 May 1977
Robert A. Frosch	21 June 1977–20 January 1981
James M. Beggs	10 July 1981–4 December 1985
James C. Fletcher	12 May 1986–8 April 1989
Richard H. Truly	14 May 1989–31 March 1992
Daniel S. Goldin	1 April 1992–17 November 2001
Sean O'Keefe	21 December 2001–

Note: Biographies of administrators and deputy administrators are available at the NASA History Office Web site at http://www.hq.nasa.gov/office/pao/History/prsnnl.htm.

Headquarters Associate Administrator for Space Science

Homer Newell	1958–1961,[a] 1963–1973[b]
Noel Hinners	1974–1979
Thomas Mutch	1979 (July)–1980
Andrew J. Stofan	1981–1982
Burton Edelson	1982–1987
Lennard Fisk	1987 (6 April)–1993
Wesley T. Huntress	1993–1998
Edward Weiler	1998–

[a] Office of Space Science
[b] Office of Space Science and Applications (OSSA) through 1993

Headquarters Life Sciences Directors

Orr Reynolds	February 1962–1970 (Director of Bioscience Programs)
Gen. J. W. Humphreys	1970–1972
Charles A. Berry	1972–1974
David Winter	April 1974–April 1979
Gerald Soffen	April 1979–1983
Arnauld Nicogossian	1983–March 1993
Harry Holloway	March 1993–April 1996
Arnauld Nicogossian	May 1996–January 2001
Kathie Olson	January 2001–July 2001
Mary Kicza	11 March 2002–

Note: The Office of Life and Microgravity Sciences was established on 8 March 1993 for the first time at the same level as the Office of Space Sciences. Joan Vernikos was the director of its Life Sciences Division from April 1993 to August 2000. On 29 September 2000 the office was restructured to become the Office of Biological and Physical Research.

Headquarters Exobiology Program Managers

Freeman Quimby	August 1963–1967
Richard S. Young[a]	1967–August 1979
Donald DeVincenzi[a]	August 1979–12 December 1986
John Rummel[a]	12 December 1986–1992
Michael Meyer[a]	1993–2002
Michael New	2002–

Note: The headquarters exobiology program managers were renamed "discipline scientists" in the mid–1980s.

[a] Also planetary protection officer. Rummel resumed the position of planetary protection officer in 1997.

Headquarters SETI Program Managers

Dick Henry	1977–1978
Jeffrey D. Rosendhal	June 1978–December 1979
Donald DeVincenzi	1979–1986
Lynn Griffiths (Harper)	1986–1988
Gary Coulter	1988–1993

NASA Research Ames Center Directors

Smith J. DeFrance	1 October 1958–15 October 1965
Harvey Julian Allen	15 October 1965–15 November 1968
Hans Mark	20 February 1969–15 August 1977
Clarence A. Syvertson	15 August 1977–13 January 1984
William F. Ballhaus Jr.	16 January 1984–1 February 1988

Dale L. Compton   15 July 1989–28 January 1994[a]
Ken K. Munechika   28 January 1994–4 March 1996
Henry McDonald   4 March 1996–19 September 2002
Scott Hubbard   19 September 2002–

[a] Compton served as acting director from 1 February 1988 to 1 February 1989.

NASA Ames Life Science Directors

Webb Haymaker   July 1961–1963
Harold P. "Chuck" Klein   January 1964–May 1984
John Billingham   1984–1991

NASA Ames Space Science Division Chiefs

David Morrison   1988–1996
Donald DeVincenzi   1996–2003

NASA Ames Exobiology Division

Harold P. "Chuck" Klein   1963
Richard S. Young   1963–1967
L. P. "Pete" Zill   1967–1974
Keith Kvenvolden   1974–1975
John Billingham   1975–1986
Sherwood Chang   1987–1998
David Blake   2000–

Note: The Exobiology Division was named the "Extraterrestrial Research Division" under Billingham. It became a branch under Life Sciences in 1986 and a branch under Space Science in 1988.

NASA Ames SETI Office Chiefs

John Billingham   1991–1993
(Bernard M. Oliver, Deputy)

Jet Propulsion Laboratory Directors

William H. Pickering   1 October 1958–31 March 1976
Bruce C. Murray   1 April 1976–30 June 1982
Lew Allen Jr.   22 July 1982–31 December 1990
Edward C. Stone   1 January 1991–30 April 2001
Charles Elachi   1 May 2001–

Appendix C: Topics at First Astrobiology Science Conference, April 2000

Topic	Number of Poster Papers
Solar system dynamics	24
Planetary detection	11
Cosmic chemistry	9
Chirality and life	8
Meteorites and organic chemistry	7
Studies of prebiotic chemistry	35
Cosmochemistry missions	2
Habitable planets	17
Europa	3
Mars geology	11
Mars climatology	8
Mars oxidants	3
Mars missions	3
Microbes and Mars	3
Mars meteorites	14
Biomarkers	22
SETI	2
Ancient Earth / geochemistry	10
Rise of oxygen on Earth	8
Snowball Earth	5
Biogeochemistry	11
Impacts	4
Evolution of the genome	14
Evolution of metabolism	11
Microbial community structure	3
Phylogeny	8
Life in extreme environments	34
Metazoan evolution	5
Life beyond the planet of origin	4
Education	14
Astrobiology programs	6

Source: From Abstracts, First Astrobiology Science Conference, 3–5 April 2000.

APPENDIX D: OBJECTIVES IN THE ASTROBIOLOGY ROADMAP (1999)

Question: How Does Life Begin and Develop?

SOURCES OF ORGANICS ON EARTH
Objective 1: Determine whether the atmosphere of the early Earth, hydrothermal systems, or exogenous matter were significant sources of organic matter.

ORIGIN OF LIFE'S CELLULAR COMPONENTS
Objective 2: Develop and test plausible pathways by which ancient counterparts of membrane systems, proteins, and nucleic acids were synthesized from simpler precursors and assembled into protocells.

MODELS FOR LIFE
Objective 3: Replicate catalytic systems capable of evolution and construct laboratory models of metabolism in primitive living systems.

GENOMIC CLUES TO EVOLUTION
Objective 4: Expand and interpret the genomic database of a select group of key microorganisms in order to reveal the history and dynamics of evolution.

LINKING PLANETARY AND BIOLOGICAL EVOLUTION
Objective 5: Describe the sequences of causes and effects associated with the development of Earth's early biosphere and the global environment.

MICROBIAL ECOLOGY
Objective 6: Define how ecophysiological processes structure microbial communities, influence their adaptation and evolution, and affect their detection on other planets.

Question: Does Life Exist Elsewhere in the Universe?

THE EXTREMES OF LIFE
Objective 7: Identify the environmental limits for life by examining biological adaptations to extremes in environmental conditions.

PAST AND PRESENT LIFE ON MARS
Objective 8: Search for evidence of ancient climates, extinct life, and potential habitats for extant life on Mars.

LIFE'S PRECURSORS AND HABITATS IN THE OUTER SOLAR SYSTEM
Objective 9: Determine the presence of life's chemical precursors and potential habitats for life in the outer solar system.

NATURAL MIGRATION OF LIFE
Objective 10: Understand the natural processes by which life can migrate from one world to another. Are we alone in the universe?

ORIGIN OF HABITABLE PLANETS
Objective 11: Determine (theoretically and empirically) the ultimate outcome of the planet-forming process around other stars, especially the habitable ones.

EFFECTS OF CLIMATE AND GEOLOGY ON HABITABILITY
Objective 12: Define climatological and geological effects upon the limits of habitable zones around the Sun and other stars to help define the frequency of habitable planets in the universe.

EXTRASOLAR BIOMARKERS
Objective 13: Define an array of astronomically detectable spectroscopic features that indicate habitable conditions and/or the presence of life on an extrasolar planet.

Question: What Is Life's Future on Earth and Beyond?

ECOSYSTEM RESPONSE TO RAPID ENVIRONMENTAL CHANGE
Objective 14: Determine the resilience of local and global ecosystems through their response to natural and human-induced disturbances.

EARTH'S FUTURE HABITABILITY
Objective 15: Model the future habitability of Earth by examining the interactions between the biosphere and the chemistry and radiation balance of the atmosphere.

BRINGING LIFE WITH US BEYOND EARTH
Objective 16: Understand the human-directed processes by which life can migrate from one world to another.

PLANETARY PROTECTION
Objective 17: Refine planetary protection guidelines and develop protection technology for human and robotic missions.

Source: Astrobiology Roadmap, issued 6 January 1999. Refined Goals and Objectives, issued in November 2002, are found at http://astrobiology.arc.nasa.gov/roadmap/goals_and_ objectives.html.

NOTES

Introduction

1. Tony Reichhardt, "NASA Lines Up for a Bigger Slice of the Biological Research Pie," *Nature* 391 (8 January 1998): 109.
2. Howard E. McCurdy, *Space and the American Imagination* (Washington, D.C.: Smithsonian Institution Press, 1997), chap. 5.

CHAPTER 1   ***The Big Picture***

1. Percival Lowell confined himself to planets in *The Evolution of Worlds* (New York: Macmillan, 1909), and George Ellery Hale dealt only with stars in *The Study of Stellar Evolution* (Chicago: University of Chicago Press, 1908). Among historians stellar evolution has been treated in David DeVorkin's work on the development of the Hertzsprung-Russell diagram, but no history of ideas of cosmic evolution exists.
2. On the natural selection of universes, see Lee Smolin, *The Life of the Cosmos* (New York: Oxford University Press, 1997). Freeman Dyson proposes *cosmic ecology,* in *Infinite in All Directions* (New York: Harper and Row, 1988), 51. It is important to note that *evolution* has general and specific meanings. When scientists speak about "cosmic evolution," they usually have a general idea of "development" in mind. When Smolin speaks of the "natural selection" of universes that may compose the multiverse, he is applying the more specific idea of Darwinian evolution to astronomy.
3. Harlow Shapley saw extraterrestrial life as one of four adjustments in humanity's view of itself since ancient Greece (*Of Stars and Men* [Boston: Beacon Press, 1958], 104). Otto Struve compared the idea of extraterrestrial life to the Copernican theory and the discovery that we were in a peripheral position in our galaxy (*The Universe* [Cambridge, Mass.: MIT Press, 1962], 157). Bernard M. Oliver and John Billingham term the idea "biocosmology" ("Project Cyclops: A Design Study of a System for Detecting Extraterrestrial Intelligence," Washington, D.C., 1971), and Steven Dick makes the case in "The Concept of Extraterrestrial Intelligence—An Emerging Cosmology," *Planetary Report* 9 (March–April 1989): 13–17; and *The Biological Universe: The Twentieth Century Extraterrestrial Life Debate and the Limits of Science* (Cambridge: Cambridge University Press: 1996), 542. See also

Dick, "Extraterrestrial Life and Our Worldview at the Turn of the Millennium" (Washington, D.C.: Smithsonian Institution, 2000).
4. The close connection between philosophical issues in terrestrial and cosmic evolution are discussed in Dick, *Biological Universe,* 378–389. Thirteen authors with diverse backgrounds explore some of the implications of the biological universe in Steven J. Dick, ed., *Many Worlds: The New Universe, Extraterrestrial Life, and the Theological Implications* (Philadelphia: Templeton Foundation Press, 2000).
5. Michael J. Crowe, *The Extraterrestrial Life Debate, 1750–1900* (Cambridge: Cambridge University Press, 1986), 224–225, 274–277, 464–465. Simon Schaffer has shown the place of the nebular hypothesis in a general "science of progress" in early Victorian Britain ("The Nebular Hypothesis and the Science of Progress," in *History, Humanity and Evolution: Essays for John C. Greene,* ed. J. R. Moore [Cambridge: Cambridge University Press, 1989], 131–164). On the role of Spencer and Fiske in nineteenth-century origin of life debates, see James Strick, *Sparks of Life: Darwinism and the Victorian Debates over Spontaneous Generation* (Cambridge, Mass.: Harvard University Press, 2000), esp. 94–95.
6. On spontaneous generation and Darwinism, see Strick, *Sparks of Life.* On Proctor and Flammarion, see Crowe, *Extraterrestrial Life Debate,* 367–386. An early case of nineteenth-century astronomical evolution which Lightman points to is the astronomer/popularizer Robert S. Ball, "The Relation of Darwinism to Other Branches of Science," *Longman's Review* 2 (November 1883): 76–92. See Bernard Lightman, "The Story of Nature's Victorian Popularizers and Scientific Narrative," *Victorian Review* 25, no. 2 (1999): 1–29.
7. On Lowell as Spencerian, and as influenced by Spencer's American disciple John Fiske, see David Strauss, *Percival Lowell: The Culture and Science of a Boston Brahmin* (Cambridge, Mass.: Harvard University Press, 2001), 97–165; W. W. Campbell, "The Daily Influences of Astronomy," *Science* 52, 10 December 1920, 540. David DeVorkin has found archival evidence that Hale's interest in cosmic evolution extended beyond the physical universe to biology and culture ("Evolutionary Thinking in American Astronomy from Lane to Russell," presented at a session on "Evolution and Twentieth Century Astronomy," History of Science Society meeting, Denver, Colo., 8 November 2001).
8. The quotation is from A. R. Wallace, "Man's Place in the Universe," *Independent,* 26 February 1903, 396. This argument was elaborated in Wallace, *Man's Place in the Universe* (1903; rpt., New York: Macmillan, 1904); the appendix is found in the London 1904 edition on 326–336. On Wallace, see Martin Fichman, *An Elusive Victorian: The Evolution of Alfred Russel Wallace* (Chicago: University of Chicago Press, 2004); also Michael Shermer, *In Darwin's Shadow: The Life and Science of Alfred Russel Wallace* (Oxford: Oxford University Press, 2002); for Wallace's "heresy" in breaking with Darwin on the matter of the evolution of the human brain, see 157–162.
9. L. J. Henderson, *The Fitness of the Environment* (New York: Macmillan, 1913), reprinted with an introduction by Harvard biologist George Wald (Gloucester, Mass.: Peter Smith, 1970), 312. The complexity of Henderson's ideas on the fitness of the environment and their connection to modern ideas on the subject are analyzed in detail in Iris Fry, "On the Biological Significance of the Properties of Matter: L. J. Henderson's Theory of the Fitness of the Environment," *Journal of the History of Biology* 29 (1996): 155–196.
10. Dick, *Biological Universe,* chap. 4.

11. Spencer Jones, *Life on Other Worlds* (New York: Macmillan, 1940), 57. On Oparin, see Iris Fry, *The Emergence of Life on Earth: A Historical and Scientific Overview* (New Brunswick, N.J.: Rutgers University Press, 2000), chap. 6; and Dick, *Biological Universe,* chap. 7. On the influence of Marxism, see Loren Graham, *Science and Philosophy in the Soviet Union* (New York: Alfred A. Knopf, 1972).
12. A. I. Oparin and V. G. Fesenkov, *Life in the Universe* (New York: Twayne Publishers, 1961); George Wald, "The Origin of Life," *Scientific American* (August 1954): 44.
13. For details of SETI history, see Dick, *Biological Universe,* chap. 8.
14. Joseph Shklovskii and Carl Sagan, *Intelligent Life in the Universe* (San Francisco: Holden-Day, 1966). In May 1964 the Armenian Academy of Sciences sponsored a meeting on extraterrestrial intelligence at Byurakan Astrophysical Observatory; the proceedings, published in 1965, are available in English in *Extraterrestrial Civilizations,* ed. G. M. Tovmasyan (Jerusalem: Israeli Program for Scientific Translations, 1967). For a list of additional Soviet meetings, see Dick, *Biological Universe,* 484–485.
15. JoAnn Palmieri has discussed the popularization of the idea of cosmic evolution in "Popular and Pedagogical Uses of Cosmic Evolution," presented at a session on "Evolution and Twentieth Century Astronomy," History of Science Society meeting, Denver, Colo., 8 November 2001.
16. Otto Struve, "Life on Other Worlds," *Sky and Telescope* 14 (February 1955): 137–146; Joshua Lederberg, "Exobiology: Experimental Approaches to Life beyond the Earth," in *Science in Space,* ed. Lloyd V. Berkner and Hugh Odishaw (New York: McGraw-Hill, 1961), 407–425; John Billingham, *Life in the Universe* (Cambridge, Mass.: MIT Press, 1981), ix.
17. In 1960 the NSF's John Wilson looked forward to funding space biology. But NASA took an early dominant lead, which it has continued to hold. By 1963 NASA's life sciences expenditures (including exobiology) had already reached $17.5 million. Toby Appel, *Shaping Biology: The National Science Foundation and American Biological Research, 1952–1975* (Baltimore: Johns Hopkins University Press, 2000), 132.
18. George Gaylord Simpson, "The Non-Prevalence of Humanoids," *Science* 143, 21 February 1964, 769–775.

CHAPTER 2 *Organizing Exobiology*

1. Joshua Lederberg, "Sputnik + 30," *Journal of Genetics* (India), 66 (December 1987): 217. Much of Lederberg's early involvement in exobiology is discussed here. Lederberg repeated this story in interviews with numerous historians, including ourselves and Audra Wolfe.
2. Lest one think such ideas only wild fancies worthy of the height of the Cold War, it is worth noting that Carl Sagan was hired in 1958 by the Department of Defense (through the Armour Research Foundation) for Project A119, to make calculations for what would result from detonating nuclear bombs on the moon. Keay Davidson, *Carl Sagan: A Life* (New York: Wiley, 1999), 93–95.
3. Joshua Lederberg and Dean B. Cowie, "Moondust," *Science* 127, 27 June 1958, 1473–1475.
4. See, e.g., Lederberg to Harrison Brown, 19 January 1961, Lederberg papers, NLM; also "Earthlike Life Unlikely on Moon or Planets, Scientist Contends," Phil Abelson

being the scientist quoted in *Washington Post,* 11 December 1961, 1–2. Abelson had been one of the first to replicate and extend Miller's results experimentally, in 1956, and since that time had been a key participant at scientific meetings on the origin of life.
5. Phil Abelson, "Extra-terrestrial Life," *Proceedings of the National Academy of Sciences (PNAS)* 47 (1961): 575–581.
6. Sagan to Lederberg, 20 February 1959. Lederberg papers, NLM.
7. William Poundstone, *Carl Sagan: A Life in the Cosmos* (New York: Henry Holt, 1999).
8. Lederberg to Harry Eagle, March 1958; also Lederberg to Robert Jastrow, 4 March 1959, Lederberg papers, NLM.
9. Lederberg notes: "The irony of advocating a parochial approach to cosmic questions has not escaped me. But I was exhausted from traveling!" Lederberg to Strick, personal communication, 22 December 2002.
10. Edward C. Ezell and Linda N. Ezell, *On Mars: Exploration of the Red Planet, 1958–1978,* NASA SP-4212 (Washington, D.C.: NASA, 1984), 63–64. The minutes of this meeting and an informal preliminary meeting of 4 December 1958 (Richard Davies's copies) can be found in JPL Archives, fld 2-1067a and 1067c.
11. Susan Spath, "C. B. Van Niel and the Culture of Microbiology, 1920–1965" (Ph.D. diss., University of California–Berkeley, 1999), esp. app. 2.
12. Stanley Miller, "A Production of Amino Acids under Possible Primitive Earth Conditions," *Science* 117, 15 May 1953, 528–529; reprinted in David W. Deamer and Gail R. Fleischaker, eds., *Origins of Life: The Central Concepts* (Boston: Jones and Bartlett, 1994), 147–148. For a somewhat historical assessment of the Miller-Urey experiment after fifty years, see Jeff Bada and Antonio Lazcano, "Prebiotic Soup: Revisiting the Miller Experiment," *Science* 300, 2 May 2003, 745–746.
13. Sidney Fox, "Evolution of Protein Molecules and Thermal Synthesis of Biochemical Substances," *American Scientist* 44 (October 1956): 347–359. For contemporary biographies, photos, and research interests of Fox, Miller, Abelson, Calvin, Lilly, Orr Reynolds, Carl Sagan, Harold Urey, and Wolf Vishniac, see Shirley Thomas, *Men of Space: Profiles of the Leaders in Space Research, Development, and Exploration,* vol. 6 (Philadelphia: Chilton, 1963).
14. Alexander Oparin, *The Origin of Life,* English trans. S. Morgulis (New York: Macmillan, 1938). By 1957 he had significantly developed and expanded it, especially the role for dialectical materialist thinking: see Oparin, *The Origin of Life on the Earth,* 3d rev. ed., English trans. Ann Synge (Edinburgh: Oliver and Boyd, 1957).
15. Stanley Miller, J. William Schopf, and Antonio Lazcano, "Oparin's *Origin of Life:* Sixty Years Later," *Journal of Molecular Evol*ution 44 (1997): 351–353; see also A. Lazcano, "Chemical Evolution and the Primitive Soup: Did Oparin Get It All Right?" *Journal of Theoretical Biology* 184 (1997): 219–223. On the prehistory and early days of exobiology, see Steven J. Dick, *The Biological Universe* (Cambridge: Cambridge University Press, 1996); see also Iris Fry, *The Emergence of Life on Earth: A Historical and Scientific Overview* (New Brunswick: Rutgers University Press, 2000); and Audra Wolfe, "Germs in Space: Joshua Lederberg, Exobiology, and the Public Imagination, 1958–1964," *Isis* 93 (June 2002): 183–205.
16. Miller OHI, 2 February 1997, 2; note that all dollar amounts throughout this book are given in the contemporary figures of the period in question, not adjusted to current dollar values. I would like to thank Dr. Miller for sharing with me an un-

published note on the subject of funding. In it he laments that the unofficial "bootlegging" procedure, to support "almost all really original work," is now untenable in an age of intensely scrutinized review of how grant dollars are spent. However small NSF's initial investment in Miller (related to earlier papers he had published, suggesting his promise as a chemist), in testimony before Congress in late 1955 NSF program officer Alan T. Waterman was happy to cite support for Miller as a good example that NSF had invested wisely in its research, stating that "this work had been listed in a *Fortune* magazine article as one of ten major discoveries in basic research in the past year." See Toby Appel, *Shaping Biology: The National Science Foundation and American Biological Research, 1945–1975* (Baltimore: Johns Hopkins University Press, 2000), 104–105.

17. The proceedings of this conference were published as F. Clark and R. L. Synge, eds., *Proceedings of the First International Symposium on the Origin of Life on the Earth* (New York: Pergamon Press, 1959). Another participant in the 1957 conference was the soon-to-be discredited Lysenkoist biologist Olga Lepeschinskaya. On her origin of life claims, see L. N. Zhinkin and V. P. Mikhailov, "On 'the New Cell Theory,'" *Science* 128 (1958): 182–186; see also L. J. Rather, *Addison and the White Corpuscles* (London: Wellcome Institute, 1972), 218–219; Valery Soyfer, *Lysenko and the Tragedy of Soviet Science* (New Brunswick: Rutgers University Press, 1994); and, most recently, Larisa Shumeiko, "Der *lebende Stoff* und die Umwandlung der Arten Die "neue" Zellentheorie von Ol'ga Borisovna Lepešinskaja (1871–1963)," in Uwe Hoßfeld and Rainer Brömer, eds., *Darwinismus und/als Ideologie* (Berlin: VWB-Verlag, 2001), 213–228.

18. Loren Graham, *Science, Philosophy, and Human Behavior in the Soviet Union* (New York: Columbia University Press, 1987), esp. chap. 3.

19. Erwin Chargaff, *The Heraclitean Fire* (New York: Rockefeller University Press, 1978), 142–144; See also Chargaff OHI, 1–3.

20. Miller OHI, 18 February 1997 and 23 February 1999.

21. Lederberg had missed the Moscow meeting because of the opportunity to work at a lab in Australia; Lederberg to Horowitz, 4 March 1958, Horowitz papers 4.3, California Institute of Technology Archives (hereafter Horowitz papers).

22. Lederberg, "Exobiology: Approaches to Life beyond the Earth," *Science* 132, 12 August 1960, 393–400. Lederberg's first published use of *exobiology* is in his paper delivered at the first meeting of the international Council on Space Research (COSPAR) in Nice, France, 11–16 January 1960; "Exobiology: Experimental Approaches to Life beyond the Earth," in *Space Research: Proceedings of the First International Space Science Symposium,* ed. H. K. Kallmann Bijl (Amsterdam: North-Holland, 1960), 1153–1170.

23. NASA Third Semiannual Report to Congress, 1 October 1959–31 March 1960, 90, 158–159; NASA Fifth Semiannual Report to Congress, 1 October 1960–30 June 1961, 133–135.

24. There is extensive literature on scientific discipline formation. For just two useful examples, see Robert Kohler, *From Medical Chemistry to Biochemistry* (Cambridge: Cambridge University Press, 1981); and David Edge and Michael Mulkay, *Astronomy Transformed: The Emergence of Radio Astronomy in Britain* (New York: Wiley, 1976).

25. See, e.g., George Gaylord Simpson, "The Nonprevalence of Humanoids," *Science* 143, 21 February 1964, 769–775.

26. See E. O. Wilson, *Naturalist* (Washington, D.C.: Island Press, 1994), chap. 12: "The Molecular Wars." The tensions between these biologists and the "molecularizing" group are explored thoroughly by Michael Dietrich in "Paradox and Persuasion: Negotiating the Place of Molecular Evolution within Evolutionary Biology," *Journal of the History of Biology* 31 (1998): 85–111.
27. Donald DeVincenzi OHI, 11 February 1997; see also Dick, *Biological Universe;* and James Strick, "The Cambrian Explosion (of Books on the Origin of Life)," *Journal of the History of Biology* 33 (2000): 371–384.
28. Lederberg diary entry, 29 July 1959; courtesy of Joshua Lederberg.
29. See Wolfe, "Germs in Space," 190. This fear was borne out on more than one occasion. An early example occurred in August 1967, when Congress completely canceled funding of the *Voyager* Mars mission scheduled for the late 1960s. After the race riots ravaged many U.S. cities that summer, spending on exploring Mars suddenly appeared politically inexpedient. See Ezells, *On Mars,* 110–118.
30. On the relegating of science to a far-back seat in Project Mercury, see, e.g., Tom Wolfe, *The Right Stuff* (New York: Farrar, Straus and Giroux, 1979), chap. 13.
31. Horowitz to Lederberg, 16 and 19 May 1960, Horowitz papers 11.3.
32. NASA Second Semiannual Report to the Congress, 1 April–30 September 1959, 209.
33. Vishniac, "Space Flights and Biology," *Science* 144, 17 April 1964, 245–246; Fox, "Humanoids and Proteinoids," *Science* 144, 22 May 1964, 954.
34. Gilbert Levin, "Significance and Status of Exobiology," *BioScience,* 15 (January 1965): 17–20 (a paper presented at the American Institute of Biological Sciences annual meeting, 26 August 1964, 17).
35. Ibid., 18–19.
36. Isaac Asimov, "A Science in Search of a Subject," *New York Times Magazine,* 23 May 1965, 52–58.
37. Dana Hedgpeth, "The Man Who Wants to Return to Mars," *Washington Post,* 1 December 2000, A1, 10–11.
38. NASA Fifth Semiannual Report to Congress, 1 October 1960–30 June 1961, 204; NASA Ninth Semiannual Report, 198; Eleventh Report, 220.
39. NASA Fifth Report, 202, 206, 212, 214.
40. NASA Sixth Semiannual Report to Congress, July–December 1962, 174, 180.
41. See M. Scott Blois, "Random Polymers as a Matrix for Chemical Evolution," in *The Origins of Prebiological Systems and Their Molecular Matrices,* ed. Sidney Fox (New York: Academic Press, 1965), 19–38.
42. NASA Sixth Semiannual Report to Congress, 1 July–31 December 1961, 173, 174, 180–181; see Charles R. Phillips and Robert K. Hoffman, "Sterilization of Interplanetary Vehicles," *Science* 132, 14 October 1960, 991–995. For more on the exobiology sterilization / Fort Detrick germ warfare connection, see Wolfe, "Germs in Space," 199–203; Wolfe has written persuasively here on the cultural implications, in a Cold War context, of rhetoric about "contamination" and "containment" of such contaminants.
43. NASA Seventh Semiannual Report to Congress.
44. Henry S. F. Cooper, *The Search for Life on Mars* (New York: Holt, Rinehart and Winston, 1980), 96.
45. Hedgpeth, "Man Who"; Gilbert Levin OHI.
46. Horowitz to Lederberg, 18 May 1962, Horowitz papers 16.5.
47. See the winter 1963 issue of *Stanford Today,* including articles on exobiology by

Lederberg and on Multivator and "exobiology at Stanford" by Levinthal (the cover story). The Multivator project is also discussed in Wolfe, "Germs in Space," 194–195.
48. NASA Tenth Semiannual Report to Congress, 216; NASA Eleventh Semiannual Report to Congress, 216.
49. Appel, *Shaping Biology,* 145, table 5.2a.
50. NASA Twelfth Semiannual Report to Congress, 149.
51. NASA Ninth Semiannual Report to Congress, 198; Eleventh Report, 223; Sidney Fox OHI , 1 February 1993, 33.
52. NASA Eleventh Semiannual Report to Congress, January–June 1964, 238.
53. Lederberg to Pittendrigh, 5 April 1962, Lederberg papers; in 1968 Pittendrigh moved to Stanford, joining Lederberg's exobiology group there.
54. NASA Eleventh Report, 238; Slepecky to Strick, 12 November 1999; NASA Ninth Semiannual Report, 198.
55. Richard S. Young, Paul H. Deal, Joan Bell, and Judith L. Allen, "Bacteria under Simulated Martian Conditions," in *Life Sciences and Space Research,* ed. M. Florkin and A. Dollfus, 2 (1964): 105–111; see also Young, Deal, and O. Whitfield, "The Response of Spore-Forming vs. Nonspore-Forming Bacteria to Diurnal Freezing and Thawing," *Space Life Sciences* 1 (1968): 113–117.
56. NASA Ninth Semiannual Report, January–June 1963, 193.
57. NASA Twelfth Semiannual Report, July–December 1964, 64.
58. NASA Tenth Semiannual Report, July–December 1963, 219.
59. DeVincenzi OHI, 21 January 1997.
60. By 30 June 1961 the NASA Space Sciences Steering Committee chair was Homer Newell; BioScience Subcommittee chair was Quimby; the secretary was Richard Young; members were Siegfried Gerathewohl, George J. Jacobs, Jack Posner, and G. D. Smith; with consultants Abelson, Calvin, Fox, Horowitz, Linschitz, Colin Pittendrigh, Ernest Pollard, and Carl Sagan. By early 1962 Orr Reynolds (physiologist and head of research at the Office of Defense Research and Engineering) "came to NASA to take charge of the biology division in the new Office of Space Sciences." Homer Newell, *Beyond the Atmosphere: Early Years of Space Science* (Washington, D.C.: NASA, 198), 276.
61. See Lederberg to Young, 21 December 1961, Lederberg papers. On the history of Ames Research Center overall, see Elizabeth Muenger, *Searching the Horizon: A History of Ames Research Center, 1940–1976,* NASA SP-4304 (Washington, D.C.:NASA, 1985); see also the more recent history by Glenn Bugos, *Atmosphere of Freedom: Sixty Years at the NASA Ames Research Center,* NASA SP-4314 (Washington, D.C.: NASA, 2000).
62. Richard Young OHI, 3.
63. NRC postdocs at Ames included, after Ponnamperuma, Henry Speer, Duane Rohlfing, Klaus Dose, Janos Lanyi, Linda Caren, Alan Schwartz, Ellen Weaver, Don DeVincenzi, Akiva Bar-nun, William Bonner, Clair Folsome, Richard Turco, Rivers Singleton, George Yuen, Carleton Moore, Norm Gabel, Noam Lahav, Jill Tarter, Owen Toon, James Ferris, Thomas Ackerman, Lelia Coyne, Adolph Smith, Neal Blair, James Kasting, Amos Banin, Chris McKay, Louis Allamandola, Kim Wedeking, Friedemann Freund, John Rummel, Kevin Zahnle, David Blake, Lynn Rothschild, Chris Chyba, Jack Farmer, and LuAnn Becker, to name only a few whose names continued to be recognized in exobiology, many still today; see *NRC*

Directory of Resident Research Associates, 1959–1995 (Washington, D.C.: NAS Press, 1996), 14–261.
64. Ponnamperuma OHI, 24 May 1982, 2.
65. Ibid., 3.
66. Klein, *A Personal History* (Mountain View, Calif.: privately printed, 1998), chap. 18.
67. Katherine Pering OHI, 8 January 2002; Keith Kvenvolden OHI, 4 January 2002. When friction soon developed between Kvenvolden and Ponnamperuma, they seem to have had differing perceptions of what Kvenvolden had actually been hired to do. Kvenvolden believed he had been hired to head up and supervise the lunar sample lab. Ponnamperuma seems to have thought he had hired Kvenvolden as just one more staff scientist in the Chemical Evolution branch, all, including the sample lab, under Ponnamperuma's control.
68. DeVincenzi OHI, 4 February 1997; Kvenvolden OHI.
69. John Jungck OHI, 10 November 1999; Alan Schwartz to Strick, personal communication, 21 February 1999.
70. See report by Richard Young and Cyril Ponnamperuma of NASA Ames Research Center, "Life: Origin and Evolution," *Science* 143, 24 January 1964, 385–388; see also Young OHI, 7–8. The proceedings were published as Sidney Fox, ed., *The Origins of Prebiological Systems and of Their Molecular Matrices* (New York: Academic Press, 1965).
71. Sidney Fox, *The Emergence of Life* (New York: Basic Books, 1988), 10–11; Fox OHI, 27 January 1993.
72. Sidney Fox, Kaoru Harada, and Jean Kendrick, "Production of Spherules from Synthetic Proteinoid and Hot Water," *Science* 129, 1 May 1959, 1221–1223.
73. For progressively stronger claims, see Sidney Fox, "How Did Life Begin?" *Science* 132, 22 July 1960, 200–208; "A Theory of Macromolecular and Cellular Origins," *Nature* 205, 23 January 1965, 328–340; "Spontaneous Generation, the Origin of Life, and Self-Assembly," *Current Modern Biology* 2 (1968): 235–240; "The Proteinoid Theory of the Origin of Life and Competing Ideas," *American Biology Teacher* 36 (March 1974): 161–172, 181; S. W. Fox and Klaus Dose, *Molecular Evolution and the Origin of Life* (San Francisco: W. H. Freeman, 1972).
74. See, e.g. Fox, *Emergence*, 173–176.
75. Stanley L. Miller and Harold C. Urey, "Organic Compound Synthesis on the Primitive Earth," *Science* 130, 31 July 1959, 245–251.
76. Sidney Fox, "Origin of Life," *Science* 130, 11 December 1959, 1622–1623.
77. S. L. Miller and H. C. Urey, "Reply," *Science* 130, 11 December 1959, 1623–1624.
78. This attitude, the dominant paradigm in the immediate aftermath of the Miller-Urey experiment, was well portrayed in George Wald's article "The Origin of Life," *Scientific American* 192 (August 1954): 44–53.
79. Norman Horowitz OHI, 15 January 1999, 8; Miller OHI, 9–10.
80. Stanley Miller and Leslie Orgel, *The Origins of Life on the Earth* (Englewood Cliffs, N.J.: Prentice-Hall, 1973), 145. Horowitz to Miller, 7 June 1972, Horowitz papers, 4.34, CalTech archives.
81. See, e.g., S. W. Fox, K. Harada, P. E. Hare, G. Hinsch, G. Mueller, "Bio-organic Compounds and Glassy Particles in Lunar Fines and Other Materials," *Science* 167, 30 January 1970, 767–770; Kvenvolden OHI, 4 January 2002.
82. Fox, "Proteinoid Theory."

83. See S. W. Fox and Aristotel Pappelis, "Synthetic Molecular Evolution and Protocells," *Quarterly Review of Biology* 68 (March 1993): 79–82; see also Alan Schwartz's obituary of his former mentor, "Sidney W. Fox, 1912–1998," *Origins of Life and Evolution of the Biosphere* 29 (1999): 1–3.
84. William Day, *Genesis on Planet Earth* (Ann Arbor, Mich.: Talos, 1979). Day's book was reviewed prominently and favorably by Lynn Margulis, who also helped arrange the publication of a second, revised edition by Yale University Press in 1984. See also, more recently, W. Day, *How Life Began* (Cambridge, Mass.: Foundation for New Directions, 2002).
85. OHI with Miller, Horowitz, DeVincenzi, William Day, William Stillwell, Adolph Smith, John Jungck, and Lynn Margulis (Jungck, Day, and Stillwell were doctoral students or postdocs in Fox's lab); James P. Ferris, "Review of Fox's *Emergence of Life*," *Nature* 337, 16 February 1989, 609–610; William Hagan, "Review of Fox's *Emergence of Life*," *Isis* 80 (1989): 162–163. See also Andre Brack, "Review of *Chemical Evolution: Physics of the Origins and Evolution of Life*," *Origins of Life and Evolution of the Biosphere* 29 (1999): 110.
86. Albert Lehninger, *Biochemistry*, 2d ed. (New York: Worth, 1975), chap. 37: "The Origin of Life."
87. Christopher Wills and Jeffrey Bada (a doctoral student of Stanley Miller's), e.g., call Fox an "excellent self-promoter" and strongly imply, without saying outright, that he duped NASA officials into thinking his work was important for the origin of life. See their book *The Spark of Life: Darwin and the Primeval Soup* (Cambridge, Mass.: Perseus, 2000), 52–55.
88. See DeVincenzi, "NASA's Exobiology Program," *Origins of Life* 14 (1984): 796.
89. Ponnamperuma OHI, 7.
90. See David Buhl and Cyril Ponnamperuma, "Interstellar Molecules and the Origin of Life," *Space Life Sciences* 3 (Fall 1971): 157–164. See also Richard Young, "The Beginning of Comparative Planetology," lecture delivered at Special Symposium on Photochemistry and the Origin of Life, August 1972, *Origins of Life* 4 (1973): 505–515.
91. Jan Sapp, *Beyond the Gene: Cytoplasmic Inheritance and the Struggle for Authority in Genetics* (New York: Oxford University Press, 1987). On the development of Margulis's SET, see Sapp, *Genesis: The Evolution of Biology* (London: Oxford University Press, 2003), chap. 19.
92. Lynn Margulis OHI, 23–24 June 1998, 1.
93. Ibid.
94. By 1986 Margulis's yearly funding had reached $87,000; for 1987 $87,891; for 1988 $90,000; for 1989 $89,850; and for 1990 $89,850 (of $125,000 requested), Margulis papers. In 1977 Woese was already requesting $73,000; by the early 1990s he was requesting $125,000 and was receiving most of this; Woese to Strick, personal communication, 14 January 2002.
95. See J. William Schopf, *Cradle of Life: The Discovery of Earth's Earliest Fossils* (Princeton, N.J.: Princeton University Press, 1999), chap. 2.
96. Lovelock, *Gaia: A New Look at Life on Earth* (New York: Oxford University Press, 1979), 10. Lovelock had found most of the older scientists at the meeting, especially Preston Cloud, unsympathetic to his ideas; see Margulis OHI, 23 June 1998, 6; see also Lovelock, *Homage to Gaia: The Life of an Independent Scientist* (New

York: Oxford University Press, 2000), 239–240. See chaps. 4–5 for more on the later fortunes of the Gaia hypothesis and its relations with other developments in exobiology.
97. John Rummel OHI, 2 September 1998, 8.
98. Homer Newell, *Beyond the Atmosphere: Early Years of Space Science,* NASA SP-4211 (Washington, D.C.: NASA), chap. 16.
99. Ibid., 278–282. The NAS report was edited by Bentley Glass, *Life Sciences in Space: Report from the Space Science Board* (Washington, D.C.: NAS, 1970).
100. DeVincenzi OHI, 21 January 1997, 1–2; Carleton Moore OHI, 9 January 2002.
101. Moore OHI.
102. DeVincenzi OHI, 21 January 1997, 1–2; also Young OHI, 8–9.
103. DeVincenzi OHI, 1–2.
104. Fry, *Emergence,* 88. The point is similarly made, Fry says, in Manfred Eigen's *Steps toward Life* (Oxford: Oxford University Press, 1992), 31–32.
105. Quoted in Ezells, *On Mars,* 235.
106. Ibid.
107. Horowitz to Lederberg, 18 May 1962, Horowitz papers 16.5; emphasis added.
108. DeVincenzi OHI, 21 January 1997, 2–3; Rummel OHI, 15; Young OHI, 5–7.
109. See, e.g., Mamikunian and Michael Briggs, "'Organized Elements' in Carbonaceous Meteorites," *Science* 139, 8 March 1963, 873.
110. Ponnamperuma OHI, 4–5.
111. Ibid., 5.
112. See, e.g., Oró to Horowitz, 27 October 1972, Horowitz papers 5.11.
113. N. H. Horowitz and Jerry S. Hubbard, "The Origin of Life," *Annual Reviews of Genetics* 8 (1974): 393.
114. Horowitz to Elso Barghoorn, 19 July 1972; Horowitz to Frank Drake, 19 July 1972; and Horowitz to Barghoorn, 2 August 1972, Horowitz papers, 4.34, CalTech Archives.
115. Miller received his first NASA grant only at the time of *Mariner 4,* for which he helped design an instrument. Miller OHI, 18 February 1997, 14.
116. Horowitz to Paul D. Boyer, 27 July 1972, Horowitz papers.
117. Simpson was interviewed for "Life in Darwin's Universe" by Gene Bylinsky published in *Omni* (September 1979): 63–66, 116.

CHAPTER 3 *Exobiology, Planetary Protection, and the Origins of Life*

1. See, e.g., Ponnamperuma, Richard Lemmon, Ruth Mariner, and Melvin Calvin, "Formation of Adenine by Electron Irradiation of Methane, Ammonia and Water," *Proceedings of the National Academy of Sciences* 49, 15 May 1963, 737–740; Ponnamperuma, Ruth Mariner, and Carl Sagan, "Formation of Adenosine by Ultraviolet Irradiation of a Solution of Adenine and Ribose," *Nature* 198, 22 June 1963, 1199–1200; Ponnamperuma, Sagan, and Mariner, "Synthesis of ATP under Possible Primitive Earth Conditions," *Nature* 199, 20 July 1963, 222–226; Ponnamperuma, "Abiological Synthesis of Some Nucleic Acid Constituents," in *The Origins of Prebiological Systems,* ed. Sidney Fox (New York: Academic Press, 1965), 221–242; "Primordial Organic Chemistry and the Origin of Life," *Quarterly Review of Biophysics* 4 (January 1971): 77–106; *The Origins of Life* (New York: E. P. Dutton, 1972), a popular work; Ponnamperuma, ed., *Exobiology* (Amsterdam: Elsevier, 1972).

2. This resonates with a criticism still being made today, now of astrobiology. Michael Drake and Bruce Jakosky are still urging that the field remember not to become so focused on those bodies where it is hoped to find life, lest it fail to develop the comparative knowledge of other worlds needed to fully understand the effect of differing environmental conditions. See "Narrow Horizons in Astrobiology," *Nature* 415, 14 February 2002, 733–734.
3. Daniel S. Greenberg, "Soviet Space Feat," *Science* 137, 24 August 1962, 590–592.
4. Carl Bruch, "Instrumentation for the Detection of Extraterrestrial Life," in *Biology and the Exploration of Mars*, ed. Colin Pittendrigh, Wolf Vishniac, and J. P. T. Pearman (Washington, D.C.: NAS Press, 1966), 487–502. For a more popular account, see "Exobiology: The Search for Life on Mars," *Time*, 7 May 1965, 80–81.
5. Bruch, "Instrumentation," 494–495. For photos, see NASA Tenth Semiannual Report to Congress, July–December 1963, 84–85; see also E. Levinthal, "Payload to Mars," *Stanford Today*, ser. 1, no.7 (Winter 1963): 10–15.
6. Pittendrigh et al., *Biology*.
7. "Exobiology: The Search for Life on Mars," *Time*, 7 May 1965, 80.
8. The "one gene, one enzyme" hypothesis was developed by George Beadle and Edward Tatum while Horowitz was a young researcher in Beadle's group. This was the belief that genes controlled the nature of an organism by coding on a one-for-one basis for the enzymes that control the metabolism. It was later modified to "one gene, one polypeptide" to include nonenzyme proteins important to the organism in other ways.
9. Lederberg to Horowitz, January 1963, Horowitz papers 4.3.
10. Horowitz to Lederberg, 23 Jan. 1963, Horowitz papers 4.3.
11. Charles R. Phillips, *The Planetary Quarantine Program: Origins and Achievements*, NASA SP-4902 (Washington, D.C.: NASA, 1974), 5. See also Morton Werber, *Objectives and Models of the Planetary Quarantine Program*, NASA SP-344 (Washington, D.C.: NASA, 1975). A more recent update is Space Studies Board, National Research Council (NRC), *Biological Contamination of Mars* (Washington, D.C.: NRC, 1992); see also Donald DeVincenzi, Margaret Race, and Harold Klein, "Planetary Protection, Sample Return Missions, and Mars Exploration: History, Status and Future Needs," *Journal of Geophysical Research* 103 (November 1998): 28577–28585. On Sagan and Lederberg's basic stance, see Carl Sagan OHI.
12. NASA Tenth Semiannual Report to Congress, July–December 1963, 84–85.
13. Brown, "'Back Contamination' and Quarantine: Problems and Perspectives," in Pittendrigh et al., *Biology*, 443–445.
14. Horowitz, memo, "Back-contamination and the Goals of Exobiological Research," attached to Horowitz letter to Lederberg, 6 February 1960, Horowitz papers 11.3.
15. Lederberg to Horowitz, 8 February 1960, Horowitz papers 11.3.
16. Aaron Novick memo to WESTEX 5-C, 19 February 1959 [*sic*] (actually 1960); attached to letter from Novick to Lederberg, 19 February 1960, Horowitz papers 11.3.
17. Horowitz to Strick, personal communication, 10 February 2002.
18. DeVincenzi and Rummel OHI.
19. Morowitz, "Requirements of q Minimum Free Living Replicating System," in Marcel Florkin, ed., *Life Sciences and Space Research* 3 (1965): 149–153.
20. Morowitz to Strick, personal communication, 31 March 2002.
21. Ibid., 3 April 2002.

22. Clair Folsome and Harold Morowitz, "Prebiological Membranes: Synthesis and Properties," *Space Life Sciences* 1 (1969): 538–544.
23. Morowitz, *Energy Flow in Biology: Biological Organization as a Problem in Thermal Physics* (1968; rpt., New Haven, Conn.: OxBow Press 1993).
24. Morowitz to Strick, personal communication, 3 April 2002.
25. Woese to Morowitz, 12 August 1977, Morowitz papers, fld 5.20, George Mason University, Fairfax, Va. (hereafter Morowitz papers).
26. Lederberg to Young, 18 November 1976, Lederberg papers, NLM.
27. Woese to Morowitz, 12 August 1977, Morowitz papers, fld. 5.20.
28. Clair Edwin Folsome, *The Origin of Life: A Warm Little Pond* (San Francisco: W. H. Freeman, 1979), 73; see also John Allen, *Biosphere 2: The Human Experiment* (New York: Penguin, 1990), 13–14. Folsome collaborated with Adolph Smith and Krishna Bahadur on "autotrophs first" origin of life experiments; see nn. 77–78.
29. See, e.g., Bruce Weber, "Emergence of Life and Biological Selection from the Perspective of Complex Systems Dynamics," in *Evolutionary Systems,* ed. G. Van de Vijver et al. (Dordrecht: Kluwer, 1998), 59–66; "Closure in the Emergence and Evolution of Life: Multiple Discourses or One?" *Annals of the New York Academy of Sciences* 401 (2000): 132–138.
30. Morowitz, B. Heinz, and D. W. Deamer, "The Chemical Logic of a Minimal Protocell," *Origins of Life and Evolution of the Biosphere* 18 (1988): 281.
31. Morowitz, "Phase Separation, Charge Separation, and Biogenesis," *Biosystems* 14 (1981): 41–47. He had published in support of Mitchell's chemiosmotic ideas since at least 1978. See J. F. Nagle and H. J. Morowitz, "Molecular Mechanisms for Proton Transport in Membranes," *Proceedings of the National Academy of Sciences* 75 (January 1978): 298–302.
32. Morowitz, *Mayonnaise and the Origin of Life* (New York: Scribners, 1985). The title essay (27–30) explains the properties of amphiphilic molecules such as lecithin which make them essential for living membranes as well as for the emulsified nature of salad dressing.
33. Morowitz, *The Beginnings of Cellular Life* (New Haven, Conn.: Yale University Press, 1992).
34. See, e.g., Morowitz, Jennifer Kostelnik, Jeremy Yang, and George Cody, "The Origin of Intermediary Metabolism," *Proceedings of the National Academy of Sciences* 97, 5 July 2000, 7704–7708.
35. Morowitz to Strick, personal communication, 3 April 2002.
36. Danielli had been the initial Ph.D. advisor to Peter Mitchell, before he left Cambridge University; by Quastler, see "Introduction to Symposium on Theoretical Radiobiology," *American Naturalist* 94 (January–February 1960): 57–58.
37. Minutes of Meeting of Theoretical Biology, Nassau Inn, Princeton, N.J., 30 October 1962, Morowitz papers, fld 5. 4a.
38. NASA Eleventh Semiannual Report to Congress, 242.
39. Ernest C. Pollard, "The Fine Structure of the Bacterial Cell and the Possibility of Its Artificial Synthesis," *American Scientist* 53 (1965): 437–463.
40. Yčas is credited with priority for this idea by Antonio Lazcano and Stanley Miller in "On the Origin of Metabolic Pathways," *Journal of Molecular Evolution* 49 (1999): 424–431. For Yčas's original papers, see "A Note on the Origin of Life,"

Proceedings of the National Academy of Sciences 41, 15 October 1955, 714–716; and "On the Earlier States of the Biochemical System," *Journal of Theoretical Biology* 44 (1974): 145–160.

41. On the planning of the 1968 summer institute, see 12 October 1967, meeting of Committee on Theoretical Biology, Cosmos Club, Washington D.C. (minutes in Morowitz papers, fld 5.4a). Present at the meeting were: D. R. Beem, Danielli, Engelberg, Gregg, H. Hollister, G. Jacobs, Jehle, Morowitz, J. R. Olive, Pollard, and W. D. Taylor.
42. Elsasser, *The Physical Foundation of Biology* (New York: Pergamon Press, 1958); *Atom and Organism: A New Approach to Theoretical Biology* (Princeton: Princeton University Press, 1966); *Reflections on a Theory of Organisms* (Baltimore: Johns Hopkins University Press, 1998).
43. Morowitz to Strick, personal communication, 31 March 2002.
44. Elso Barghoorn and Stanley Tyler, "Microorganisms from the Gunflint Chert," *Science* 147 (February 1965): 563–577; J. W. Schopf had also done work on this paper; Preston Cloud, "Significance of the Gunflint (Precambrian) Microflora," *Science* 148, 2 April 1965, 27–35.
45. Thomas Brock, "Life at High Temperatures," *Science* 158, 24 November 1967, 1012–1019. Brock was anticipated in the hypothesis of a high-temperature origin of life by earlier Yellowstone researcher R. B. Harvey, in "Enzymes of Thermal Algae," *Science* 60, 21 November 1924, 481–482. Brock wrote an update of the subject under the same title in 1985, when undersea hydrothermal vent discoveries, Woese's revelations about the Archaea, and proliferating theories about the relevance of hyperthermophiles for the origin of life produced a surge of new data and interest in the subject. See Brock, "Life at High Temperatures," *Science* 230, 11 October 1985, 132–138.
46. Brock, "Life (1967)," 1017.
47. Brock to Strick, personal communication, 4 February 1999; Brock OHI.
48. For the Surtsey research, see Cyril Ponnamperuma, Richard Young, and Linda Caren, "Some Chemical and Microbiological Studies of Surtsey," *Surtsey Research Project Report* 3 (1968): 70–80; my thanks to Dr. Caren for a copy of the paper and for the story about Ponnamperuma's injury.
49. Malcolm Walter, J. Bauld, and Thomas Brock, "Siliceous Algal and Bacterial Stromatolites in Hot Spring and Geyser Effluents of Yellowstone National Park," *Science* 178, 27 October 1972, 402–405; see also M. Walter, "A Hot Spring Analog for the Depositional Environment of Precambrian Iron Formations of the Lake Superior Region," *Economic Geology* 67 (1972): 965–972. His doctoral dissertation was entitled "Stromatolites and the Biostratigraphy of the Australian Precambrian" (University of Adelaide, 1970).
50. Malcolm Walter, ed., *Stromatolites* (Amsterdam: Elsevier, 1976).
51. Malcolm Walter, "What Do Stromatolites Tell Us, if Anything?" Paper delivered at Gordon Research Conference on OOL, 23 February 1999, Ventura, Calif.
52. Walter to Strick, personal communication, 23 February 1999.
53. The proceedings volume is Gregory R. Bock and Jamie A, Goode, eds., *Evolution of Hydrothermal Ecosystems on Earth (and Mars?)* (New York: John Wiley, 1996). See also Walter's more popular treatment, *The Search for Life on Mars* (Cambridge, Mass.: Perseus Books, 1999).

54. See Iris Fry, *The Emergence of Life on Earth: A Historical and Scientific Overview* (New Brunswick, N.J.: Rutgers University Press, 2000), for a detailed and philosophically astute discussion of this debate up to the present.
55. Freeman Dyson, *Origins of Life,* 2d ed. (Cambridge: Cambridge University Press, 1999); John Maynard Smith and Eörs Szathmáry, *The Origins of Life: from the Birth of Life to the Origin of Language* (Oxford: Oxford University Press, 1999).
56. Arguing over a precise definition of "when life begins" has tended to be unhelpful, as it has caused many investigators to overlook the dual-origin process as one of the likely transitional stages. See John Farley, *The Spontaneous Generation Controversy from Descartes to Oparin* (Baltimore: Johns Hopkins University Press, 1977), 179–183; see also Harmke Kamminga, "The Problem of the Origin of Life in the Context of Developments in Biology," *Origins of Life and Evolution of the Biosphere* 18 (1988): 1–11.
57. Norman W. Pirie actually suggested a "dual" or "multiple origins" hypothesis at least as far back as 1957, though analogy with Margulis's cell symbiosis argument has certainly attracted much more attention to the idea. See Pirie, "Chemical Diversity and the Origins of Life," in *Proceedings of the First International Symposium on the Origin of Life on the Earth, Moscow, 19–24 August 1957,* ed. F. Clark and R. L. M. Synge (New York: Pergamon Press, 1959), 76–83, esp. 78–79. See also Carl Lindegren, *The Cold War in Biology* (Ann Arbor, Mich.: Planarian Press, 1966), 82.
58. Dyson, *Origins of Life,* 6–7.
59. Ibid., 7–8.
60. Ibid., 6.
61. See William Hagan, "Review of Fox's *The Emergence of Life,*" *Isis* 80 (1989): 162–163; see also Andre Brack, "Review of *Chemical Evolution: Physics of the Origins and Evolution of Life,*" *Origins of Life and Evolution of the Biosphere* 29 (1999): 110.
62. See, e.g., Sidney Fox, "The Proteinoid Theory of the Origin of Life and Competing Ideas," *American Biology Teacher* 36 (1974): 161–172, 181.
63. Note that since the mid-1960s most researchers have preferred to use the plural, *origins* of life, to emphasize their belief that the process may have occurred multiple times, or in multiple steps.
64. Maynard Smith and Szathmáry, *Origins of Life,* 18–25.
65. Ibid, 12.
66. See, e.g., pp. 35 and 37, on which they say "catch–22 of the origin of life," but they mean the origin of replication.
67. Evelyn Fox Keller's *Refiguring Life* (New York: Columbia University Press, 1995) is one of several recent analyses that begin to examine this topic, as is Lily Kay's *Who Wrote the Book of Life?* (Stanford, Calif.: Stanford University Press, 2000). Donald Fleming also raised this issue in his classic paper "Émigré Physicists and the Biological Revolution," in *The Intellectual Migration,* ed. Fleming and Bernard Bailyn (Cambridge, Mass.: Harvard University Press, 1969).
68. Maynard Smith and Szathmáry, *Origins of Life,* 2.
69. Ibid., 12–13.
70. John Farley, *Spontaneous Generation,* 159–179. Equally important, but often historically overlooked, is Charles B. Lipman, "The Origin of Life," *Scientific Monthly*

19 (October 1924): 357–367, a cogent statement of the heterotroph hypothesis, contemporaneous with Oparin and preceding Haldane.
71. Maynard Smith and Szathmáry, *Origins of Life,* 36–37.
72. Leslie Orgel, "The Origin of Life: A Review of Facts and Speculations," *Trends in Biochemical Sciences* 23 (1998): 491–495; see also Stephen Freeland, Robin Knight, and Laura Landweber, "Do Proteins Predate DNA?" *Science* 286 (1999): 690–692; Gerald F. Joyce, "The Rise and Fall of the RNA World," *New Biologist* 3 (1991): 399–407. This, along with many other critical papers in the field, are reprinted in an invaluable collection, David Deamer and Gail Fleischaker, eds., *Origins of Life: The Central Concepts* (Boston: Jones and Bartlett, 1994).
73. The field underwent a similar burst of growth following the Miller-Urey experiment of 1953. But it was found that here, too, one of the most basic quandaries proved more difficult than expected. This problem is discussed further in chapter 5.
74. Fox OHI, 9–10; see Young, "Prebiological Evolution: The Constructionist Approach to the Origin of Life," in *Molecular Evolution and Protobiology,* ed. K. Matsuno et al. (New York: Plenum, 1984), 45–54.
75. Miller to Horowitz, 14 June 1972, Horowitz papers 4.34. Oparin had criticized Herrera's work in the 1957 edition of his *Origin of Life on Earth,* 3d ed. (Edinburgh: Oliver and Boyd, 1957). For more on Herrera, see Fox, *The Emergence of Life* (New York: Basic Books, 1988); see also Ismael Ledesma-Mateos and Ana Barahona, "The Institutionalization of Biology in Mexico in the Early Twentieth Century: The Conflict between Alfonso Luis Herrera and Isaac Ochoterena," *Journal of the History of Biology* 36 (2003): 285–307; Alicia Negrón-Mendoza, "Alfonso L. Herrera: A Mexican Pioneer in the Study of Chemical Evolution," *Journal of Biological Physics* 20 (1994): 11–15.
76. Horowitz to Miller, 21 June 1972, Horowitz papers 4.34.
77. See Krishna Bahadur, "Photosynthesis of Amino Acids from Paraformaldehyde and Potassium Nitrate," *Nature* 173, 12 June 1954, 1141; "The Reactions Involved in the Formation of Compounds Preliminary to the Synthesis of Protoplasm and Other Materials of Biological Importance," in Clark and Synge, *Proceedings,* 140–150; *Synthesis of Jeewanu: The Protocell* (Allahabad, India: Ram Narain Lal Beni Prasad, 1966); Bahadur, S. Ranganayaki, and L. Santamaria, "Photosynthesis of Amino Acids from Paraformaldehyde Involving the Fixation of Nitrogen in the Presence of Colloidal Molybdenum Oxide as Catalyst," *Nature* 182 (1958): 1668. Most recently, see Bahadur and S. Ranganayaki, *Origin of Life: A Functional Approach* (Allahabad, India: Ram Narain Lal Beni Prasad, 1981); Adolph E. Smith, Clair Folsome, and Krishna Bahadur, "Nitrogenase Activity of Organo-Molybdenum Microstructures," *Experientia* 37 (1981): 357–359. Bahadur's work was severely criticized in Linda Caren and Cyril Ponnamperuma, "A Review of Experiments on the Synthesis of 'Jeewanu,'" NASA Technical Memorandum X-1439, 1 September 1967. Unlike Fox, he became persona non grata in the NASA exobiology network soon thereafter (Sidney Fox OHI, 35–36; A. E. Smith to Strick, personal communication, 7 February 1993). My thanks to Dr. Adolph Smith, a sometime collaborator of Bahadur's, for the loan of some Bahadur letters and a film of a Jeewanu preparation made in 1969.
78. See A. E. Smith and F. T. Bellware, "Dehydration and Rehydration in a Prebiological System," *Science* 152, 15 April 1966, 362–363; Smith, Bellware, and J. J. Silver,

"Formation of Nucleic Acid Coacervates by Dehydration and Rehydration," *Nature* 214, 3 June 1967, 1038–1040; Smith, Silver, and Gary Steinman, "Cell-like Structures from Simple Molecules under Simulated Primitive Earth Conditions," *Experientia* 24, 15 January 1968, 36–38; see also Smith and Dean Kenyon, "Is Life Originating De Novo?" *Perspectives in Biological Medicine* 15 (August 1972): 529–542. A good overall survey of the field during the late 1960s is Dean H. Kenyon and Gary Steinman, *Biochemical Predestination* (New York: McGraw Hill, 1969). Both Smith and Sol Kramer credited their interest in a "synthetic" approach to their study of the work of Wilhelm Reich, e.g., *The Bion Experiments on the Origin of Life* (New York: Farrar, Straus and Giroux, 1979).

79. See, e.g., Carl R. Woese, D. H. Dugre, W. C. Saxinger, and S. A. Dugre, "The Molecular Basis for the Genetic Code," *PNAS* 55 (1966): 966–974; Woese, *The Genetic Code: The Molecular Basis for Genetic Expression* (New York: Harper and Row, 1967). See also Leslie Orgel, "Evolution of the Genetic Apparatus," *Journal of Molecular Biology* 38 (1968): 381–393; Orgel, *The Origins of Life: Molecules and Natural Selection* (New York: Wiley, 1973).

80. John Oró, "Synthesis of Adenine from Ammonium Cyanide," *Biochemical and Biophysical Research Communications* 2 (1960): 407–412; "Comets and the Formation of Biochemical Compounds on the Primitive Earth," *Nature* 190, 29 April 1961, 389–390; John Oró and A. P. Kimball, "Synthesis of Purines under Possible Primitive Earth Conditions, I. Adenine from Hydrogen Cyanide," *Archives of Biochemistry and Biophysics* 94 (1961): 217–227; Oró and Kimball, "Synthesis of Purines under Possible Primitive Earth Conditions, II. Purine Intermediates from Hydrogen Cyanide," *Arch. Biochem. Biophysics* 96 (1962): 293–313; Oró, "Stages and Mechanisms of Prebiological Organic Synthesis," in *The Origins of Prebiological Systems*, ed. Sidney Fox (New York: Academic Press, 1965), 137–171; Oró, Stanley Miller, Richard Young, and Cyril Ponnamperuma, eds., *Cosmochemical Evolution and the Origins of Life: Proceedings of the Fourth International Conference on the Origins of Life and the First Meeting of ISSOL, Barcelona, June 25–28, 1973* (Dordrecht: Reidel, 1984), 2 vols. See also Juan Oró OHI.

81. Lynn Margulis, "Review of Cairns-Smith's *The Life Puzzle*," *Origins of Life* 4 (1973): 516.

82. A. Graham Cairns Smith, *The Life Puzzle: On Crystals and Organisms and on the Possibility of a Crystal as an Ancestor* (Toronto: University of Toronto Press, 1972).

83. His first ideas are contained in "The Structure of the Primitive Gene and the Prospect of Generating Life" (MS, dated October 1964); my thanks to Dr. Cairns-Smith for a copy of this ms., which was submitted to (and rejected by) *Nature* then *Science* and then sent to Melvin Calvin, who suggested that the author submit it to the *Journal of Theoretical Biology*, which published a longer version. The first published version was Cairns-Smith, "The Origin of Life and the Nature of the Primitive Gene," *Journal of Theoretical Biology* 10 (1966): 53–88. As early as 29 February 1968, the theory received very favorable public notice in an article by "gene-first" advocate C. H. Waddington, "That's Life," *New York Review of Books*, 19–22.

84. John Desmond Bernal, "The Physical Basis of Life," *Proceedings of the Physics Society of London* 62 (September 1949): 537–558; later revised and expanded as a monograph (London: Routledge and Kegan Paul, 1951).

85. "The Case for an Alien Ancestry," *Proceedings of the Royal Society (London) B*

189, 6 May 1975, 249–272; Lovelock's paper was "Thermodynamics and the Recognition of Alien Biospheres," ibid., 167–181.
86. Cairns-Smith, *Genetic Takeover and the Mineral Origins of Life* (Cambridge: Cambridge University Press, 1982).
87. Mella Paecht-Horowitz, J. Berger, and A. Katchalsky, "Prebiotic Synthesis of Polypeptides by Heterogeneous Polycondensation of Amino Acid Adenylates," *Nature* 228, 14 November 1970, 636–639.
88. Hartman to Strick, personal communication, 3 February 2002.
89. The proceedings were published as Cairns-Smith and Hyman Hartman, eds., *Clay Minerals and the Origin of Life* (Cambridge: Cambridge University Press, 1986).
90. See, e.g., Cairns-Smith, "The First Organisms," *Scientific American* 252 (June 1985): 90–100; see also James Gleick, "Quiet Clay Revealed as Vibrant and Primal," *New York Times,* 5 May 1987, C1, 5; And, most recently, Cairns-Smith, "The Origin of Life: Clays," in *Frontiers of Life,* ed. David Baltimore, Renato Dulbecco, Francois Jacob, and Rita Levi-Montalcini (New York: Academic Press, 2001), 1:169–192.
91. DeVincenzi OHI, 4 February 1997, 36.
92. Oró OHI, 28 January 1997.
93. On the episode of "organized elements" in the Orgueil meteorite, see Steven Dick, *The Biological Universe* (Cambridge: Cambridge University Press, 1996).
94. Kvenvolden OHI.
95. Sidney Fox, Kaoru Harada, P. Edgar Hare, G. Hinsch, and Georg Mueller, "Bio-Organic Compounds and Glassy Microparticles in Lunar Fines and Other Materials," *Science* 167, 30 January 1970, 767–770.
96. Gordon Hodgson, Edward Bunnenberg, Berthold Halpern, Etta Peterson, Keith Kvenvolden, and Cyril Ponnamperuma, "Carbon Compounds in Lunar Fines from Mare Tranquilitatis, II. Search for Porphyrins," *Proceedings of the Apollo 11 Lunar Science Conference* 2 (1970): 1829–1844.
97. Kvenvolden OHI.
98. Harold Morowitz to Thomas Paine, June 1969, fld 5.2, Morowitz papers.
99. Harold Morowitz OHI.
100. Quoted in James Lawless, Clair Folsome, and Keith Kvenvolden, "Organic Matter in Meteorites," *Scientific American* 226 (June 1972): 38–46.
101. Carleton Moore OHI.
102. Kvenvolden OHI, 6.
103. Katherine Pering to Strick, personal communication, 9 January 2002.
104. Kvenvolden OHI, 5–8; Keith Kvenvolden, James Lawless, Katherine Pering, Etta Peterson, Jose Flores, Cyril Ponnamperuma, Ian R. Kaplan, and Carleton Moore, "Evidence for Extraterrestrial Amino Acids and Hydrocarbons in the Murchison Meteorite," *Nature* 228, 5 December 1970, 923–926. Kvenvolden's story, with his editorial supervision, was also told in Christopher Wills and Jeff Bada, *The Spark of Life* (Cambridge, Mass.: Perseus, 2000), 89–90.
105. Pering to Strick, personal communication, 9 January 2002; Ponnamperuma is dead, and thus far no other participants have been located to interview with whom accounts can be compared. Because the crucial encounter took place face to face between Kvenvolden and Ponnamperuma with nobody else present, the point may never be resolved with 100 percent certainty. Katherine Pering does not recall a conflict about the results after the time of the hospital encounter; she does say professional

relations between the two men had been strained for a long time prior to this episode.
106. Kvevolden OHI, 6.
107. See, e.g., John Cronin and Carleton B. Moore, "Amino Acid Analysis of the Murchison, Murray, and Allende Carbonaceous Chondrites," *Science* 172, 25 June 1971, 1327–1329; Cronin, "Acid-Labile Amino Acid Precursors in the Murchison Meteorite," *Origins of Life* 7 (October 1976): 337–342; Cronin, Sandra Pizzarello, and Carleton B. Moore, "Amino Acids in an Antarctic Carbonaceous Chondrite," *Science* 206 (1979): 335–337.
108. Cronin and Pizzarello, "Enantiomeric Excesses in Meteoritic Amino Acids," *Science* 275, 14 February 1997, 951–955.
109. A concise, up-to-date summary of knowledge in origin of life research is J. William Schopf, ed., *Life's Origin: The Beginnings of Biological Evolution* (Berkeley: University of California Press, 2002). See also A. Lazcano, "The Never-Ending Story," *American Scientist* 91 (September–October 2003): 452–455.

CHAPTER 4 **Vikings *to Mars***

1. Richard S. Young, "The Origin and Evolution of the Viking Mission to Mars," *Origins of Life* 7 (July 1976): 271–272.
2. Norman Horowitz, *To Utopia and Back: The Search for Life in the Solar System* (San Francisco: W. H. Freeman, 1986), 146.
3. See, e.g., Carl Sagan, "Life," *Encyclopedia Britannica,* 15th ed. (London: 1974), 10:893–911.
4. In an interesting historical irony Sagan, like Horowitz, was proclaiming by the early 1980s that the fragility of life on Earth was one of the important lessons derived from planetary exploration because of his work on the "nuclear winter" theory (see chap. 5). Although he might never share Horowitz's hardboiled negativity about the rarity of life, this represented a dramatic shift from Sagan's deep basic optimism up to this time, that life must be spread throughout the cosmos and must, thus, be fairly tenacious.
5. MS, JPL Archives, Richard Davies papers, fld 5–1189.
6. See "Exobiology Program at the Jet Propulsion Laboratory" (MS, JPL, Pasadena, Calif., 1972).
7. See Clayton R. Koppes, *JPL and the American Space Program: A History of the Jet Propulsion Laboratory* (New Haven, Conn.: Yale University Press, 1982), chap. 8.
8. One of Lovelock's discoveries with the ECD was the rising concentration of chlorofluorocarbons in the atmosphere, even far from population centers and industrial areas. Thus, he made a seminal contribution to what soon became the ozone depletion debates of the 1970s (about spray can propellants as well as supersonic transport planes). See Lydia Dotto and Harold Schiff, *The Ozone War* (New York: Doubleday, 1978); see also Lovelock, "The Independent Practice of Science," *New Scientist* 83, 6 September 1979, 716–717.
9. Silverstein to Lovelock, 9 May 1961, Lovelock papers. My thanks to Dr. Lovelock for giving me access to this material.
10. NASA, Sixth Semiannual Report to Congress, 1 July–31 December 1961 (Washington, D.C.: NASA, 1962), 181.

11. James Lovelock, *Homage to Gaia: The Life of an Independent Scientist* (London: Oxford University Press, 2000), 137–145, 227–264.
12. Carl Bruch, "Instrumentation for the Detection of Extraterrestrial Life," in *Biology and the Exploration of Mars,* ed. C. S. Pittendrigh et al. (Washington, D.C.: NAS, 1966), 488–489.
13. Ibid., 142–145.
14. James Lovelock, *Gaia: A New Look at Life on Earth* (Oxford: Oxford University Press, 1979), 1. Stapledon had enormous influence on several generations of origin of life and exobiology researchers, most notably J. B. S. Haldane. See Mark B. Adams, "Last Judgment: The Visionary Biology of J. B. S. Haldane," *Journal of the History of Biology* 33 (2000): 457–491.
15. For a survey of the strategies being considered at this time, see Bruch, "Instrumentation."
16. It is worth noting that this basic insight of Lovelock's, seen as so challenging in 1965, has since become the new paradigm in exobiology and astrobiology. See, e.g., Pamela Conrad and Kenneth Nealson (both at JPL), "A Non-Earthcentric Approach to Life Detection," *Astrobiology* 1 (2001): 15–24; see also Stephen Schneider, "A Goddess of Earth or the Imagination of a Man?" *Science* 291, 9 March 2001, 1906–1907, a well-balanced assessment of Lovelock's fundamental contributions.
17. Edward C. Ezell and Linda N. Ezell, *On Mars: Exploration of the Red Planet, 1958–1978,* NASA-SP 4212 (Washington, D.C.: NASA, 1984), 107.
18. James Lovelock, "A Physical Basis for Life-Detection Experiments," *Nature* 207, 7 August 1965, 568–570.
19. Lovelock, *Homage to Gaia,* 237–239.
20. Dian R. Hitchcock and James E. Lovelock. "Life Detection by Atmospheric Analysis," *Icarus* 7 (1967): 149–159.
21. J. E. Lovelock and C. E. Giffin, "Planetary Atmospheres: Compositional and Other Changes Associated with the Presence of Life," in *Advanced Space Experiments,* vol. 25, ed. O. L. Tiffany and E. Zaitzeff (Washington, D.C.: American Astronautical Society, 1968), 179–193.
22. Lovelock, *Homage to Gaia,* 239; Lovelock OHI; see also Horowitz to Orgel, 3 February 1968, Horowitz papers 5.10.
23. Lovelock OHI; see also *Gaia,* 10.
24. Margulis OHI.
25. J. E. Lovelock, "Geophysiology: A New Look at Earth Science," *Bulletin of the American Meteorological Society* 67 (April 1986): 392.
26. Robert J. Charlson, James Lovelock, Meinrat Andreae, and Stephen Warren, "Oceanic Phytoplankton, Atmospheric Sulphur, Cloud Albedo and Climate," *Nature* 326, 16 April 1987, 655–661.
27. It was so much the norm that a psychiatrist (Frank Fremont-Smith) and an ethologist (Sol Kramer) were reinvited (having been at the 1967 meeting as well), and there was much talk of the origin of life being an epistemological problem as much as a scientific one, invoking Marshall McLuhan's slogan that "the medium *is* the message." Kramer described first getting interested in the origin of life problem while enrolled in a course on cancer, taught by famed psychoanalyst turned natural scientist Wilhelm Reich. See Lynn Margulis, ed., *Origins of Life II* (New York: Gordon and Breach, 1970), 8–13.
28. Lovelock, *Homage to Gaia,* 239; see also Horowitz, R. P. Sharp, and R. W. Davies,

"Planetary Contamination I: The Problem and the Agreements," *Science* 155, 24 March 1967, 1501–1505. Horowitz was opposed in this opinion by Carl Sagan, Elliott Levinthal, and Joshua Lederberg, "Contamination of Mars," *Science* 159, 15 March 1968, 1191–1196; see also Sagan OHI.
29. Lovelock, "Independent Practice," 715.
30. Lyndon B. Johnson, "Remarks upon Viewing New Mariner 4 Pictures from Mars," 29 July 1965, *Public Papers of the Presidents of the United States,* 806; cited in Howard McCurdy, *Space and the American Imagination* (Washington, D.C.: Smithsonian Institution Press, 1997), 122.
31 Steven D. Kilston, Robert R. Drummond, and Carl Sagan, "A Search for Life on Earth at Kilometer Resolution," *Icarus* 5, (January 1966): 79–98. Interestingly, when the *Galileo* spacecraft tested out this proposition by observing Earth in 1993, Sagan was proved wrong. He admitted as much in "A Search for Life on Earth from the Galileo Spacecraft," *Nature* 365 (1993): 715–721.
32. Horowitz to Strick, personal communication, 16 January 2002.
33. Henry S. F. Cooper, *The Search for Life on Mars* (New York: Holt, Rinehart and Winston, 1980), 69.
34. For an excellent description of these Antarctic Dry Valleys, see Stephen Pyne, *The Ice: A Journey to Antarctica* (Iowa City: University of Iowa Press, 1986), 226–233, 312–316. In a stunning stroke of historical irony, these valleys make a spectacular reappearance in the exobiology story after *Viking,* as a source of meteorites, some later determined to be from Mars, most notably EETA79001 from Elephant Moraine and ALH84001 from the Allan Hills (see chap. 8).
35. Norman Horowitz, Roy E. Cameron, and Jerry S. Hubbard. "Microbiology of the Dry Valleys of Antarctica," *Science* 176, 21 April 1972, 242–245; see also Roy E. Cameron, "Properties of Desert Soils," in Pittendrigh et al., *Biology,* 164–186; and Ezells, *On Mars,* 235–237, including errata sheet.
36. See 1 May 1966, JPL press release "Can Exploration of a Chilean Desert Assist in the Search for Life on Mars?" "JPL scientists Richard Davies, Roy E. Cameron, and Roy Brereton will leave 2 May on a six-week exploration trip in Chile's Atacama Desert." NASA History Office, Exobiology files.
37. Ezells, *On Mars,* 235–237; Levin to Horowitz, letter, 27 June 1972, Horowitz papers, 4.18.
38. Brock to Strick, personal communication, 5 February 1999.
39. Cooper, *Life on Mars,* 65–69, 71–80. Brown University geologist Tim Mutch was the head of the *Viking* lander imaging team; he and Sagan conducted tests on a lander model in the Colorado desert, to determine the camera's capabilities.
40. Carl Sagan, "Life," in *Encyclopedia Britannica.* Sagan's interest in definitions and terminology in the discussion was also reflected in an exchange he had with Dean Kenyon and N. W. Pirie in the journal *Origins of Life;* see Sagan, "On the Terms 'Biogenesis' and 'Abiogenesis,'" *Origins of Life,* 5 (October 1974): 529.
41. N. H. Horowitz, "The Search for Extraterrestrial Life," *Science* 151, 18 February 1966, 790.
42. Frank Herbert, *Dune* (New York: Ace Books, 1965).
43. Horowitz, "Search," 789.
44. Ibid., 790.
45. Ibid., 792.

46. On the earlier history of *Gulliver*, see Gilbert Levin, A. H. Heim, J. R. Clendenning, and M. F. Thompson, "Gulliver: A Quest for Life on Mars," *Science* 138, 12 October 1962, 114–119; see also Gilbert Levin, A. H. Heim, M. F. Thompson, D. R. Beem, and N. H. Horowitz, "'Gulliver': An Experiment for Extraterrestrial Life Detection and Analysis," in *Life Sciences and Space Research*, ed. M. Florkin and A. Dollfus, 2 (1964): 124–132.
47. Cooper, *Life on Mars*, 100.
48. Ibid.
49. Cooper, *Life on Mars*, 94.
50. Horowitz to Lederberg, 4 December 1973, Horowitz papers 4.3.
51. Harold P. Klein, *A Personal History* (Mountain View, Calif.: privately printed, 1998), 203; see also 202–3, 218, 269–282, 287–292, on Klein's experience with the Biology team throughout the mission. On the Biology Committee, see also Cooper, *Life on Mars*, 94–106; and Ezells, *On Mars*, 229–242.
52. Klaus Biemann, "Detection and Identification of Biologically Significant Compounds by Mass Spectrometry," in *Life Sciences and Space Research*, ed. M. Florkin, 3 (1965): 77–85. See Helge Kragh, "The Chemistry of the Universe: Historical Roots of Modern Cosmochemistry," *Annals of Science* 57 (2000): 353–368.
53. Klaus Biemann, Juan Oró, Priestly Toulmin III, Leslie Orgel, A. O. Nier, D. M. Anderson, P. G. Simmonds, D. Flory, A. V. Diaz, D. R. Rushneck, J. E. Biller, and Arthur K. LaFleur. "The Search for Organic Substances and Inorganic Volatile Compounds in the Surface of Mars," *Journal of Geophysical Research* 82, 30 September 1977, 4641. Note that this is their construction of their reasoning *after* the data have come in, in a way that took everyone by surprise.
54. Jerry Hubbard, James P. Hardy, and N. H. Horowitz, "Photocatalytic Production of Organic Compounds from CO and H_2O in a Simulated Martian Atmosphere," *PNAS* 68 (March 1971): 574–578.
55. Horowitz to Miller, 21 June 1972, Horowitz papers 4.34. This work was published as Jerry Hubbard, J. P. Hardy, G. E. Voecks, and Ellis E. Golub, "Photocatalytic Synthesis of Organic Compounds from CO and Water: Involvement of Surfaces in the Formation and Stabilization of Products," *Journal of Molecular Evolution* 2 (1973): 149–166.
56. Horowitz to Orgel, 10 April 1974, Horowitz papers 5.10. On the general state of the mission planning by the summer of 1972, see Richard S. Young, "The Beginning of Comparative Planetology," lecture to August 1972 Special Symposium on Photochemistry and the Origin of Life, *Origins of Life* 4 (Summer 1973): 505–515. On the biology package, see Harold P. Klein, Joshua Lederberg, and Alex Rich, "Biological Experiments: The Viking Mars Lander," *Icarus* 16 (1972): 139–146; Gilbert V. Levin, "Detection of Metabolically Produced Labeled Gas: The Viking Mars Lander," *Icarus* 16 (1972): 153–166. In the same issue of *Icarus*, on the GCMS experiment, see D. M. Anderson et al., "Mass Spectrometric Analysis of Organic Compounds, Water, and Volatile Constituents in the Atmosphere and Surface of Mars," *Icarus* 16 (1972): 111–138.
57. Ezells, *On Mars*, 231.
58. Ibid., 229.
59. Ibid., 232.
60. Ibid. (By October 1972 Lederberg wrote to NASA administrator John Naugle about

future Mars missions, already imagining that budget constraints could postpone the next one until 1979; Lederberg papers. The next Mars mission was not until the *Mars Pathfinder,* which landed on 4 July 1997.)
61. Ezells, *On Mars,* 232.
62. Ibid., 233; see also Klein autobiography, 270; and Lederberg to Richard Young, 15 March 1972, Lederberg papers.
63. Ezells, *On Mars,* 234–235.
64. Friedmann (at Florida State University from 1968 to 2001) first met Vishniac at the annual American Society for Microbiology meeting in 1973; the first samples were brought back after Vishniac's death, given to Friedmann by his widow, Helen. Since that time Friedmann himself became much more active in Antarctic research, in some ways picking up in Vishniac's stead. See *Antarctic Cryptoendolithic Microbial Ecosystem (ACME) Research Group Newsletter,* no. 8 (May 1986), wherein all of Friedmann's correspondence with Vishniac and Helen Vishniac is reproduced.
65. Lederberg was apparently one of the first to get the news of Vishniac's death; see Lederberg to VanNiel, 12 December 1973, Lederberg papers.
66. Friedmann to Strick, personal communication, 27 May 2002.
67. Chris McKay, "Relevance of Antarctic Microbial Ecosystems to Exobiology," in *Antarctic Microbiology,* ed. E. Imre Friedmann (New York: Wiley-Liss, 1993), 593–601; see also McKay, "The Search for Life on Mars," *Origins of Life and Evolution of the Biosphere* 27 (1997): 273–275.
68. Friedmann to Strick, personal communication, 27 May 2002. He explained: "I have been supported by different NSF programs (e.g. Systematic Biology) and the following remarks refer only to my experience with DPP (Division of Polar Programs), later OPP (Office of Polar Programs). In contrast to other NSF programs, the 'managers' of Polar Programs are not professors serving for a limited number of years in a temporary position, but permanent federal employees. In this, they are similar to NASA program directors. But the same system produced, in the two agencies, quite different 'cultures.'"
69. Harold P. Klein, Joshua Lederberg, Alex Rich, Norman Horowitz, Vance Oyama, and Gilbert Levin "The Viking Mission Search for Life on Mars," *Nature* 262, (July 1976): 24–27.
70. Richard S. Young, "The Origin and Evolution of the Viking Mission to Mars," *Origins of Life* 7 (July 1976): 271–272.
71. Harold P. Klein, "General Constraints on the Viking Biology Investigation," *Origins of Life* 7 (July 1976): 273–279. At this time Klein wrote numerous general information articles on the biology experiments, including "Life on Mars?" *Trends in Biochemical Sciences* 1 (1976): 174–176; and "Microbiology on Mars?" *ASM [American Society for Microbiology] News* 42 (April 1976): 207–214.
72. Gilbert Levin and Patricia A. Straat, "Labeled Release: An Experiment in Radiorespirometry," *Origins of Life* 7 (July 1976): 293–311.
73. Jerry S. Hubbard, "The Pyrolytic Release Experiment: Measurement of Carbon Assimilation," *Origins of Life* 7 (July 1976): 281–292.
74. Vance Oyama, Bonnie J. Berdahl, G. C. Carle, M. E. Lehwalt, and H. S. Ginoza, "The Search for Life on Mars: Viking 1976 Gas Changes as Indicators of Biological Activity," *Origins of Life* 7 (July 1976): 313–333.
75. These included (before the mission) *Icarus* 16 (1972) and *Origins of Life* 5 (1974); then, reporting of the "preliminary results" in the 17 December 1976 issue of *Sci-*

ence; definitive descriptions of the experiment and the data set in the 30 September 1977 issue of *Journal of Geophysical Research;* a further discussion of the ambiguous biology package results in a special issue of *Origins of Life* (9) and of *Icarus* (34), both in 1978; and "completion" of the experiments (compared with simulations of them run in Earth labs trying to duplicate the Mars results) reported in *Journal of Molecular Evolution* 14 (1979).

76. Ezells, *On Mars,* 384. For accounts that capture a more popular sense of the mission, see Timothy Ferris, "The Odyssey and the Ecstasy," *Rolling Stone,* 7 April 1977; Anon., "One Man's Mars, No Martians," *Science News* 111, 5 March 1977, 149; David L. Chandler, "Life on Mars," *Atlantic* 242 (June 1977): 34–49; see also, and with more scientific content, Norman Horowitz, "The Search for Life on Mars," *Scientific American* 237 (November 1977): 57–68.
77. Benton Clark et al., "Inorganic Analyses of Martian Surface Samples at the Viking Landing Sites," *Science* 194, 17 December 1976, 1283–1288.
78. McKay, "Search for Life," 264.
79. Ezells, *On Mars,* 403.
80. Harold P. Klein, "Did Viking Discover Life on Mars?" *Origins of Life and Evolution of the Biosphere* 29 (1999): 628.
81. Ibid.
82. Gilbert V. Levin and Patricia A. Straat, "Viking Labeled Release Biology Experiment: Interim Results," *Science* 194, 17 December 1976, 1322–1329; see also, with all the additional control experiments over many months reported, Levin and Straat, "Recent Results from the Viking Labeled Release Experiment on Mars," *Journal of Geophysical Research* 82, 30 September 1977, 4663–4667.
83. Levin OHI; Ezells, *On Mars,* 403; see also Levin, "The Issue of Life on Mars and Its Implications on Science and Philosophy," talk at Philosophical Society of Washington, Cosmos Club, 9 February 2001, Washington, D.C.
84. Horowitz in 7 August 1976 *Viking* press conference at JPL, quoted in Ezells, *On Mars,* 405.
85. Ibid., 407.
86. Biemann et al., "Search." Carl Sagan was among those floored by the GCMS results: "That really knocked me for a loop," he said (Sagan OHI, 5).
87. Ezells, *On Mars,* 408.
88. McKay, "Search," 264.
89. Oró OHI, 4–6; Levin OHI, 5–8.
90. Ibid.
91. Oró OHI, 5.
92. Ibid.; see also Oró and G. Holzer, *Journal of Molecular Evoltion* 14 (1979): 153–160.
93. Ponnamperuma, A. Shimoyama, M. Yamada, T. Hobo, and R. Pal, *Science* 197 (1977): 455–461.
94. Oyama, Bonnie Berdahl, Fritz Woeller, and M. E. Lehwalt. "The Chemical Activities of the Viking Biology Experiments and the Arguments for the Presence of Superoxide, Peroxides, γFe_2O_3, and Carbon Suboxide Polymer in Martian Soil," in *COSPAR Life Sciences and Space Research,* ed. R. Holmquist and A. C. Strickland, 16 (Oxford: Pergamon Press, 1978), 3–8.
95. Levin OHI.
96. Ibid.

97. See, e.g., every paper other than Levin and Straat's in the special issue of *Journal of Molecular Evolution* in which all the experiments were summed up and described as "concluded," including Harold P. Klein, "Simulation of Viking Biology Experiments: An Overview," *Journal of Molecular Evolution* 14 (1979): 161–165; see also Oró and Holzer, 153–160.
98. Levin and Straat, "Reappraisal of Life on Mars," in Reiber, *NASA Mars Conference*, 187–192.
99. Levin OHI, 40.
100. Sagan OHI; Levin OHI.
101. Barry DiGregorio, *Mars: The Living Planet* (Berkeley, Calif.: Frog, Ltd., 1997).
102. Harold P. Klein, "Did Viking Discover Life on Mars?" *Origins of Life and Evolution of the Biosphere* 29 (1999): 625–631; see also Klein to Strick, personal communication, 22 February 2000; Klein OHI, 28 November 2000.
103. Klein, "Did Viking Discover Life," 630.
104. Ibid., 627–629.
105. Ibid., 629.
106. Ibid., 630.
107. Sagan OHI, 7–8; the paper referred to is Joshua Lederberg and Carl Sagan, "Microenvironments for Life on Mars," *PNAS* 48, 15 September 1962, 1473–1475.
108. Dana Hedgpeth, "The Man Who Wants to Return to Mars," *Washington Post*, 1 December 2000, A1, 10–11.
109. Steven Benner, Kevin Devine, Lidia Matveeva, and David Powell, "The Missing Organic Molecules on Mars," *PNAS* 97, 14 March 2000, 2425–2430.

CHAPTER 5   **The Post-Viking Revolutions**

1. Iris Fry, *The Emergence of Life on Earth: A Historical and Scientific Overview* (New Brunswick, N.J.: Rutgers University Press, 2000), 112–113. For a survey of this creationist literature, see Charles B. Thaxton, Walter L. Bradley, and R. L. Olsen, *The Mystery of Life's Origin* (1984; rpt., Dallas, Tex.: Lewis and Stanley, 1992); Percival Davis and Dean H. Kenyon, *Of Pandas and People: The Central Question of Biological Origins*, 2d ed. (Dallas, Tex.: Haughton, 1993); see also 1994 video interviews with Charles Thaxton and Dean H. Kenyon, available from Access Research Network, Colorado Springs, Colo., <http://www.arn.org>.
2. The beginning stages of much of this ferment and reconceptualization process can be seen in the volume by the Space Studies Board, National Research Council/National Academy of Sciences, *The Search for Life's Origins: Progress and Future Directions in Planetary Biology and Chemical Evolution* (Washington, D.C.: NAS Press, 1990). One of the authors of this volume, Hyman Hartman, has said that origin of life work seemed to fall into strikingly different pre-*Viking* and post-*Viking* phases. Hartman to Strick, personal communication, 3 February 2002.
3. See William K. Hartmann and Donald R. Davis. "Satellite-Sized Planetesimals and Lunar Origin," *Icarus* 24 (1975): 504–515; see also Donald Wilhelms, *To a Rocky Moon* (Tucson: University of Arizona Press, 1992), 353.
4. See Stephen Jay Gould, "Toward the Vindication of Punctuational Change," in *Catastrophes and Earth History: The New Uniformitarianism*, ed. W. A. Berggren and J. A. Van Couvering (Princeton, N.J.: Princeton University Press, 1984), 9–34. Gould had also been making such arguments in his influential and widely read

monthly column "This View of Life" in *Natural History* since 1975. See also Stephen Brush, *Fruitful Encounters* (Cambridge, UK: Cambridge University Press, 1996); and William Broad, "Apollo Opened Window on Moon's Violent Birth," *New York Times,* 20 July 1999, F1–2.
5. See, e.g., John B. Corliss and R. D. Ballard, "Oases of Life in the Cold Abyss," *National Geographic* 152 (October 1977): 441–453; Holger Jannasch and C. O. Wirsen, "Microbial Life in the Deep Sea," *Scientific American* 236 (1977): 42–52.
6. See, e.g., John B. Corliss et al., "Submarine Thermal Springs on the Galápagos Rift," *Science* 203, 16 March 1979, 1073–1083; see also Holger Jannasch and M. J. Mottl, "Geomicrobiology of Deep-Sea Hydrothermal Vents," *Science* 229, 23 August 1985, 717–725. For Jannasch's reminiscences, see "Adventures Discovering Microbes Changing the Planet," in *Many Faces, Many Microbes,* ed. Ronald M. Atlas (Washington, D.C.: American Society of Microbiology Press, 2000), 71–76; see also "Small Is Powerful: Recollections of a Microbiologist and Oceanographer," *Annual Reviews of Microbiology* 51 (1997): 1–45.
7. Woese, "Bacterial Evolution," *Microbiology Review* 51 (1987): 221–271; Woese, O. Kandler, and M. L. Wheelis, "Towards a Natural System of Organisms: Proposal for the Domains Archaea, Bacteria, and Eucarya," *Proceedings of the National Academy of Sciences(PNAS)* 87 (1990): 4576–4579.
8. Woese, "The Universal Ancestor," *PNAS* 95, 9 June 1998, 6854–6859.
9. See, e.g., Virgina Morrell, "Microbiology's Scarred Revolutionary," *Science* 276, 2 May 1997, 699–702; Woese to Strick, personal communication, 10 June 1997; see also Woese, "There Must Be a Prokaryote Somewhere: Microbiology's Search for Itself," *Microbiology Reviews* 58 (March 1994): 1–9; see also Sherrie Lyons, "Thomas Kuhn Is Alive and Well," *Perspectives in Biology and Medicine* 45 (Summer 2002): 359–376. Lyons describes Radhey Gupta's alternative "monoderm/diderm" classification, which shows that alternative schemas to Woese's are also possible, outside the previous prokayote/eukaryote "paradigm."
10. See, e.g., Joseph Fruton, *Eighty Years* (New Haven, Conn.: Epikouros Press, 1994), 146–149; by 1960 Pollard, thoroughly miffed at Yale, had relocated to Penn State. My thanks to Nicolas Rasmussen for this reference.
11. MacNab to Strick, personal communication, 25 May 1982; see also J. E. Strick, "Swimming against the Tide: Adrianus Pijper and the Debate over Bacterial Flagella, 1946–1956," *Isis* 87 (June 1996): 274–305.
12. Woese to Strick, personal communication, 20 December 2001.
13. Ernst Mayr, "Two Empires or Three?" PNAS 95 (1998): 9720–9723; see also Woese's reply, "Default Taxonomy: Ernst Mayr's View of the Microbial World," *PNAS* 95 (1998): 11043–11046. See Jan Sapp, *Genesis: The Evolution of Biology* (London: Oxford University Press, 2003), chap. 18.
14. Margulis OHI.
15. See Benton Clark, "Sulfur: The Fountainhead of Life in the Universe?" in *Life in the Universe,* ed. John Billingham (Cambridge, Mass.: MIT Press, 1981). See also John B. Corliss, John Baross, and S. E. Hoffman, "An Hypothesis Concerning the Relationship between Submarine Hot Springs and the Origin of Life," *Oceanologica Acta,* No. Sp. (1981): 59–69; see also John A. Baross, and S. E. Hoffman. "Submarine Hydrothermal Vents and Associated Gradient Environments as Sites for the Origin of Life," *Origins of Life and Evolution of the Biosphere* 15 (1985): 327–345. For the general recognition that the temperature at which life could survive

was higher than anyone had thought, see Thomas Brock, "Life at High Temperatures," *Science* 230, 11 October 1985, 132–138.
16. See, e.g., a special issue of the journal *Cell* in June 1996, including Patrick Forterre, "A Hot Topic: The Origin of Hyperthermophiles," *Cell* 85, 14 June 1996, 789–792; see also Antonio Lazcano and Stanley Miller, "The Origin and Early Evolution of Life: Prebiotic Chemistry, the Pre-RNA World, and Time," ibid., 793–798.
17. Sarah Simpson, "Life's First Scalding Steps," *Science News* 155, 9 January 1999, 24–26; see also M. Balter, "Did Life Begin in Hot Water?" *Science* 280, 3 April 1998, 31.
18. William W. Rubey, "Geologic History of Sea Water," *Geological Society of America Bulletin* 62 (September 1951), 1111–1147; see also idem., "Development of the Hydrosphere and Atmosphere, with Special Reference to Probable Composition of the Early Atmosphere," *Geological Society Special Paper* 62 (1955): 631–650.
19. Cronin to Strick, personal communication, 20 December 2001. The papers referred to are: T. C. Chamberlin and R. T. Chamberlin, "Early Terrestrial Conditions That May Have Favored Organic Synthesis," *Science* 28, 25 December 1908, 897–911, reprinted in Deamer and Fleischaker, *Origins,* 15–29; Harrison Brown, "Rare Gases and the Form of the Earth's Atmosphere," in *The Atmospheres of the Earth and the Planets,* ed. G. P. Kuiper (Chicago: University of Chicago Press, 1949); Hans Suess, "Die Häufigkeit der Edelgase auf der Erde und im Kosmos," *Journal of Geology* 57 (1949): 600–607; Heinrich D. Holland, "Model for the Evolution of the Earth's Atmosphere," in *Petrologic Studies: A Volume to Honor A. F. Buddington* (Washington, D.C.: Geological Society of America, 1962), 447–477, reprinted in *Geochemistry and the Origin of Life,* ed. Keith Kvenvolden (Stroudsburg, Pa.: Dowden, Hutchinson and Ross, 1974), 210–240.
20. Kasting to Strick, personal communication, 21 December 2001. Walker's book is *The Evolution of the Atmosphere* (New York: Macmillan, 1977); Brack's book is *The Molecular Origins of Life* (New York: Academic Press, 1998). A good review article (except for the caveats in the quoted passage) is James Kasting, "Earth's Early Atmosphere," *Science* 259, 12 February 1993, 920–926.
21. Philip Abelson, "Chemical Events on the Primitive Earth," *PNAS* 55, 15 June 1966, 1365–1372, reprinted in Kvenvolden, *Geochemistry and the Origin of Life,* 48–55. Rubey and Abelson are discussed in Horowitz to Miller, 6 December 1973 (Horowitz papers, 4.34).
22. Richard Kerr, "Origin of Life: New Ingredients Suggested," *Science* 210, 3 October 1980, 42–43.
23. Stanley Tyler and Elso Barghoorn, "Occurrence of Structurally Preserved Plants in Pre-Cambrian Rocks of the Canadian Shield," *Science* 119, 30 April 1954, 606–608.
24. Elso Barghoorn and Stanley Tyler, "Microorganisms from the Gunflint Chert," *Science* 147 (February 1965): 563–577; Preston Cloud, "Significance of the Gunflint (Precambrian) Microflora," *Science* 148, 2 April 1965, 27–35; Elso Barghoorn and J. William Schopf, "Microorganisms Three Billion Years Old from the Precambrian of South Africa," *Science* 152 (1966): 758–763; Elso Barghoorn and J. W. Schopf, "Alga-like Fossils from the Early Precambrian of South Africa," *Science* 156 (1967): 508–512. For the inside story on many of these discoveries and of how tensions over priority were negotiated between Barghoorn and Cloud in 1965, see J. Wil-

liam Schopf, *Cradle of Life: The Discovery of Earth's Earliest Fossils* (Princeton, N.J.: Princeton University Press, 1999), 56–61.
25. Schopf to Strick, personal communication, 5 May 2002. Barghoorn and Schopf both attended the 1967 NASA/Smithsonian OOL meeting in Princeton. It may be there that they first made the contacts leading to their first (1967–1969) Exobiology grant.
26. A. H. Knoll and E. S. Barghoorn, "Archean Microfossils Showing Cell Division from the Swaziland System of South Africa," *Science* 198 (1977): 396–398.
27. Stephen Jay Gould, "An Early Start," *Natural History* 87 (February 1978): 10–24, reprinted in *The Panda's Thumb* (New York: W. W. Norton, 1980), 217–226. Quote from that edition, 221.
28. Schwartz to Strick, personal communication, 3 February 2002.
29. George Wald, "The Origin of Life," *Scientific American* 192 (August 1954): 44–53.
30. Antonio Lazcano and Stanley Miller, "How Long Did It Take for Life to Begin and Evolve to Cyanobacteria?" *Journal of Molecular Evolution* 39 (1994): 546–554.
31. J. William Schopf, ed., *Earth's Earliest Biosphere: Its Origin and Evolution* (Princeton, N.J.: Princeton University Press, 1983), xxi.
32. Ibid.
33. J. W. Schopf and Cornelis Klein, eds., *The Proterozoic Biosphere* (New York: Cambridge University Press, 1992).
34. J. William Schopf, "Microfossils of the Early Archean Apex Chert: New Evidence of the Antiquity of Life," *Science* 260, 30 April 1993, 640–646.
35. Schopf to Strick, personal communication, 5 May 2002.
36. Christopher Wills and Jeffrey Bada, *The Spark of Life: Darwin and the Primeval Soup* (New York: Perseus, 2000), 198.
37. The proceedings were published as John Billingham, ed., *Life in the Universe* (Cambridge, Mass.: MIT Press, 1981).
38. Stephen Schneider and Randi Londer, *The CoEvolution of Climate and Life* (San Francisco: Sierra Club, 1984).
39. The proceedings of this conference were published as Stephen Schneider and Penelope Boston, eds., *Scientists on Gaia* (Cambridge, Mass.: MIT Press, 1991).
40. David Milne, David Raup, John Billingham, Karl Niklas, and Kevin Padian, eds., *The Evolution of Complex and Higher Organisms (ECHO)*, NASA SP-478 (Moffet Field, Calif.: NASA Ames, 1985), 24.
41. See Robert M. Young, "Darwin's Metaphor: Does Nature Select?" *Darwin's Metaphor* (Cambridge: Cambridge University Press, 1985), 79–125.
42. The first sharp critique in this vein was W. Ford Doolittle, "Is Nature Really Motherly?" *CoEvolution Quarterly* (Spring 1981): 58–63, with replies by Lovelock (62–63) and Margulis (63–65); then came Richard Dawkins's *The Blind Watchmaker* (1984). Quite a bit of the criticism is summed up in Charles Mann, "Lynn Margulis: Science's Unruly Earth Mother," *Science* 252, 19 April 1991, 378–381. The tone used by critics is rather more harsh and dismissive than is typical for a scholarly scientific exchange.
43. The book was *Gaia: A New Look at Life on Earth* (Oxford: Oxford University Press, 1979).
44. James Lovelock and A. J. Watson. "The Regulation of Carbon Dioxide and Climate: Gaia or Geochemistry," *Planetary and Space Sciences* 30 (1982): 795–802;

see also A. J. Watson and J. E. Lovelock, "Biological Homeostasis of the Global Environment: The Parable of Daisyworld," *Tellus* 35B (1983): 284–289. For a balanced retrospective on the entire controversy, see Stephen Schneider, "A Goddess of Earth or the Imagination of a Man?" *Science* 291, 9 March 2001, 1906–1907.

45. Milne et al., *ECHO,* 154.
46. See, e.g., Wills and Bada, *Spark,* 81–83, for the initial negative reaction of scientists to Gaia based on its "Earth Mother" aspects and for Lovelock's response.
47. See, e.g., Pamela Conrad and Kenneth Nealson (both currently at JPL), "A Non-Earthcentric Approach to Life Detection," *Astrobiology* 1 (2001): 15–24. Nealson has also written an excellent review of new discoveries and changed thinking in microbiology since *Viking* which are relevant to exobiology and astrobiology: "Post-Viking Microbiology: New Approaches, New Data, New Insights," *Origins of Life and Evolution of the Biosphere* 29 (1999): 73–93.
48. Morowitz, *Beginnings of Cellular Life* (New Haven: Yale University Press, 1992), 5–6. For Morowitz the "systems approach" of Gaia must have had inherent appeal early on.
49. James Lovelock, "A Way of Life for Agnostics?" *Skeptical Inquirer* 25 (September–October 2001): 40–42; Lovelock OHI. Margulis has replied in numerous articles, several of them in her recent collection with Dorion Sagan, *Slanted Truths* (New York: Springer Verlag, 1997); see esp. "Big Trouble in Biology," 265–282.
50. Lovelock to Strick, personal communication, 11 March 2002. See Midgeley, *Science and Poetry* (London: Routledge, 2002).
51. See Lovelock, "On Being an Independent Scientist," *New Scientist,* 6 September 1979, 714–717; Lovelock, *The Ages of Gaia,* 2d ed. (New York: W. W. Norton, 1995), xvi–xvii; he develops the discussion much further as the central focus of his autobiography, *Homage to Gaia: The Life of an Independent Scientist* (Oxford: Oxford University Press, 2000).
52. Lovelock to Strick, personal communication, 10 June 2002. The article he mentions is Timothy Lenton, "Gaia and Natural Selection," *Nature* 394 (1998): 439–447. The Lawton article is "Earth System Science," *Science* 292, 15 June 2001, 1965–1966. Lovelock's *Geophysiology of Amazonia* paper was republished as "Geophysiology: A New Look at Earth Science," *Bulletin of the American Meteorological Society* 67 (April 1986): 392–397.
53. Lawton, "Earth System," 1965.
54. Thomas Kuhn, *The Structure of Scientific Revolutions,* 2d ed. (Chicago: University of Chicago Press, 1970).
55. Lovelock to Strick, personal communication, 6 June 2002. The article to which he refers is Dennis Overbye, "NASA Presses Its Search for Extraterrestrial Life," *New York Times,* 4 June 2002, D1, 4, specifically to a quote about Gaia by Kevin Zahnle.
56. See Luis W. Alvarez, *Alvarez: Adventures of a Physicist* (New York: Basic Books, 1987), chap. 15.
57. Luis W. Alvarez, Walter Alvarez, Frank Asaro, and Helen V. Michel, "Extraterrestrial Cause for the Cretaceous-Tertiary Extinction," *Science* 208, 6 June 1980, 1095–1108.
58. DeVincenzi OHI, 28 January 1997.
59. David M. Raup and Joseph J. Sepkoski Jr., "Mass Extinctions in the Marine Fossil Record," *Science* 215 (March 1982): 1501–1503.
60. Milne, et al., *ECHO,* xix.

61. D. Raup and J. J. Sepkoski, "Periodicity of Extinctions in the Geologic Past," *PNAS* 81, 1 February 1984, 801–805.
62. Ibid.
63. Raup to Strick, personal communication, 18 March 2002; DeVincenzi OHI, 28 January 1997.
64. Ibid.
65. Richard Kerr, "Periodic Impacts and Extinctions Reported," *Science* 223, 23 March 1984, 1277.
66. Ibid.; see also Alvarez, *Alvarez,* 265–267.
67. Alvarez, *Alvarez,* 266–267.
68. Raup to Strick, personal communication, 8 May 2002; see also David Raup, "Periodicity of Extinction: A Review," in *Controversies in Modern Geology,* ed. D. W. Muller et al. (New York: Academic Press, 1991), 193–208.
69. Alvarez, *Alvarez,* 282–283.
70. William Poundstone, *Carl Sagan: A Life in the Cosmos* (New York: Holt, 1999), 297; see also Keay Davidson, *Carl Sagan: A Life* (New York: Wiley, 1999). The entire special issue of *Ambio* was reprinted by the Royal Swedish Academy of Sciences as a monograph: *Aftermath: The Human and Ecological Consequences of Nuclear War,* ed. Jeannie Peterson (New York: Pantheon, 1983); see Paul Crutzen and John W. Birks, "The Atmosphere after a Nuclear War: Twilight at Noon," 73–96. Among those credited with advice and commentary on early drafts of the paper were James Lovelock and Steven Schneider. Small wonder, then, that it quickly came to the attention of Sagan and others in NASA circles.
71. Poundstone, *Sagan,* 297–298.
72. Ibid., 301–303.
73. Richard P. Turco, Owen B. Toon, Thomas P. Ackerman, James B. Pollack, and Carl Sagan, "Nuclear Winter: Global Consequences of Multiple Nuclear Explosions," *Science* 222, 23 December 1983, 1283–1292.
74. Donald DeVincenzi, "NASA's Exobiology Program," *Origins of Life* 14 (1984): 793–799; see 797.
75. DeVincenzi OHI, 28 January 1997.
76. Richard P. Turco, Owen B. Toon, Thomas P. Ackerman, James B. Pollack, and Carl Sagan, "Climate and Smoke: An Appraisal of Nuclear Winter," *Science* 247, 12 January 1990, 166–176.
77. Richard Turco, "Carl Sagan and Nuclear Winter," in *Carl Sagan's Universe,* ed. Yervant Terzian and Elizabeth Bilson (Cambridge: Cambridge University Press, 1997), 239–246.
78. DeVincenzi, "NASA's Exobiology Program," 798–799.
79. NAS Space Studies Board, *The Search for Life's Origins* (Washington, D.C.: NAS Press, 1990).
80. Ibid., 101.
81. Stanley Miller and Chris Chyba, "Whence Came Life?" *Sky and Telescope* (June 1992): 604–605; quote on 605.
82. Poundstone, *Carl Sagan,* 329–330.
83. Rummel OHI, 9. A concise version of the contrasting views of the Miller group with Chyba and hydrothermal vent researchers is contained in point-counterpoint fashion in Miller and Chyba, "Whence Came Life?"
84. Jon Cohen, "Novel Center Seeks to Add Spark to Origins of Life," *Science* 270,

22 December 1995, 1925–1926. Note the pun by which Cohen makes clear that the NSCORT group pushes a strong Miller-Urey agenda.
85. Table courtesy of John Rummel.
86. Cohen, "Novel Center," 1925.
87. Wills and Bada, *Spark*. For their critique of exogenous delivery of organics, see 92–94; of "ventists," see esp. 96–101.
88. Cohen, "Novel Center," 1925; Wills and Bada, *Spark*, 101–105. For more recent work on a wider rule for clays, see M. M. Hanczyc, S. M. Fujikawa, and J. W. Szostak, "Experimental Models of Primitive Cellular Compartments: Encapsulation, Growth, and Division," *Science* 302, 24 October 2003, 618–621.
89. Kruger et al., *Cell* 31 (1982): 147–157.
90. Cech to Strick, personal communication, 29 May 1997.
91. Schopf to Strick, personal communication, 14 May 2002.
92. See, e.g., Walter Gilbert, "The RNA World," *Nature* 319 (1986): 618, reprinted in Deamer and Fleischaker, *Origins*, 375; see also Thomas Cech, "RNA as an Enzyme," *Scientific American* 255 (November 1986): 64–75.
93. Gerald Joyce, "The Rise and Fall of the RNA World," *New Biologist* 3 (1991): 399–407.
94. Leslie Orgel, "The Origin of Life—A Review of Facts and Speculations," *Trends in Biochemical Sciences* 23 (1998): 491–495.
95. Cronin to Strick, personal communication, 9 February 2000.
96. Ibid.
97. Cohen, "Novel Center," 1926.
98. Carl Woese, "On the Evolution of Cells," *PNAS* 99, 25 June 2002, 8742–8747.

CHAPTER 6   *The Search for Extraterrestrial Intelligence*

1. Portions of this chapter are based on Steven J. Dick, "The Search for Extraterrestrial Intelligence and the NASA High Resolution Microwave Survey (HRMS): Historical Perspectives," *Space Science Reviews* 64 (1993): 93–139.
2. G. Cocconi and P. Morrison, *Nature* 184 (1959): 844–846. The proceedings of the JPL meeting are in G. Mamikunian and M. H. Briggs, eds., *Current Aspects of Exobiology* (Oxford: Pergamon Press, 1964). On Ozma and the Green Bank meeting, see S. Dick, *The Biological Universe* (Cambridge: Cambridge University Press, 1996), 414–431.
3. Billingham OHI, 12 September 1990, 1–5; Swift, *SETI Pioneers* (Tucson: University of Arizona Press), 247–278.
4. Billingham OHI, 6–7.
5. C. Ponnamperuma and A. G. W. Cameron, eds., *Interstellar Communication: Scientific Perspectives* (Boston: Houghton Mifflin: 1974). The mini-study consisted of only four people at Ames: David Black on planetary systems, Ponnamperuma on origin of life, Dale Dunn on communications, and Billingham.
6. Oliver, OHI, 1–6; Swift, *SETI Pioneers*, 86–115. Many of Oliver's published and unpublished papers on SETI are collected in *The Selected Papers of Bernard M. Oliver* (Palo Alto, Calif.: Hewlett-Packard, 1997).
7. B. Oliver and J. Billingham, *Project Cyclops: A Design Study of a System for Detecting Extraterrestrial Intelligence,* NASA CR 114445 (Washington, D.C.: NASA, 1972).

8. The first meeting of the Interstellar Communication Committee took place on 1 December 1972. Detailed minutes for all the meetings are in the SETI Institute Archives. The original members of the Interstellar Communication Committee included Billingham as chief, J. Wolfe as deputy chief, and D. Black, E. Duckworth, R. Eddy, M. Hansen, H. Hornby, R. Johnson, and D. Lumb as members. Vera Buescher soon joined and became a key member of the SETI team for the next thirty years. Mark kept NASA administrator James Fletcher apprised of progress and sought his support and advice. Mark to Fletcher, personal communication, 3 October 1972, SETI Institute Archives.
9. These studies are found in the SETI Institute Archives. The Fletcher quote is in Richard Berenzden, ed., *Life beyond Earth and the Mind of Man*, NASA SP-328 (Washington, D.C.: NASA, 1973). In order to concentrate on the interstellar communication plans, in October Billingham obtained from his immediate boss, Chuck Klein (who was simultaneously skeptical and supportive), and Hans Mark a sabbatical from his position as Biotechnology Division chief.
10. P. Morrison, J. Billingham, and J. Wolfe, *The Search for Extraterrestrial Intelligence*, NASA SP-419 (Washington, D.C.: NASA, 1977). On Billingham's remark, and for a succinct overview of his role in NASA SETI, see Billingham, "SETI in NASA," presented at the conference commemorating Frank Drake's seventieth birthday and forty years of SETI, held at Harvard and Boston Universities, 6–7 May 2000.
11. Ibid., 20. For the early discussions of the bimodal approach, see C. Seeger, in Morrison et al., *Search for Extraterrestrial Intelligence*, 77–92; and S. Gulkis, E. Olsen, and J. Tarter, in M. Papagiannis, ed., *Strategies in the Search for Life in the Universe* (Dordrecht: Reidel, 1980), 93–105.
12. R. E. Edelson, *Mercury* 6, no. 4 (1977): 8–12; B. Murray, S. Gulkis, and R. E. Edelson, *Science* 199 (1978): 485–492; D. Black et al., *Mercury* 6, no. 4 (1977): 4–7.
13. NASA, *Outlook for Space: Report to the NASA Administrator by the Outlook for Space Study Group* (Washington, D.C.: NASA, 1976), 38, 1435–149.
14. J. Wolfe et al., in J. Billingham, ed., *Life in the Universe* (Cambridge, Mass.: MIT Press, 1981), 391–417.
15. Ibid., 391–417. The detectability of terrestrial transmitters at interstellar distances, and the implications for detection of leakage radiation from extraterrestrial civilizations, was studied by Woodruff T. Sullivan III and his colleagues in Sullivan, S. Brown, and C. Wetherill, *Science* 199 (1978): 377–388, and reviewed in Billingham, *Life in the Universe*, 377–390.
16. F. Drake, J. Wolfe, and C. Seeger, SETI Science Working Group Report, NASA Technical Paper 2244 (1983).
17. Michael Hart, *Quarterly Journal of the Royal Astronomical Society* (*QJRAS*) 16 (1975): 128–135; D. Viewing, *Journal of the British Interplanetary Society* (*JBIS*) 28 (1975): 735–744.
18. M. Hart and B. Zuckerman, *Extraterrestrials: Where Are They?* (New York: Pergamon, 1982; 2d ed. with new chapters published by Cambridge University Press, 1995); F. Tipler, *QJRAS* 21 (1980): 267–281 and 22 (1981): 133–145 and 279–292; J. Barrow and F. Tipler, *The Anthropic Cosmological Principle* (Oxford: Oxford University Press, 1986); D. Brin, *QJRAS* 24 (1983): 283–309.
19. Senator William Proxmire, press release, 16 February 1978; U.S. Congress, *Extraterrestrial Intelligence Research,* Hearings before the Space Science and Applications Subcommmittee of the Committee on Science and Technology, U.S. House

of Representatives, 95th Cong., 2d sess., 19–20 September 1978. For congressional action related to SETI, I am indebted to Vera Buescher's unpublished compilation, "A Brief History of Congressional Actions Regarding the Search for Extraterrestrial Intelligence (SETI)," SETI Institute, October 1995.

20. Introduction, Billingham, *Life in the Universe*. This volume is the proceedings of a meeting held at NASA Ames in 1979.
21. *Congressional Record*, Senate, 30 July 1981, S 8812; *Congressional Record*, House, 11 September 1981, H 6156; Frank Drake, "Putting the Cosmos on Hold," *Cosmic Search* (1982): 8–9.
22. National Research Council (NRC), *Astronomy and Astrophysics for the 1980s* (Field report) (Washington, D.C.: NRC, 1982).
23. The technical details of the system implemented in 1992 have been described by the participants elsewhere. See G. R. Coulter, M. J. Klein, P. R. Backus, and J. D. Rummel, "Searching for Intelligent Life in the Universe: NASA's High Resolution Microwave Survey," *Space Biology and Medicine* 3 (1993).
24. On the MCSA 1.0, see A. M. Peterson, K. S. Chen, and I. R. Linscott, "The Multichannel Spectrum Analyzer," in *The Search for Extraterrestrial Life: Recent Developments*, ed. M. D. Papagiannis (Dordrecht: Reidel, 1985), 373–383.
25. I. R. Linscott, J. Duluk, J. Burr, and A. Peterson, in *Bioastronomy: The Next Steps*, ed. G. Marx (Dordrecht: Reidel, 1988), 319–335.
26. On the software algorithms see Cullers, "Software Implementation of Detection Algorithms for the MCSA," in Pagiagiannis, *Search for Extraterrestrial Life*, 385–390.
27. On the WBSA, see M. P. Quirk, M. F. Garyantes, H. C. Wilck, and M. J. Grimm, *IEEE Transactions on Acoustical Speech Signal Processing* 36 (1988): 1854–1861. On the sky survey signal processing and data acquisition, see E. T. Olsen, A. Lokshin, and S. Gulkis, "An Analysis of the Elements of an All Sky Survey," in Papagiannis, *Search for Extraterrestrial Life*, 405–410.
28. NASA, Program Plan for the Search for Extraterrestrial Intelligence, MS, NASA 1987.
29. Alan Boss, *Looking for Earths* (New York: Wiley and Sons, 1998), 117–118.
30. Tarter, OHI, 2; Swift, *SETI Pioneers*, 346–377.
31. Tarter, OHI, 18–19; Pierson OHI.
32. *Congressional Record*, House, H4356–4359, 28 June 1990.
33. Senate Report 101-474, to accompany H.R. 5158, 10 September 1990.
34. A summary of the Lederberg workshop was included in the 1977 landmark Morrison volume, *Search for Extraterrestrial Intelligence*. The CASETI proceedings were eventually published as J. Billingham et al., *Social Implications of the Detection of an Extraterrestrial Civilization* (Mountain View, Calif.: SETI Press, 1999).
35. News bulletin, Richard Bryan, U.S. senator, State of Nevada, 103d Cong., "Bryan Amendment Passes to Cut Expensive Search for "Martians"—Great Martian Chase to End?" 22 September 1993. The debate is found in *Congressional Record*, Senate, 22 September 1993, S 12151–12153. On Bryan's previous action on 14 May 1991 (for FY 1992), see news bulletin, Richard Bryan, "Bryan Eliminates Government Waste, Cuts $14.5 Million Martian Hunt," 14 May 1991; news bulletin, Richard Bryan, "Senate Committee Votes to Cut Alien Search Funding," 16 June 1992.

36. Bryan, 22 September 1993 press release; and George Johnson, "E.T., Don't Call Us, We'll Call You. Someday," *New York Times,* 10 October 1993, 4:2.
37. Stephen J. Garber, "Searching for Good Science: The Cancellation of NASA's SETI Program," *JBIS* 52 (1999): 3–12.
38. Keay Davidson, "'Giggle Factor' Helps Kill Project to Contact Aliens," *Washington Times,* 10 October 1993, D8; Christopher Anderson and Jeffrey Mervis, "Congress Boosts NSF, NASA Budgets," *Science* 262, 8 October 1993, 173. For an interview with the headquarters manager for SETI as these events were occurring, see Coulter OHI, 30 September 1993.
39. Wesley T. Huntress to Dale Compton and Ed Stone, letter, 12 October 1993, SETI Institute Archives, folder marked "U.S. Congress, NASA HRMS Termination"; "Toward Other Planetary Systems, High Resolution Microwave Survey, Project Termination Report," 31 March 1994 (Ames Research Center and JPL).
40. Jill Tarter, "Past and Future Observing Plans: The Fate of the NASA HRMS, Soon to Be Reborn as Project Phoenix," *SETI News* 3, no. 1 (first quarter, 1994): 1.

CHAPTER 7 **The Search for Planetary Systems**

1. Oliver and Billingham, *Project Cyclops: A Design Study of a System for Detecting Extraterrestrial Intelligent Life,* NASA CR 11445 (Washington, D.C.: NASA, 1971), 13–15; A. G. W. Cameron, "Planetary Systems in the Galaxy," in *Interstellar Communication: Scientific Perspectives,* ed. Cyril Ponnamperuma and A. G. W. Cameron (Boston: Houghton Mifflin, 1974), 26–44.
2. The agenda and attendee list for the workshop is in *The Search for Extraterrestrial Intelligence,* ed. Philip Morrison, John Billingham, and John Wolfe, NASA SP-419 (Washington, D.C.: NASA, 1977), 269 ff. Detailed minutes of the workshops are located at the SETI Institute archives.
3. Black OHI, 1, 4; Oliver and Billingham, *Project Cyclops,* 14–15; Alan Boss, *Looking for Earths: The Race to Find New Solar Systems* (New York: Wiley and Sons, 1998), 30. The Second Workshop on Extrasolar Planetary Detection was held at NASA Ames Research Center, 20–21 May 1976.
4. Morrison et al., *Search for Extraterrestrial Intelligence,* 57–58; Boss, *Looking for Earths,* 79. Bracewell had written *The Galactic Club* (San Francisco: W. H. Freeman, 1974) and in the aftermath of the NASA meetings wrote "Detecting Nonsolar Planets by Spinning Infrared Interferometer," *Nature* 274 (1978): 780.
5. Minutes of the NASA-Ames Astrometric Conference, U.S. Naval Observatory, Washington, D.C., 10–11 May 1976, SETI Institute Archives.
6. Jesse Greenstein and David Black, "Detection of Other Planetary Systems," in Morrison et al., *Search for Extraterrestrial Intelligence,* 55–60.
7. Black, OHI, 5; David Black, ed., *Project Orion: A Design Study of a System for Detecting Extrasolar Planets* (Washington, D.C.: NASA, 1980). This book is based on the 1976 NASA/ASEE-Stanford Summer Faculty Workshop in Engineering Systems Design, 14 June–20 August 1976. The appendix includes discussion of Space Telescope capability to detect planets.
8. David C. Black and William E. Brunk, eds., *An Assessment of Ground-Based Techniques for Detecting Other Planetary Systems,* NASA CP-2124 (Washington, D.C.: NASA 1980), 1:4–8. Volume 1 is an overview, and volume 2 presents position papers.
9. Black OHI, 8.

10. Black OHI, 5–9; Black, "In Search of Planetary Systems," *Space Science Reviews* 25 (January 1980): 35–81.
11. Foreword to Black and Brunk.
12. David C. Black, "Prospects for Detecting Other Planetary Systems," in *Life in the Universe,* ed. John Billingham (Cambridge, Mass.: MIT Press, 1981). The meeting was held at Ames on 19–20 June 1979. Black reviewed the prospects in more detail in his article in "In Search of Planetary Systems."
13. For a single telescope the case of "speckle interferometry" for planet detection was put forth at the Ames meeting by Simon P. Worden, "Detecting Planets in Binary Systems with Speckle Interferometry." It had been put forward earlier by one of the pioneers in the technique, H. A. McAlister, in "Speckle Interferometry as a Method for Detecting Nearby Extrasolar Planets," *Icarus* 30 (1977): 789–792. Although the technique proved extremely useful for binary star observations, by the end of the century it had not yet detected any planets.
14. *Life in the Universe,* ed. John Billingham (Cambridge, Mass.: MIT Press, 1981), xiv.
15. Bernard F. Burke, ed., *TOPS: Toward Other Planetary Systems: A Report by the Solar System Exploration Division* (Washington, D.C.: NASA, 1992), preface, vii.
16. *Planetary Exploration through the Year 2000: A Core Program,* SSEC Report, May 1983. The Space Science Board and COMPLEX reports were: *Report on Space Science 1975* (1976); *Strategy for Exploration of the Inner Planets: 1977–1987* (1978); and *Strategy for Exploration of Primitive Solar System Bodies* (1980). Pages 4–7 of the latter report placed the origin of the solar system in the broader context of star formation in the galaxy.
17. On the long-standing "love-hate" relationship between NASA and the Space Science Board, see Homer Newell, *Beyond the Atmosphere,* 205–214; Boss, *Looking for Earths,* 82–83; and Black OHI, 9–10.
18. *Strategy for the Detection and Study of Other Planetary Systems and Extrasolar Planetary Materials: 1990–2000* (Washington, D.C.: National Academy Press, 1990), 1–3. It is notable that the chairman of COMPLEX from 1985 to 1988 was Robert Pepin, who had been Black's thesis advisor at the University of Minnesota.
19. Astronomy and Astrophysics Survey Committee, National Academy of Sciences, National Research Council, *The Decade of Discovery in Astronomy and Astrophysics* (Washington, D.C.: National Academy Press, 1991), 30–31.
20. *Planetary Exploration through the Year 2000: An Augmented Program* (Washington, D.C.: NASA, 1986), 15.
21. H. H. Aumann et al., "Discovery of a Shell around Alpha Lyrae," *Astrophysical Journal* 278, 1 March 1984, L23–L27; front page of the *Washington Post,* 10 August 1983. The story of the "Vega Phenomenon" is told in Ken Crosswell, *Planet Quest: The Epic Discovery of Alien Solar Systems* (New York: Free Press, 1997), 100–113.
22. *Planetary Exploration,* 22, 183–184.
23. Ibid., 205; Black OHI, 10–13.
24. *Other Worlds from Earth: The Future of Planetary Astronomy* (Washington, D.C.: NASA, 1989), 9, 21–31.
25. Ibid., 21–31, 68–69, 76–81, 90–91, 95. The ATF is described and pictured on pages 68–69 and the CIT on pages 78–79. Their strengths and weaknesses are described on 91.

26. *TOPS*. The work of the PSSWG is colorfully described by one of its members in Boss, *Looking for Earths,* esp. 83–86, 127–132.
27. *TOPS,* viii.
28. Ibid., 59–66.
29. Ibid., 99–110, on the Keck telescopes for planet searches; Boss, *Looking for Earths,* 97–98. On NASA funding for Keck at the rate of 6.8 million per year from 1994 to 2000, see Boss, *Looking for Earths,* 128.
30. R. A. Brown and C. J. Burrows, "On the Feasibility of Detecting Extrasolar Planets by Reflected Starlight Using the Hubble Space Telescope," *Icarus* 87 (1990): 484; Black, OHI, 16; Boss, *Looking for Earths,* 85–86.
31. Boss, *Looking for Earths,* 104–106.
32. *TOPS,* 59, 114–117.
33. Boss, *Looking for Earths,* 122–124, 128.
34. TOPS, xviii, and p. 1.
35. Paul Butler to S. Dick, personal communication, 13 February 2002. Butler's master's thesis was entitled "A Precision Astronomical Instrument to Measure Doppler Shifts."
36. Butler to Dick, personal communication, 13 February 2002.
37. The discoveries of the first planets around solar-type stars by the Swiss team of Michel Mayor and Didier Queloz and the American team of Marcy and Butler have been described many times; see especially Crosswell, *Planet Quest;* Michael D. Lemonick, *Other Worlds: The Search for Life in the Universe* (New York: Simon and Schuster, 1998); and Donald Goldsmith, *Worlds Unnumbered: The Search for Extrasolar Planets* (Sausalito, Calif.: University Science Books, 1997).
38. Steven Beckwith and Anneila Sargent review progress in "Circumstellar Disks and the Search for Neighbouring Planetary Systems," *Nature* 383, 12 September 1996, 139–144. The "proplyds" of Orion are announced in C. R. Odell and Z. Wen, *Astrophysical Journal* 387 (1994): 194–202.
39. *The Space Interferometry Mission: Taking the Measure of the Universe,* Final Report of the Space Interferometry Science Working Group, 5 April 1996. NASA/JPL published a more popular version under the same title in March 1999. On the twists and turns of the committee's goals and actions, see "History of the Working Group" in the 1996 document.
40. *HST and Beyond—Exploration and the Search for Origins: A Vision for Ultraviolet-Optical-Infrared Space Astronomy* (Washington, D.C.: AURA, 1996).
41. Charles Beichman, ed., *Roadmap for the Exploration of Neighboring Planetary Systems* (August 1996).
42. R. N. Bracewell, *Nature* 274 (1978): 780; Roger Angel, in *Next Generation Space Telescope,* ed. P. Bely, C. Burrows, and J. G. Illingworth (1990), 81–94; M. Shao, in same volume, 160.
43. Beichman, *Roadmap,* 10–16.
44. Lemonick, *Other Worlds,* 161.
45. Harley Thronson, "Our Cosmic Origins: NASA's Origins Theme and the Search for Earth-like Planets," in *Planets beyond the Solar System and the Next Generation of Space Missions,* ed. D. R. Soderblom (San Francisco: Astronomical Society of the Pacific, 1997).
46. *Origins: Roadmap for the Office of Space Science Origins Theme* (Washington, D.C.: NASA, 1997). The original forty-eight-page publication was updated in April 2000 with a ninety-four-page publication.

47. C. A. Beichman, N. J. Woolf, and C. A. Lindensmith, *The Terrestrial Planet Finder (TPF): A NASA Origins Program to Search for Habitable Planets*, JPL Publication 99-3 (May 1999). The European Space Agency also proposed a space infrared interferometer known as "Darwin."

CHAPTER 8 **The Mars Rock**

1. John Noble Wilford, "Clues in Meteorite Seem to Show Signs of Life on Mars Long Ago," *New York Times*, 7 August 1996, A1, A10.
2. John Noble Wilford, "Mars and Its Meteorites Targets of New Research," *New York Times*, 13 August 1996, C1, C8.
3. William J. Broad, "Jupiter's Moon Europa Could Be Habitat for Life," *New York Times*, 13 August 1996, C1, C7. See also Broad, "Scientists Widen the Hunt for Alien Life," *New York Times*, 6 May 1997, Arizona ed., B9, 15.
4. John Noble Wilford, "Plotting a Mission to Retrieve Rocks from Mars," *New York Times*, 10 September 1996, Arizona ed., B5, 9. By January 1999 the sample return was being planned for 2008, after an ambitious series of preparatory missions. See William Broad, "Spacecraft Speed to Mars, High Hopes on Board," *New York Times*, 5 January 1999, D5.
5. Chris Romanek OHI and David McKay OHI; Everett Gibson OHI; the published article was David S. McKay et al., "Search for Past Life on Mars: Possible Relic Biogenic Activity in Martian Meteorite ALH84001," *Science* 273, 16 August 1996, 924–930.
6. Richard Kerr, "A Lunar Meteorite and Maybe Some from Mars," *Science* 220, 15 April 1983, 288–289.
7. Bogard and Johnson, "Martian Gases in an Antarctic Meteorite," *Science* 221, 12 August 1983, 651–654.
8. Brian Mason, "A Lode of Meteorites," *Natural History* 90 (April 1981): 62–67.
9. Each of the rocks resembled most closely one of the three meteorites first found in this group, Shergotty, Nakhla, and Chassigny. Nakhla fell near El Nakhla el Baharia, Egypt, on 28 June 1911.
10. Kerr, "Lunar Meteorite," 288. The age of Shergotty was later found to be only 165 to 300 million years since crystallization, indicating that Mars must have had at least intermittent volcanism until fairly recently.
11. Ibid., 289.
12. Ibid.
13. Christopher Wills and Jeffrey Bada, *The Spark of Life: Darwin and the Primeval Soup* (Cambridge, Mass.: Perseus, 2000), 237.
14. Richard Kerr, "Martian Meteorites Are Arriving," *Science* 237, 14 August 1987, 721.
15. Ibid.
16. See <http://www.jpl.nasa.gov/snc/> for the latest update on new SNC meteorites.
17. Everett Gibson et al., "Life on Mars: Evaluation of the Evidence within Martian Meteorites ALH84001, Nakhla, and Shergotty," *Precambrian Research* 106, 1 February 2001, 16. They cite D. Bogard and D. Garrison, "Noble Gas Abundances in SNC Meteorites," *Meteoritics and Planetary Science* 33 (1998): A19.
18. Ian P. Wright, Monica M. Grady, and Colin T. Pillinger, "Organic Materials in a Martian Meteorite," *Nature* 340 (1989): 220–222. This team had used an indirect

method that did not attempt to characterize the organic matter; hence, the ALH84001 team could claim to have made the first direct measurements of organic molecules in a Martian meteorite. See Romanek OHI, 21–22. For a somewhat ironic look at the history of "life on meteorite" claims, see Colin and J. M. Pillinger, "A Brief History of Exobiology, or There's Nothing New in Science," *Meteoritics and Planetary Science* 32 (1997): 443–445.

19. Wills and Bada, *Spark,* 236.
20. Romanek OHI, 4.
21. The length of time in space was determined by isotopic changes produced by cosmic ray exposure there. See McKay et al., "Possible Relic," 924.
22. Romanek OHI, 3.
23. Romanek OHI, 4–5; Romanek et al., *Nature* 372 (1994): 655–659.
24. McKay et al., "Possible Relic," 924. Harvey and McSween later withdrew their high-temperature model for the carbonates in ALH84001.
25. Romanek OHI, 6.
26. Ibid., 6–7.
27. David McKay OHI; McKay's funding from NASA Exobiology began in 1990 for analysis of cosmic dust for carbon.
28. Romanek OHI, 7–8.
29. Ibid., 9. For Schopf's account of his January 1995 visit and his opinion at that time, see J. William Schopf, *Cradle of Life: The Discovery of Earth's Earliest Fossils* (Princeton, N.J.: Princeton University Press, 1999), 304–305.
30. Romanek OHI, 9–10; see also McKay et al., "Possible Relic." The team believed the PAHs toward the outside of the meteorite were burned off, vaporized as the rock entered Earth's atmosphere and was heated.
31. McKay et al., "Possible Relic," 929.
32. Romanek OHI, 15. Four of the five anonymous peer reviewers for *Science* identified themselves afterward to the team. Gibson later discovered that Carl Sagan was the fifth. In the end Sagan and the others were satisfied with the changes and recommended publication; see Gibson OHI, 6.
33. For a thorough and philosophically astute analysis of the argument in the 1996 paper and of the first responses, through late 1998, see Iris Fry, *The Emergence of Life on Earth* (New Brunswick: Rutgers University Press, 2000), 222–235.
34. McKay et al., "Possible Relic," 924.
35. Gibson et al., "ALH84001, Nakhla and Shergotty," 16.
36. McKay et al., "Possible Relic," 927.
37. Ibid., 925.
38. Ibid., 928. The 1996 paper used the spelling *nannobacteria,* apparently after Folk's original usage. This was later (by late 1997) corrected to a standardized spelling of *nanobacteria,* analogous to metric terms beginning with the prefix *nano-*.
39. Romanek OHI.
40. McKay OHI, 11.
41. Some scientists insisted that James D. Watson was unique in 1952–1953 in the degree to which he described being driven by desire for priority in discovering the structure of DNA (in his 1968 memoir *The Double Helix*). By 1996 such behavior seems to have become more common—or at least more commonly acknowledged.
42. Gibson OHI.
43. Schopf, *Cradle,* 306–309.

44. Gibson OHI, 5. It was the press recognition that this otherwise unknown woman had an inside track to highly sensitive information from the White House itself which led shortly afterward to the revelation that she had a personal relationship with Dick Morris.
45. Meyer OHI, 12.
46. Gibson OHI; Romanek and McKay OHI. See Romanek OHI.
47. Michael Meyer OHI, 12. A history of early disputes and standards is Ron Westrum, "Science and Social Intelligence about Anomalies: The Case of Meteorites," *Social Studies of Science* 8 (1978): 461–493.
48. H. P. Klein OHI; Donald DeVincenzi OHI.
49. Richard Kerr, "Requiem for Life on Mars? Support for Microbes Fades," *Science* 282, 20 November 1998, 1398. For more of Kerr's coverage of the ongoing debate for *Science*, see also "Martian Rocks Tell Divergent Stories," *Science* 274, 8 November 1996, 918–919; "Martian 'Microbes' Cover Their Tracks," *Science* 276, 4 April 1997, 30–31; "Putative Martian Microbes Called Microscopy Artifacts," *Science* 278, 5 December 1997, 1706–1707; "Geologists Take a Trip to the Red Planet," *Science* 282, 4 December 1998, 1807–1809; "Are Martian 'Pearl Chains' Signs of Life?" *Science* 291, 9 March 2001, 1875–1876; "Rethinking Water on Mars and the Origin of Life," *Science* 292, 6 April 2001, 39–40; "Reversals Reveal Pitfalls in Spotting Ancient and E.T. Life," *Science* 296, 24 May 2002, 1384–1385.
50. Meyer OHI, 13–14.
51. Schopf, *Cradle*, 308.
52. Ibid., chap. 12; quotes on 306.
53. Oró OHI, 13–14.
54. Quoted in Schopf, *Cradle*, 304.
55. Romanek OHI, 9.
56. On the enthusiasm generated for planetary exploration, see Dava Sobel, "Among Planets," *New Yorker*, 9 December 1996, 84–90. The first round of scientific criticism appeared in *Science* 273, 20 September 1996, under the title "Past Life on Mars?"; it included Frank Von Hippel and Ted Von Hippel (1639), Harold Morowitz (1639–1640), Louis DeTolla (1640), with a reply by McKay, Gibson, and Thomas-Keprta (1640). A second, more detailed round of criticisms appeared in the 20 December 1996 issue of *Science*, including: Jeffrey Bell, "Evaluating Evidence for Past Life on Mars," 2121–2122; Edward Anders, 2119–2120.
57. Kerr, "Martian 'Microbes,'" 30–31.
58. Gibson et al., "ALH84001, Nakhla, and Shergotty," 16–18.
59. Kerr, "Martian 'Microbes,'" 31.
60. Ibid.
61. J. P. Bradley, R. P. Harvey, and H. Y. McSween Jr., "No 'Nanofossils' in Martian Meteorite," *Nature* 390, 4 December 1997, 454.
62. David McKay, Everett Gibson, Kathie Thomas-Keprta, and Hojatollah Vali, "No 'Nanofossils' in Martian Meteorite: A Reply," *Nature*, 4 December 1997, 455–456.
63. Kerr, "Martian 'Meteorites,'" 31.
64. Jack Maniloff, Kenneth H. Nealson, Roland Psenner, Maria Loferer, and Robert Folk, "Nannobacteria: Size Limits and Evidence," *Science* 276 (June 1997): 1776–1777; Nicholas Wade, "Mars Meteorite Fuels Debate on Life on Earth," *New York Times*, 29 July 1997, C1, 3; Schopf, *Cradle*, 316–321. Note, again, that the *nanobacteria* spelling only became standardized usage in late 1997.

65. Morowitz, "Past Life on Mars?" *Science* 273, 20 September 1996, 1639.
66. DeTolla, "Past Life on Mars?" 1640; E. Olavi Kajander, I. Kuronen, and N. Ciftcioglu, "Fetal Bovine Serum: Discovery of Nanobacteria," *Molecular Biology of the Cell*, 7, 3007 (supp. S); E. O. Kajander, I. Kuronen, K. Akerman, A. Pelttari, and N. Ciftcioglu, "Nanobacteria from Blood, the Smallest Culturable Autonomously Replicating Agent on Earth," *Proceedings of the Society for Optical Engineering (SPIE)*, 3111 (1997), 420–428; Milton Wainwright, "Nanobacteria and Associated 'Elementary Bodies' in Human Disease and Cancer," *Microbiology Today* 145 (October 1999): 2623–2624.
67. Wade, "Meteorite Fuels Debate"; Philippa Uwins, Richard I. Webb, and Anthony P. Taylor, "Novel Nano-Organisms from Australian Sandstones," *American Mineralogist* 83 (1998): 1541–1550; William J. Broad, "Scientists Find Smallest Form of Life, if It Lives," *New York Times*, 18 January 2000, Arizona ed., D1, 4.
68. Kajander, "Nanobacteria from Blood," 420.
69. Gretchen Vogel, "Finding Life's Limits," *Science* 282, 20 November 1998, 1399.
70. Kerr, "Requiem," 1398. A 100 nm sphere, though half the diameter of a 200 nm sphere, it should be noted, has only one-eighth the volume.
71. Ibid.
72. Broad, "Scientists Find Smallest Form of Life," D4.
73. Ibid.
74. Ibid.
75. Derek Sears and William Hartmann, "Conference on Early Mars, Houston, Texas, 24–27 April 1997," *Meteoritics and Planetary Science* 32 (1997): 445–446.
76. Jan Toporski, Andrew Steele, Frances Westall, Kathie Thomas-Keprta, and David McKay, "The Simulated Silicification of Bacteria—New Clues to the Modes and Timing of Bacterial Preservation and Implications for the Search for Extraterrestrial Microfossils," *Astrobiology* 2 (Spring 2002): 1–26.
77. Frances Westall, "The Nature of Fossil Bacteria: A Guide to the Search for Extraterrestrial Life," *Journal of Geophysical Research* 104, no. E7, 25 July 1999, 16437–16451.
78. Kerr, "Requiem," 1400.
79. Buseck to Strick, personal communication, 30 May 2002. The grant, for the period from August 2001 through June 2003, was for $401,673.
80. Ibid.; this grant, for the period from August 2002 through July 2005, totals $21,141. Buseck has numerous other NASA grants and has been a grantee of NASA Cosmochemistry since 1978.
81. Sears and Hartmann, "Conference on Early Mars," 445–446.
82. Luann Becker, Daniel P. Glavin, and Jeffrey L. Bada, "Polycyclic Aromatic Hydrocarbons (PAHs) in Antarctic Martian Meteorites, Carbonaceous Chondrites, and Polar Ice," *Geochimica et Cosmochimica Acta* 61 (1997): 475–481; see also A. J. T. Jull, C. Courtney, D. A. Jeffrey, and J. W. Beck, "Isotopic Evidence for a Terrestrial Source of Organic Compounds Found in Martian Meteorites Allan Hills 84001 and Elephant Moraine 79001," *Science* 279, 16 January 1998, 366–369.
83. Romanek OHI, 11.
84. Kerr, "Requiem," 1400.
85. Jeffrey L. Bada, Daniel P. Glavin, Gene D. McDonald, and Luann Becker, "A Search for Endogenous Amino Acids in the Martian Meteorite ALH84001," *Science* 279, 16 January 1998, 362–365; see also Daniel Glavin, Jeffrey Bada, Karen Brinton,

and Gene McDonald, "Amino Acids in the Martian Meteorite Nakhla," *PNAS* 96 (August 1999): 8835–8838.
86. Glavin et al., "Amino Acids, 8835.
87. Jeffrey Bada and Gene McDonald, "Detecting Amino Acids on Mars," *Analytical Chemistry* 68 (1996): 674A. Naturally, the line of reasoning employed here foregrounds the role of Miller-Urey synthesis, notwithstanding the recent consensus of skepticism about the relevance of a reducing atmosphere on early Earth, let alone Mars.
88. Kerridge, "Life on Mars? A Critique," talk presented in Geology Department, Arizona State University, 26 March 1997. Kerridge had trained under J. D. Bernal and Alan Mackay at Birkbeck College, University of London. He first became connected with the NASA Exobiology community through an NRC postdoc at Ames, then he worked on isotope geochemistry with David DesMarais, under Ian Kaplan at UCLA, 1975–76. He has received exobiology grant support fairly steadily since the 1970s. John Kerridge OHI.
89. Ibid., 23.
90. DeVincenzi OHI, 12 May 1997, 19.
91. Ibid.
92. Kerr, "Requiem."
93. Imre Friedmann, Jacek Wierzchos, Carmen Ascaso, and Michael Winklhofer, "Chains of Magnetite Crystals in the Meteorite ALH84001: Evidence of Biological Origin," *PNAS* 98, 27 February 2001, 2176–2181.
94. K. L. Thomas-Keprta et al., "Truncated Hexa-octahedral Magnetite Crystals in ALH84001: Presumptive Biosignatures," *PNAS* 98, 27 February 2001, 2164; see also Thomas-Keprta et al., "Elongated Prismatic Magnetite Crystals in ALH84001 Carbonate Globules: Potential Martian Magnetofossils," *Geochimica et Cosmochimica Acta* 64 (December 2000): 4049–4081.
95. See Kathy Sawyer, "New Findings Energize Case for Life on Mars," *Washington Post*, 27 February 2001, A3, 24.
96. Richard Kerr, "Are Martian 'Pearl Chains' Signs of Life?" *Science* 291, 9 March 2001, 1875–1876.
97. Cronin to Strick, personal communication, 15 December 2001.
98. Buseck to Strick, personal communication, 30 May 2002.
99. Fry, *Emergence*, 221.
100. Martin Brasier et al., "Questioning the Evidence for Earth's Oldest Fossils," *Nature* 416, 7 March 2002, 76–81.
101. J. William Schopf, Anatoliy Kudryavtsev, David Agresti, Thomas Wdowiak, and Andrew Czaja, "Laser-Raman Imagery of Earth's Earliest Fossils," *Nature* 416, 7 March 2002, 73–76. See also, from that issue, Henry Gee, "That's Life?" 28.
102. David Tenenbaum, "Ancient Fossils—or Just Plain Rocks," *Astrobiology News*. 6 (January 2003), <http://www.astrobio.net/news/print.php?sid=350>.
103. See, e.g., Kenneth Chang, "Oldest Bacteria Fossils? Or Are They Merely Tiny Rock Flaws?" *New York Times*, 12 March 2002, D4.
104. Two very different accounts have appeared; see Richard Kerr, "Reversals Reveal Pitfalls," *Science* 1384–1385; see also Rex Dalton, "Microfossils: Squaring Up over Ancient Life," *Nature* 417, 21 June 2002, 782–784. The *Nature* account is much more supportive of the British team and highly accusatory of Schopf's behavior as a scientist.
105. S. J. Mojzsis, Gustaf Arrhenius, K. D. McKeegan, T. M. Harrison, A. P. Nutman,

and C. R. L. Friend, "Evidence for Life on Earth before 3800 Million Years Ago," *Nature* 384, 7 November 1996, 55–59.
106. See the exchange between both sides in the debate in *Science* 298, 1 November 2002, 917a ("Technical Comments" section) and 961–962.
107. Kerr, "Reversals Reveal Pitfalls." Schopf must surely be chagrined at inclusion in such company, particularly given the publicity he received for taking the McKay team to task for errors parallel to those he now stands accused of.
108. Ibid., 1385. See also the comments of paleontologist Roger Buick in a 7 January 2003 press release, <http://www.spaceref.com/news/viewpr.html?pid=10315>.
109. Gibson to Strick, personal communication, 7 June 2002.

CHAPTER 9 **Renaissance**

1. Glenn E. Bugos, *Atmosphere of Freedom: Sixty Years at the NASA Ames Research Center* (Washington, D.C.: NASA History Office, 2000), 224–225.
2. "A Budget Reduction Strategy" (MS, 2 February 1995, NASA Ames files).
3. Harper OHI, 17 January 2001, 19. The ECHO report is David Milne et al., *The Evolution of Complex and Higher Organisms: A Report Prepared by the Participants of Workshops Held at NASA Ames Research Center, Moffett Field, California, July 1981, January 1982, and May 1982* (Washington, D.C.: NASA, 1985).
4. DeVincenzi OHI 12 May 1997, 11; Harper OHI, 13 May 1997, 13; Harper OHI, 17 January 2001, 13.
5. Among those who had crucial input to the late March meeting were Lynn Harper and Kathleen Connell. Outside NASA the word *astrobiology* actually predates Joshua Lederberg's coining of the term *exobiology* in 1961. For example, the American astronomer Otto Struve pondered the use of *astrobiology* to apply to the broad study of life beyond Earth in "Life on Other Worlds," *Sky and Telescope* 14 (February 1955): 137–146. But until 1995 the *exobiology* terminology was used almost exclusively among biologists, while *bioastronomy* was used among astronomers.
6. Dear Colleague letter, 30 May 1995, by Wesley T. Huntress.
7. Harper, OHI, 13 May, 1997, 14. *Astrobiology* was mentioned three places in NASA's 1996 Strategic Plan: in the Human Exploration and Development (HEDS) section, in which a map showed astrobiology assigned to Ames; in the Space Science section, in which Ames is identified with astrobiology in a diagram showing primary NASA center missions and roles; and in the glossary, in which *astrobiology* was defined as "the study of the living universe" in the terms used earlier.
8. DeVinczenzi, OHI, 12.
9. Harper, OHI 17 January 2001, 3.
10. Meyer OHI, 4 February 1997, 2–4. The exobiology budget at Headquarters had been 9.4 million, but 1 million was transferred to the Planetary Instruments Definition and Development Program, a result of centralization of planetary instrument development in the Space Science Division.
11. Tony Reichhardt, "NASA Lines Up for a Bigger Slice of the Biological Research Pie," *Nature* 391 (1998): 109; Morrison OHI, 3; Harper OHI.
12. Harper OHI, 17 January 2001, 17–18.
13. Ibid., 5.
14. D. Devincenzi, ed., *Astrobiology Workshop Final Report: Leadership in Astrobiology,*

Proceedings of a Workshop Held at NASA Ames Research Center, 9–11 September 1996, NASA Conference Publication 10153, (Ames Research Center, NASA, 1996).
15. G. Soffen, "Astrobiology: A Program Plan," 30 June 1997, stamped "Draft," NASA Ames files. An annotation dated 4 July 1997 indicated the draft plan had been seen by Huntress and forwarded to Goldin.
16. Astrobiology Development Plan, NASA Ames Research Center, 7 May 1997, revision 2; Harper OHI, 17 January 2001, 5.
17. Astrobiology Development Plan; "Ames First Astrobiology Mission Studies Leonid Firestorm over Okinawa," *Ames Astrogram,* 27 November 1998, 1, 4.
18. Astrobiology Development Plan.
19. Carl Sagan, *Pale Blue Dot: A Vision of the Human Future in Space* (New York: Random House, 1994). The first Pale Blue Dot Workshop was held at Ames on 27–28 June 1996 and the second at Ames on 19–21 May 1999. The reports of these workshops, and those described subsequently, are available at <http://www.astrobiology.arc.nasa.gov/workshops/index.html>.
20. Sara E. Acevedo, Donald L. DeVincenzi, and Sherwood Chang, eds., *Sixth Symposium on Chemical Evolution and the Origin and Evolution of Life,* sponsored by Michael Meyer, NASA CP-1998-10156 (Washington, D.C.: NASA, 1998).
21. Morrison OHI, 7–8. Harper OHI, 13 May 1997, 21.
22. Draft Cooperative Agreement Notice (CAN), sec. 1, intro.
23. NASA News Release, "NASA Selects Initial Members of New Virtual Astrobiology Institute," 19 May 1998; Andrew Lawler, "Astrobiology Institute Picks Partners," *Science* 280, 29 May 1998, 1338; Harry McDonald, e-mail to Ames staff, 19 May 1998.
24. Lawler, "Astrobiology Institute Picks Partners," 1338.
25. NASA News Release 99-33AR, "Nobel Prize Winner to Lead NASA Astrobiology Institute," 18 May 1999. Blumberg would remain director until 14 October 2002; UCLA professor Bruce Runnegar was named as his successor.
26. The Genomics/Station workshop grew out of Astrobiology's "biology beyond the planet of origin," which became "terrestrial life into space" in the roadmap. This element later became the centerpiece for the Generations Initiative approved for the 2003 budget. Harper was the originator of the initiative and did the feasibility and conceptual studies. Greg Schmidt and Kathleen Connnell were instrumental in selling it. And Mel Averner was the primary sponsor and champion. Blumberg again played a pivotal role. This completed the second-to-last piece of the Astrobiology Roadmap findings to obtain national approval. The one remaining was in Earth Sciences "co-evolution of life in the environment" theme.
27. "Remarks of NASA Administrator Daniel S. Goldin," NASA Ames Research Center Astrobiology Institute, 18 May 1999; Michael Mecham, "Astrobiology Team Taking Shape at Ames," *Aviation Week and Space Technology* 150, 14 June 14 1999, 211–212; *Washington Post,* 18 August 1999; Rebecca Rawls, "Fledgling Astrobiology Institute Aims to Foster Collaboration in Study of the Origin and Future of Life," *Chemical and Engineering News,* 20 December 1999, 25–28.
28. David Morrison to Astrobiology Workshop Participants, 2 June 1998.
29. Astrobiology Roadmap; David Morrison, "The NASA Astrobiology Program," *Astrobiology* 1 (2001): 3–13.

Epilogue

1. Henry McDonald to Astrobiology Science Conference attendees, 29 March 2000, in First Astrobiology Science Conference, abstract, 3–5 April 2000. Bruce Jakosky, a planetary scientist at the University of Colorado, was the chair of the Scientific Organizing Committee, and Lynn Rothschild of Ames served as the chair of the Local Organizing Committee.
2. Baruch Blumberg and Keith Cowing, "Astrobiology at T + 5 Years," *Ad Astra* (January–February 2002): 10–11.
3. An excellent historical treatment of origin of life research is Iris Fry, *The Emergence of Life on Earth: A Historical and Scientific Overview* (New Brunswick: Rutgers University Press, 2000).
4. The abstracts for the topics discussed here are in the abstract book from the First Astrobiology Science Conference, 3–5 April 2000. The abstracts for the second Astrobiology Science Conference are published in *International Journal of Astrobiology* 1 (April 2002): 87–176.
5. National Research Council, *A Science Strategy for the Exploration of Europa* (Washington, D.C.: National Academy Press, 1999); National Research Council, *Preventing the Forward Contamination of Europa* (Washington, D.C.: National Academy Press, 2000); NASA, Publication AO 99-OSS-04, *Deep Space Systems: Europa Orbiter Mission* (Washington, D.C.: NASA, 1999).
6. The report of the Workshop on Societal Implications of Astrobiology, 16–19 November 1999 at NASA Ames, is available as a NASA Technical Memorandum at <http://astrobiology.arc.nasa.gov/workshops/societal/>, revised 20 January 2001.
7. Steven J. Dick, "SETI and the Postbiological Universe," published as "Cultural Evolution, the Postbiological Universe, and SETI," *International Journal of Astrobiology* 2 (2003): 65–74; Dick, "Cultural Implications of Astrobiology: A Preliminary Reconnaissance at the Turn of the Millennium," in *Bioastronomy '99: A New Era in Bioastronomy*, ed. G. Lemarchand and K. Meech (San Francisco: Astronomical Society of the Pacific, 2000), 649–659.
8. "Pioneering the Future," address by Sean O'Keefe, Syracuse University, 12 April 2002.
9. DeVincenzi OHI, 12 May 1997, 18.
10. Michael J. Drake and Bruce M. Jakosky, "Narrow Horizons in Astrobiology," *Nature* 415 (2002): 733–734. In 2003 the National Research Council of the National Academies issued a seminal report on astrobiology programs, *Life in the Universe: An Assessment of U.S. and International Programs in Astrobiology* (Washington, D.C., National Academies Press, 2003). The study was conducted by the Committee on the Origins and Evolution of Life (COEL) of the Space Studies Board/Board of Life Sciences, and was co-chaired by Jonathan Lunine and John Baross.

Selected Bibliography

Alvarez, Luis W., Walter Alvarez, Frank Asaro, and Helen V. Michel. "Extraterrestrial Cause for the Cretaceous-Tertiary Extinction." *Science* 208, 6 June 1980, 1095–1108.

Appel, Toby. *Shaping Biology: The National Science Foundation and American Biological Research, 1952–1975*. Baltimore: Johns Hopkins University Press, 2000.

Association of Universities for Research in Astronomy (AURA). *HST and Beyond— Exploration and the Search for Origins: A Vision for Ultraviolet-Optical-Infrared Space Astronomy*. Washington, D.C.: AURA, 1996.

Baross, John A., and S. E. Hoffman. "Submarine Hydrothermal Vents and Associated Gradient Environments as Sites for the Origin of Life." *Origins of Life and Evolution of the Biosphere* 15 (1985): 327–345.

Barrow, John, and Frank Tipler. *The Anthropic Cosmological Principle*. Oxford: Oxford University Press, 1986.

Beichman, Charles, N. J. Woolf, and C. A. Lindensmith. *The Terrestrial Planet Finder (TPF): A NASA Origins Program to Search for Habitable Planets*. Publication 99-3. Pasadena, Calif.: JPL, 1999.

Beichman, Charles, ed. *Roadmap for the Exploration of Neighboring Planetary Systems*. Pasadena, Calif.: Jet Propulsion Laboratory (JPL), 1996.

Berenzden, Richard, ed. *Life beyond Earth and the Mind of Man*. SP-328. Washington, D.C.: NASA, 1973.

Bernal, John D. *The Origin of Life*. New York: World Publishing, 1967.

———. "The Physical Basis of Life." 1949. Reprint. London: Routledge and Kegan Paul, 1951.

Billingham, John, et al. *Social Implications of the Detection of an Extraterrestrial Civilization*. Mountain View, Calif.: SETI Press, 1999.

Billingham, John, ed. *Life in the Universe*. Cambridge, Mass.: MIT Press, 1981.

Black, David. "In Search of Planetary Systems." *Space Science Reviews* 25 (January 1980): 35–81.

———, ed. *Project Orion: A Design Study of a System for Detecting Extrasolar Planets*. SP-436. Washington, D.C.: NASA, 1980.

Black, David, and William E. Brunk, eds. *An Assessment of Ground-Based Techniques for Detecting Other Planetary Systems*. CP-2124. Washington, D.C.: NASA, 1980.

Bock, Gregory R., and Jamie A, Goode, eds. *Evolution of Hydrothermal Ecosystems on Earth (and Mars?)* New York: John Wiley and Sons, 1996.

Boss, Alan. *Looking for Earths*. New York: John Wiley and Sons, 1998.
Bracewell, Ronald. *The Galactic Club*. San Francisco: W. H. Freeman, 1974.
Brock, Thomas. "Life at High Temperatures." *Science* 158, 24 November 1967, 1012–1019.
Bugos, Glenn E. *Atmosphere of Freedom: Sixty Years at the NASA Ames Research Center*. Washington, D.C.: NASA History Office, 2000.
Burke, Bernard F., ed. *TOPS—Toward Other Planetary Systems: A Report by the Solar System Exploration Division*. Washington, D.C.: NASA, 1992.
Buvet, Rene, and Cyril Ponnamperuma, eds. *Chemical Evolution and the Origin of Life: Proceedings of the Third International Conference on the Origin of Life, Pont-a-Mousson, France, April 1970*. Amsterdam: Elsevier, 1971.
Cairns-Smith, A. Graham. *Genetic Takeover and the Mineral Origins of Life*. Cambridge: Cambridge University Press, 1982.
Cairns-Smith, A. Graham, and Hyman Hartman, eds. *Clay Minerals and the Origin of Life*. Cambridge: Cambridge University Press, 1986.
Calvin, Melvin. *Chemical Evolution*. London: Oxford University Press, 1969.
Chaisson, Eric. *Cosmic Evolution: The Rise of Complexity in Nature*. Cambridge, Mass.: Harvard University Press, 2001.
Clark, F., and R. L. M. Synge, eds., *Proceedings of the First International Conference on the Origin of Life, Moscow, 19–24 Aug. 1957*. New York: Pergamon Press, 1959.
Cocconi, Giuseppe, and Philip Morrison. "Searching for Interstellar Communications." *Nature* 184 (1959): 844–846.
Cronin, John, and Sandra Pizzarello. "Enantiomeric Excesses in Meteoritic Amino Acids." *Science* 275, 14 February 1997, 951–955.
Crosswell, Kenneth. *Planet Quest: The Epic Discovery of Alien Solar Systems*. New York: Free Press, 1997.
Crowe, Michael J. *The Extraterrestrial Life Debate, 1750–1900*. Cambridge: Cambridge University Press, 1986.
Day, William. *Genesis on Planet Earth: The Search for Life's Beginning*. 1979. 2d rev. ed., New Haven: Yale University Press, 1984.
Deamer, David. "The First Living Systems: A Bioenergetic Perspective." *Microbiology and Molecular Biology Reviews* 61 (1997): 239–261.
Deamer, David, and Gail R. Fleischaker, eds. *Origins of Life: The Central Concepts*. Boston: Jones and Bartlett, 1994.
Delsemme, Armand. *Our Cosmic Origins from the Big Bang to the Emergence of Life and Intelligence*. Cambridge: Cambridge University Press, 1998.
DeVincenzi, Donald, ed. *Astrobiology Workshop Final Report: Leadership in Astrobiology*. Proceedings of a Workshop Held at NASA Ames Research Center, 9–11 September 1996. NASA CP-10153.
DeVorkin, David. "Evolutionary Thinking in American Astronomy from Lane to Russell." Presented at a session on "Evolution and Twentieth Century Astronomy," History of Science Society meeting, Denver, Colo., 8 November 2001.
Dick, Steven J. *The Biological Universe: The Twentieth Century Extraterrestrial Life Debate and the Limits of Science*. Cambridge: Cambridge University Press: 1996.
———. "The Concept of Extraterrestrial Intelligence—An Emerging Cosmology." *Planetary Report* 9 (March–April 1989): 13–17.
———. *Extraterrestrial Life and Our Worldview at the Turn of the Millennium*. Washington, D.C.: Smithsonian Institution, 2000.

———. *Life on Other Worlds: The Twentieth Century Extraterrestrial Life Debate.* Cambridge: Cambridge University Press, 1998.

———, ed. *Many Worlds: The New Universe, Extraterrestrial Life and the Theological Implications.* Philadelphia: Templeton Foundation Press, 2000.

———. "The Search for Extraterrestrial Intelligence and the NASA High Resolution Microwave Survey (HRMS): Historical Perspectives." *Space Science Reviews* 64 (1993): 93–139.

Dorminey, Bruce. *Distant Wanderers: The Search for Planets beyond the Solar System.* New York: Springer-Verlag, 2002.

Dose, Klaus, S. W. Fox, G. A. Deborin, and T. E. Pavlovskaya, eds. *The Origin of Life and Evolutionary Biochemistry.* New York: Plenum, 1974.

Dyson, Freeman J. *Infinite in All Directions: Gifford Lectures Given at Aberdeen, Scotland, April–November 1985.* New York: Harper and Row, 1988.

———. *Origins of Life.* 2d ed. Cambridge: Cambridge University Press, 1999.

Ezell, Edward C., and Linda N. Ezell. *On Mars: Exploration of the Red Planet, 1958–1978.* SP-4212. Washington, D.C.: NASA, 1984.

Farley, John. *The Spontaneous Generation Controversy from Descartes to Oparin.* Baltimore: Johns Hopkins University Press, 1977.

Folsome, Clair E. *The Origin of Life: A Warm Little Pond.* San Francisco: W. H. Freeman, 1979.

Fox, Sidney. *The Emergence of Life: Darwinian Evolution from the Inside.* New York: Basic Books, 1988.

———, ed. *The Origins of Prebiological Systems and Their Molecular Matrices.* New York: Academic Press, 1965.

Fox, Sidney, and Klaus Dose. *Molecular Evolution and the Origin of Life.* San Francisco: W. H. Freeman, 1972.

Fry, Iris. *The Emergence of Life on Earth: A Historical and Scientific Overview.* New Brunswick, N.J.: Rutgers University Press, 2000.

———. "On the Biological Significance of the Properties of Matter: L. J. Henderson's Theory of the Fitness of the Environment." *Journal of the History of Biology* 29 (1996): 155–196.

Goldsmith, Donald. *Worlds Unnumbered: The Search for Extrasolar Planets.* Sausalito, Calif.: University Science Books, 1997.

Guthke, Karl S. *The Last Frontier: Imagining Other Worlds from the Copernican Revolution to Modern Science Fiction.* Ithaca: Cornell University Press, 1990.

Hart, Michael, and B. Zuckerman. *Extraterrestrials: Where Are They?* 2d ed. 1982. Reprint. Cambridge: Cambridge University Press, 1995.

Hartman, Hyman, J. G. Lawless, and Philip Morrison, eds. *Search for the Universal Ancestors.* SP-477. Washington, D.C.: NASA, 1985, and London: Blackwell Scientific, 1987.

Henderson, Lawrence J. *The Fitness of the Environment.* New York: Macmillan, 1913.

Holland, H. D. *The Chemical Evolution of the Atmosphere and Oceans.* Princeton: Princeton University Press, 1984.

Horowitz, Norman. *To Utopia and Back: The Search for Life in the Solar System.* San Francisco: W. H. Freeman, 1986.

Horowitz, Norman, and Jerry S. Hubbard. "The Origin of Life." *Annual Review of Genetics* 8 (1974): 393–410.

Kamminga, Harmke. "Life from Space: A History of Panspermia." *Vistas in Astronomy* 26 (1982): 67–86.

———. "The Problem of the Origin of Life in the Context of Developments in Biology." *Origins of Life and Evolution of the Biosphere* 18 (February 1988): 1–11.

Kasting, James. "Earth's Early Atmosphere." *Science* 259, 12 February 1993, 920–926.

Kenyon, Dean, and Gary Steinman. *Biochemical Predestination*. New York: McGraw-Hill, 1969.

Keosian, John. *The Origin of Life*. 1964. 2d rev. ed., New York: Reinhold, 1968.

Kerr, Richard A. "Origin of Life: New Ingredients Suggested." *Science* 210, 3 October 1980, 42–43.

Kragh, Helge. "The Chemistry of the Universe: Historical Roots of Modern Cosmochemistry." *Annals of Science* 57 (2000): 353–368.

Kuiper, G. P., ed. *The Atmospheres of the Earth and the Planets*. Chicago: University of Chicago Press, 1949.

Kvenvolden, Keith, ed. *Geochemistry and the Origin of Life*. Stroudsburg, Pa.: Dowden, Hutchinson, and Ross, 1974.

Kvenvolden, Keith, James Lawless, Katherine Pering, Etta Peterson, Jose Flores, Cyril Ponnamperuma, Ian R. Kaplan, and Carleton Moore. "Evidence for Extraterrestrial Amino Acids and Hydrocarbons in the Murchison Meteorite." *Nature* 228, 5 December 1970, 923–926.

Lederberg, Joshua. "Exobiology: Experimental Approaches to Life beyond the Earth." In *Science in Space*. Ed. Lloyd V. Berkner and Hugh Odishaw, 407–425. New York: McGraw-Hill, 1961.

———. "Signs of Life: Criterion System of Exobiology." *Nature* 207, 3 July 1965, 9–13. Reprinted in Colin Pittendrigh et al., eds., *Biology and the Exploration of Mars*, 127–140. Washington, D.C.: National Academy of Sciences, 1966.

Lemarchand, Guillermo, and K. Meech, eds. *Bioastronomy '99: A New Era in Bioastronomy*. San Francisco: Astronomical Society of the Pacific, 2000.

Lemonick, Michael. *Other Worlds: The Search for Life in the Universe*. New York: Simon and Schuster, 1998.

Lightman, Bernard. "The Story of Nature: Victorian Popularizers and Scientific Narrative." *Victorian Review* 25, no. 2 (1999): 1–29.

Lovelock, James E. *The Ages of Gaia*. New York: Norton, 1988.

———. *Gaia: A New Look at Life on Earth*. London: Oxford University Press, 1979.

———. *Homage to Gaia: The Life of an Independent Scientist*. Oxford: Oxford University Press, 2000.

McCurdy, Howard D. *Space and the American Imagination*. Washington, D.C.:: Smithsonian Institution Press, 1997.

McKay, David S., Everett K. Gibson, Kathie Thomas-Keprta, Hojatollah Vali, Christopher Romanek, Simon Clemett, Xavier Chillier, Claude Maechling, and Richard N. Zare. "Search for Past Life on Mars: Possible Relic Biogenic Activity in Martian Meteorite ALH84001." *Science* 273, 16 August 1996, 924–930.

McSween, Harry Y., Jr. "What We Have Learned about Mars from SNC Meteorites." *Meteoritics* 29 (November 1994): 757–779.

Mamikunian, G., and M. H. Briggs, eds. *Current Aspects of Exobiology*. Oxford: Pergamon Press, 1964.

Margulis, Lynn, ed. *Origins of Life, I*. New York: Gordon and Breach, 1970.

———, ed. *Origins of Life, II*. New York: Gordon and Breach, 1971.

Margulis, Lynn, and L. Olendzenski, eds., *Environmental Evolution: The Effects of the Origin and Evolution of Life on Planet Earth*. Cambridge, Mass.: MIT Press, 1992.

Margulis, Lynn, and Dorion Sagan. *Slanted Truths*. New York: Copernicus, 1997.
Marx, G., ed. *Bioastronomy: The Next Steps*. Dordrecht: Reidel, 1988.
Matsuno, Koichiro, Klaus Dose, Kaoru Harada, and Dwayne Rohlfing, eds. *Molecular Evolution and Protobiology*. New York: Plenum, 1984.
Maynard Smith, John, and Eörs Szathmáry. *The Origins of Life: From the Birth of Life to the Origin of Language*. Oxford: Oxford University Press, 1999.
Miller, Stanley. "A Production of Amino Acids under Possible Primitive Earth Conditions." *Science* 117, 15 May 1953, 528–529.
Miller, Stanley, and Leslie Orgel. *The Origins of Life on the Earth*. Englewood Cliffs, N.J.: Prentice-Hall, 1973.
Miller, Stanley, and Harold Urey. "Organic Compound Synthesis on the Primitive Earth." *Science* 130, 31 July 1959, 245–251.
Milne, David, et al. *The Evolution of Complex and Higher Organisms: A Report Prepared by the Participants of Workshops Held at NASA Ames Research Center, Moffett Field, California, July 1981 January 1982, and May 1982*. Washington, D.C.: NASA, 1985.
Morowitz, Harold. *Beginnings of Cellular Life*. New Haven: Yale University Press, 1992.
Morrison, Philip, J. Billingham, and J. Wolfe. *The Search for Extraterrestrial Intelligence*. SP-419. Washington, D.C.: NASA, 1977.
Muenger, Elizabeth A. *Searching the Horizon: A History of Ames Research Center, 1940–1976*. SP-4304. Washington, D.C.: NASA, 1985.
National Aeronautics and Space Administration (NASA). *Origins: Roadmap for the Office of Space Science Origins Theme*. Pasadena: NASA/JPL, 1997; rev. ed., 2000.
———. *Other Worlds from Earth: The Future of Planetary Astronomy*. Washington, D.C.: NASA, 1989.
———. *Planetary Exploration through the Year 2000: A Core Program*. Washington, D.C.: NASA, Space Science Exploration Committee, 1983.
———. *Planetary Exploration through the Year 2000: An Augmented Program*. Washington, D.C.: NASA, Space Science Exploration Committee, 1986.
———. *TOPS: Toward Other Planetary Systems: A Report*. Washington, D.C.: NASA, Space Science Exploration Committee, 1992.
Newell, Homer. *Beyond the Atmosphere: Early Years of Space Science*. Washington, D.C.: NASA, 1980.
Oliver, Bernard M., and John Billingham. *Project Cyclops: A Design Study of a System for Detecting Extraterrestrial Intelligence*. Washington, D.C.: NASA, 1971.
Oparin, A. I. *The Origin of Life*. Trans. S. Morgulis.1938. Reprint. New York: Dover, 1953.
Oparin, A. I., and V. G. Fesenkov. *Life in the Universe*. New York: Twayne Publishers, 1961.
Orgel, Leslie. *The Origins of Life: Molecules and Natural Selection*. New York: Wiley, 1973.
Papagiannis, M., ed. *The Search for Extraterrestrial Life: Recent Developments*. Dordrecht: Reidel, 1985.
———. *Strategies in the Search for Life in the Universe*. Dordrecht: Reidel, 1980.
Pirie, N. W. "The Meaninglessness of the Terms 'Life' and 'Living.'" In *Perspectives in Biochemistry*. Ed. Joseph Needham. Cambridge: Cambridge University Press, 1937.
Pittendrigh, C. S., Wolf Vishniac, and J. P. T. Pearman, eds. *Biology and the Exploration of Mars*. Washington, D.C.: National Academy of Sciences, 1966.

Ponnamperuma, Cyril, ed. *Comets and the Origin of Life: Proceeding of the Fifth College Park Colloquium on Chemical Evolution, 29–31 Oct. 1980.* Dordrecht: Reidel, 1981.

———, ed. *Cosmochemistry and the Origin of Life.* Dordrecht: Reidel, 1983.

Ponnamperuma, C., and A. G. W. Cameron, eds. *Interstellar Communication: Scientific Perspectives.* Boston: Houghton Mifflin, 1974.

Ponnamperuma, C., and Lynn Margulis, eds. *Limits of Life: Proceedings of the Fourth College Park Colloquium on Chemical Evolution, 18–20 Oct. 1978.* Dordrecht: Reidel, 1980.

Rohlfing, Duane, and A. I. Oparin, eds. *Molecular Evolution.* Festschrift in honor of S. Fox's sixtieth birthday. New York: Academic Press, 1972.

Rubey, W. W. "Geologic History of Sea Water." *Geological Society of America Bulletin* 62 (1951): 1111–1147.

Sagan, Carl. *Pale Blue Dot: A Vision of the Human Future in Space.* New York: Random House, 1994.

Sagan, Carl, and Stephen J. Pyne. *The Scientific and Historical Rationales for Solar System Exploration.* SPI 88-1. Washington, D.C.: Space Policy Institute, George Washington University, 1988.

Schopf, J. William. *Cradle of Life: The Discovery of Earth's Earliest Fossils.* Princeton: Princeton University Press, 1999.

———, ed. *Earth's Earliest Biosphere.* Princeton: Princeton University Press, 1983.

———, ed. *Life's Origin: The Beginnings of Biological Evolution.* Berkeley: University of California Press, 2002.

Shapley, Harlow. *Of Stars and Men.* Boston: Beacon Press, 1958.

Shklovskii, Joseph, and Carl Sagan. *Intelligent Life in the Universe.* San Francisco: Holden-Day, 1966.

Simpson, George Gaylord. "The Non-Prevalence of Humanoids." *Science* 143, 21 February 1964, 769–775.

Smolin, Lee. *The Life of the Cosmos.* New York: Oxford University Press, 1999.

Space Science Board, National Academy of Sciences. *Strategy for the Detection and Study of Other Planetary Systems and Extrasolar Planetary Materials: 1990–2000.* Washington: National Academy Press, 1990.

———. *Strategy for Exploration of the Inner Planets: 1977–1987.* Washington, D.C.: National Academy Press, 1978.

———. *Strategy for Exploration of Primitive Solar System Bodies.* Washington, D.C.: National Academy Press, 1980.

Space Studies Board, National Research Council / National Academy of Sciences. *The Search for Life's Origins: Progress and Future Directions in Planetary Biology and Chemical Evolution.* Washington, D.C.: National Academy of Sciences, 1990.

Strauss, David. *Percival Lowell: The Culture and Science of a Boston Brahmin.* Cambridge, Mass.: Harvard University Press, 2001.

Strick, James E. *Sparks of Life: Darwinism and the Victorian Debates over Spontaneous Generation.* Cambridge, Mass.: Harvard University Press, 2000.

Swift, David W. *SETI Pioneers.* Tucson: University of Arizona Press, 1990.

Thomas, Shirley. *Men of Space: Profiles of the Leaders in Space Research, Development, and Exploration.* Vol. 6. Philadelphia: Chilton, 1963.

Wald, George. "The Origin of Life." *Scientific American* (August 1954): 44–53.

Walker, James C. G. *The Evolution of the Atmosphere.* New York: Macmillan, 1977.
Wallace, A. R. *Man's Place in the Universe.* London: Macmillan, 1903.
Wills, Christopher, and Jeffrey Bada. *The Spark of Life: Darwin and the Primeval Soup.* New York: Perseus, 2000.
Woese, Carl, and George E. Fox. "Phylogenetic Structure of the Prokaryotic Domain: The Primary Kingdoms." *Proceedings of the National Academy of Sciences* 74, 1 November 1977, 5088–5090.
Wolfe, Audra. "Germs in Space: Joshua Lederberg, Exobiology, and the Public Imagination, 1958–1964." *Isis* 93 (June 2002): 183–205.

INDEX

Page references for figures are printed in italics.

Abelson, Philip, 24–25, 101, 110, 245–246n4, 246n13, 249n60
Ackerman, Thomas, 122–125, 249n63
Adams, James, 134
adenine, synthesis of, 72, 74
Akoyunoglou, George, 36
ALH84001. *See* Martian meteorite
Allamandola, Louis, 224, 225, 249n63
Allen, Paul, 153, 154
Altman, Sidney, 71, 129
Alvarez, Luis, 118–121
Alvarez, Walter, 118–121
American Institute for Biological Sciences (AIBS), 48, 51, 65
Ames Research Center (NASA), 18, 19, 35–39, 43, 50, 56, 66–67, 73–78, 82, 90, 93, 114–115, 119–126, 182, 249nn61–62; and astrobiology, 202–220; and life sciences, 133; and SETI, 132 ff.
amino acids, 2, 25–26, 39–40, 73; in lunar samples, 42, 75–76; in meteorites, 75–79, 196–197
"analytikers," 71–72, 128
Angel, Roger, 174
Antarctic dry valleys, 86–87, 183, 262n34
anthropocentrism: banned, 12; demise of, 15
anti-chance conception of OOL chemistry, 40–41, 52, 111–112
Apex chert microfossils, 113–114, 199–200
Apollo 11, 43, 73, 76, 95
Apollo 12, 43, 76

archaea, 49, 67, 102, 105–109, 111, 224, 226
Arrhenius, Gustaf, *74*, 127, 200
Ashley, Bill, *81*
Asimov, Isaac, 31
asteroids, 118–125
astrobiology: Academy, 219–220; definition, 1, 202, 205–213; disciplinary status, 231; and exobiology, 4–5, 206, 212, 221; journals, 20; and Origins program, 206, 214; roadmap, 217–220, 240–241; and SETI, 222; and society, 218–219, 222, 230–231; term used in 1950s, 17–18; workshops, 211–212
Astrobiology Institute, 1, 19–20, 202, 205, 207, 208, 211, 213–217; budget, 223; directors, 214; journals, 223; members, 214–215, 223; research, 224–229, 231
Astrobiology Science Conference, 1, 46, 222–223, 239
Astrometric Imaging Telescope, 169, *170*
Astrometric Interferometry Mission, 172
astronaut medicine, 30, 50, 53, 132
Aumann, Hartmut, 164
"autotrophs first" approach to OOL, 63, 72, 254n28
Averner, Mel, 208

Bada, Jeffrey, 48, 126–128, 196–197, 251n87
Bahadur, Krishna, 72–*74,* 254n28, 257n77

Bahcall, John, 169
Banin, Amos, 73–74, 249n63
Barghoorn, Elso, 48–49, 53, 65, 110–111, 113, 268–269nn24–25; NASA funding, 110–111
Barnard's star, 156
Bar-nun, Akiva, 249n63
Baross, John, 109, 194–195
Bay of Pigs, 82
Beem, Don, 141
Berdahl, Bonnie, *81*, 95, 265n94
Berger, J., 73
Bernal, John Desmond, 40, 71, 73, 128, 282n88
Berry, Bill, *81*, 203, 204
Beta Pictoris, 164–*165*
Biemann, Klaus, 84, 91–92, 97–99
Billingham, John, 18, 114–115, 119–120, 132 ff., *137*, *153*, 156, 160; and astrobiology, 203–204; Chief of SETI Office at Ames, 147; and societal impact of SETI, 219
bioastronomy, 14. *See also* astrobiology; exobiology
biocosmology, 10
biofilms, 195
biofriendly universe, 10
Biological Evolution of Mars: International Symposium (1990), 182
biological universe, 10
Biosphere 2, 254n28
Black, David, 136, 156–158, 159–161, 166, 167, 168; and Origins program, 176; and Orion project, 158–159
Blake, David, 249n63
Blanchard, Douglas, 189
Blois, M. Scott, 32
Blum, Harold, 32
Blumberg, Baruch, 208, 214–*216*, 219, 221, 222, 223
Bogard, Donald, 180–181
Bonner, William, 249n63
"bootlegging," 26, 246–247n16
Borucki, Bill, 166, 177, 229
Boss, Alan, 171
Bova, Ben, 219
Bowman, Gary, *81*
Bowyer, Stuart, 147

Boyce, Peter, *141*
Bracewell, Ronald, 133, *137*, 157, 158, 174
Bradbury, Ray, 2
Bradley, John, 193, 199
Brasier, Martin, 199–200, 225
Bremermann, Hans, 64
Briggs, Geoffrey, 167
Brin, David, 142
Brock, Thomas, 65–67, 87, 255n45, 267–268n15
Brocker, David, *151*
Brockett, H. R., *137*
Brown, Allan, 59
Brown, Fred, 99
Brown, Harrison, 31–32, 109, *137*
Brown, Robert, 169
brown dwarf, 165, *176*
Brunk, William E., 159
Bryan, Richard, 149–151, 160
Buescher, Vera, 136, *137*, 138, *153*
Buhl, David, 48–49
bureaucratic mindset, 37
Buseck, Peter, 195, 198
Burke, Bernard F., *141*, 167
Burrows, C. J., 169
Butler, Paul, 168, 169, 171–172, 175

CIA, 26–29
Cairns-Smith, A. Graham, 72–74, 128, 258n83
Calvin, Melvin, 3, 16, 18, 25, 32, 36, 61, 246n13, 249n60, 258n83
Cameron, A.G.W., 132, *137*, 156
Cameron, Roy, 86
Campbell, Bruce, 171
Campbell, W. W., 12
Caren, Linda, 249n63, 255n48, 257n77
Carle, Glenn, *81*
CASETI (Cultural Aspects of SETI), 149–150
catastrophism, new, 106, 111, 118
Cech, Thomas, 71, 128–129
CETI (Communication with Extraterrestrial Intelligence): distinguished from SETI, 136
Chaisson, Eric, 17, *141*
Chambers, Robert, 10

Chandra telescope, 174
Chang, Sherwood, 73–74, 133, 182
Chargaff, Erwin, 26
chemiosmotic coupling, 63, 254n31
chicken and egg problem, 67–71, 73, 128–130
Chun, Bill, *81*
Chyba, Chris, 126, 249n63, 271n83
Clark, Benton, 95, 109, 114, 181–182
Clarke, Arthur C., 227
clays, role in OOL, 72–73, 128, 272n88
Clayton, Robert, 182
Clemett, Simon, 185
Clinton, Bill, 179, 190
Cloud, Preston, 65–66, 74, 84, 110–113, 251n96, 268–269n24
Cocconi, Giuseppe, 16, 132
Cody, George, 200–201
Cold War, 23–30, 57–59, 82, 245n2
Committee on Planetary and Lunar Exploration (COMPLEX). *See* National Academy of Sciences
Compton, Dale, 152
Condon, Estelle, 205
Congress, U. S. *See* SETI
Connell, Kathleen, 205, 212–213, 219
Connors, Mary, 138
contamination problem, 2, 24–25, 29, 58–61, 73–76, 78–79, 85–86; in rhetorical Cold War sense, 248n42; Columbus and syphilis analogy, 59–61; "quarantine" of lunar samples, 76. *See also* Planetary Protection
Conte, Silvio, 148–149
Cordova, France, 204, 205
Corliss, John, 106, 109
cosmic evolution. *See* evolution, cosmic.
cosmic haystack, *140*
COSPAR, 44–45, 59, 82, 247n22
Coyne, Lelia, 249n63
Creationist OOL literature, 105, 266n1
Cronin, John, 77, 109, 129, 198
Crowe, Michael, 10
Crutzen Paul, 122
Cullers, Kent, 146
Cyclops project, 134–136

Dalton, Bonnie, *81*

Danielli, James, 254n36
Darwin, Charles, 11, 13, 106, 118
Davies, Richard, 82, 86
Davis, Mike, 141
Dawkins, Richard, 116–117
Day, William, 42, 251n85
Deamer, David, 63, 182, 225
Deep Space Network (DSN), 150, 153
Delbrück, Max, 69
Demarais, David, 67, 225, 282n88
Despain, Alvin, 145
Deverall, Genelle, *81*
DeVincenzi, Donald, *38*, 43, 48–52, 61, 114, 123–124, 197–198; and astrobiology, 206, 208, 209, 212, 231, 249n63
Dicke, Robert, 158
Dose, Klaus, 182, 249n63
Drake, Frank, 4, 16, 17, 132, 133, *137*, *141*, 143; *153*, 157; President of SETI Institute, 148
Drake, Michael, 181, 253n2
Drake Equation, 16, 144, 156
Dressler, Alan, 173
Dryden, Hugh, 24
dual origins hypothesis, 67–71, 256nn56–57
Dune, 87
Dworkin, Jason, 225
Dyson, Freeman, 67–71

EETA79001, 180–181
Earth System Science, 105, 117–118
EASTEX, 25, 30
ecosphere, 135
Edelson, Robert, *137*, 138, 139
Edsall, John, 54
Ehrlich, Richard, 32
Eigen, Manfred, 31, 128, 252n104
Eisenhower, Dwight, 24
Elachi, Charles, 166, 174
Elsasser, Walter, 65
endosymbiosis, 3
Engelberg, Joseph, 64
eobiology (coined by N.W. Pirie), 29
Epstein, Eugene, *137*
Europa, 4, 179, 227–228, *229*
European Space Agency, 20

298 Index

evolution, cosmic: 9–20, *11*, *19*; birth of idea, 10–14; and Chaisson, 17; Chambers, 10; 9-22; definition of, 9–10, 243n2; Drake Equation, 16; Fiske, 11; Flammarion, 11; Hale, 244n7; Henderson, 13; images of, 11, 19; Laplace, 10; NASA as chief patron, 18; and Origins program, 173–174, 177; Proctor, 11; Reeves, 17; as research program, 14–20; Sagan, 17; SETI, 10, 18–19; Shapley, 17; Shklovskii, 16; and space age, 17–18; Spencer, 11; Wallace, 12–13
evolution, cultural, 10, 16, 136
evolution, Darwinian, 9, 10
Evolution of Complex and Higher Organisms (ECHO) Report, 115, 119–121, 124, 126, 204
evolution of life. *See* life, evolution of
exobiology: and American space program, 3, 5; and astrobiology, 4–5, 204, 206, 212, 221; and biology, 3; birth of, 1, 18; as discipline, 4–5, 17–18, 43–55, 231; and Mars rock, 4; and public relations, 3; scope of, 4. *See also* astrobiology
Exploration of Neighboring Planetary Systems (ExNPS), 174, 175
extinctions. *See* mass extinctions

faint young sun paradox, 85
Fantasia, 2
Farmer, Jack, 227, 249n63
Fedo, Chris, 200
Fermi paradox, 142
Fesenkov, V., 16
Ferris, James, 249n63
Field, George, 160
Field Museum, 76
"fishbowl": working in, 97, 99
Fiske, John, 11–12
Fiske, Lennard, 168, 170
Flammarion, Camille, 11–12
Fletcher, James, 136
Flores, Jose, *75*
Florkin, Marcel, 54
Folk, Robert, 184, 194
Folsome, Clair, 62, 249n63, 257n77; as

consultant to *Biosphere 2*, 254n28
Fort Detrich germ warfare labs, 32, 248n42
Fox, George E., 125
Fox, Ronald, 69
Fox, Sidney, 3, 18, 26, 31–35, 39–44, 50–52, 54, 59, 67–69, 71–72, 246n13; Exobiology Program Directors and, 43, 71, 124, 251n87, 257n74. *See also* S. Miller, dispute with
Fraknoi, Andrew, 148
Fremont-Smith, Frank, 261n27
Friedmann, E. Imre, 93–94, 182, 198, 264n64
Frosch, Robert, 143, 161, 162
Fry, Iris, 51–52, 105, 199

GCMS (Viking), 80, 84, 91–93, 95–101, 182, 186, 265n86
Gabel, Norm, 249n63
Gagarin, Yuri, 82
Gaia hypothesis, 3, 48–49, 81–86, 102, 105, 114–118, 125, 130
Galileo, 179, 227, 262n31
Garn, Jake, 149
Gatewood, George, 157, 158, 159, 168
"gemischers," 71–72. *See also* "synthetic approach" to OOL
"gene-first" approach to OOL, 41–42, 63, 67–71. *See also* "information-first" approach
geophysiology, 84–85, 117; as "closet Gaia," 117
Gerathewol, Siegfried, 249n60
Gibson, Everett, *180*, 183–201
Gilbert, Walter, 208, 272n92
Gilbreath, Bill, *137*
Gillett, Fred, 164
Glennan, Keith, 30–31
Goddard Space Institute, 213
Goddard Spaceflight Center (GSFC), 203, 211
Gold, Thomas, 30
Goldin, Daniel, 1, 94; and Mars rock, 189–192; and astrobiology, 202–203, 204, 205, 207, 214–215, *216*–218; and Origins program, 174, 175–176
Golub, Ellis, 92

Goodwin, Brian, 65
Gore, Al, 179, 190
Gould, Stephen J., 106, 111
gradualism, 106, 111, 118
Grady, Monica, 183
Graham, Loren, 15, 26
Great Observatories, 177
Greenberg, J. Mayo, 182
Greenstein, Jesse, *137*, 156–158, 159, 160
Gregg, John, 64
Griffin, Roger, 158
Gulkis, Sam, *137*, *141*
Gulliver experiment, 31–35, 57, 83–88; Horowitz as advisor on, 32, 52, 86, 88
Guastaferro, Angelo "Gus," 160
Guerrero, Ricardo, *115*
Gunflint formation, 48, 66, 110
Gupta, Radhey, 267n9

Haddock, Fred, *137*
Haldane, J.B.S., 15, 23, 40–41, 109. *See also* Oparin-Haldane theory
Hale, George Ellery, 12
Hamilton, Paul, 76
Hamilton, William, 116–117
Harada, Kaoru, 40
Harper, Lynn, 203–204, 205–211, 213
Hart, Michael, 142
Hartline, M. Keffer, 30
Hartman, Hyman, 73–*74*, 125, 266n2
Harvey, Ralph, 183, 193, 199
Harvey, R. B., 255n45
Hayes, John, 66, 112–113
Haymaker, Webb, 37
Healy, Mylan, 171
Heisenberg, Werner, 99
Henderson, Lawrence J., 13
Herrera, Alfonso, 71–72, 257n75
"heterotrophs first" approach to OOL, 63, 72, 256–257n70. *See also* Oparin-Haldane hypothesis
Hewlett, William, 153
Heyns, Roger, 148
High Resolution Microwave Survey (HRMS), 144, 147, 150, 152
Hill, Henry, 158
Hines, John, 212

Hipparcos satellite, 159, 161
Hitchcock, Dian, 83
Hobby, George, 86
Hofmann, Hans, 112–113
Holland, Heinrich ("Dick"), 48–49, 109–110
Holloway, Harry, 204
Holton, Emily, 208
Horowitz, Norman, 3, 25, 30–31, 41–42, 57–61, 71, 83–97, 192; and scientific skepticism, 58–59, 80, 85–88; develops pyrolytic release experiment, 88ff
Howard, Rick, 208
Hoyle, Fred, 156
Huang, Su-Shu, 16, 132
Hubbard, Jerry, 86, 95
Hubbard, Scott: and astrobiology, 205, 208, 210–211, 213, 214, 218
Hubble Space Telescope (HST), 166, 169, 172, 174, 176
Huntoon, Carol, 189
Huntress, Wesley, 152, 176, 189, 204, 205, 208, 209
hydrothermal vents, undersea, 67, 102, 105–109, 112, 126–127, 199–200, 255n45

Icarus (journal), 83
Iceland, 66
"information first" approach to OOL, 67–71
Infrared Astronomy Satellite (IRAS), 160–161, 164, 172
intelligence, evolution of, 16. *See also* SETI
interdisciplinarity of exobiology, 47, 49–50, 52–53
interferometry, optical, 158, 169–178
information theory, computer metaphors, 69–70
International Astronomical Union, 20
International Society for the Study of the Origin of Life (ISSOL), 20, 43, 54
International Space Station, 210, 212, 217
Ivanov, Mikhail, 182

Jacobs, George, 63–65, 249n60
Jakosky, Bruce, 218, 222, 230, 253n2
Jannasch, Holger, 106
Jeans, James, 13, 134
"Jeewanu," 72–73,
Jehle, Herbert, 64–65
Jenniskens, Peter, 211
Jet Propulsion Laboratory (JPL), 36, 49, 82–86, 88–93, 97–100, 203; and astrobiology, 211, 213, 214; and exobiology, 132; and planetary systems search, 169, 171, 173; and SETI, 138, 141, 142, 144, 146–147, 150, 153
Johnson, Lyndon, 86
Johnson, Pratt, 180–181
Johnson, Richard, *81*
Johnson Space Center, 132, 203, 211, 213, 214, 230
Journal of Molecular Evolution, 54
Joyce, Gerald, 127, 129
Jungck, John, 251n85

K-T asteroid theory, 118–121, 124
Kajander, Olavi, 194
Kamminga, Harmke, *74*, 256n56
Kaplan, Isaac ("Ian"), 112, 259n104, 282n88
Kapor, Mitch, 153
Kasting, James, 110, 124–125, 225, 249n63
Katchalsky, Aharon Katzir, 73
Kauffman, Stuart, 52, 63
Keck Observatory, 168–170
Kennedy, John F., 57, 64
Kenyon, Dean, 258n78, 262n40, 266n1
Kepler mission, 177
Kerr, Richard, 181, 194, 196, 199–200
Kerridge, John, 197
Kirschvink, Joseph, 193, 198–199
Klein, Harold P., 37–39, 50, 53, 78, *81*, 90–91, 93, 95, 98–102, 125, 132, 138, 182, 231
Klein, Michael, 146
Knoll, Andrew, 111, 124–125, 182, 214
Kondo, Joji, *137*
Kramer, Sol, 48, 258n78, 261n27
Kuhn, Thomas, 42, 108, 118, 267n9

Kuiper, Gerard, 17
Kvenvolden, Keith, 37–39, 66, *75–78*, 110; dispute with Ponnamperuma, 77–78, 250n67, 259–260nn104–105

Lacey, James, 40
Lahav, Noam, 73–*74*, 249n63
Lanyi, Janos, 249n63
Laplace, Pierre Simon, 10
Laser-Raman spectroscopy, 199
Lawless, James, 73–*74*, 75
Lawrence, Jeffrey, 195
Lawton, John, 118
Lazcano, Antonio, 111, 182
Lederberg, Joshua, and exobiology, 3, 18, 23–35, 53, 57–62, 81, 90, 93, 101–102, 263-264n60
Lehninger, Albert, 42–43
Lehwalt, Marjorie, *81*, 265n94
Lenton, Tim, 117
Lepeschinskaya, Olga, 247n17
Levin, Gilbert, 31–35, 57, 81, 83, 86–*89*, 90, 95–102, 182
Levinthal, Elliott, 32–35, 262n28
Levy, Gerald, *137*
Lewis Research Center, 213
LF. *See* Life Finder
life, classification of, 3; definition of, 4, 62, 67–71, 81, 83, 87, 262n40; evolution of, 224; future of, 230; origin (OOL) of, 2, 43–47, 53–54, 61–79, 105–130, 195, 224–227
Life Finder, 177
Lightman, Bernard, 12
Lilly, John C., 16, 35, 246n13
Linscott, Ivan, 145
Lipman, Charles B., 256–257n70
Lipmann, Fritz, 30
Lockyer, Norman, 12
Lovelace, W. Randolph, 53
Lovelock, James, 3, 32, 48–49, 57, 81–85, 114–118, *115*, 271n70; and CFCs, 260n8
Lowell, Percival, 2, 12, 90, 131
Luria, Salvador, 69
Lyell, Charles, 106, 118
Lysenko, Trofim D., 26, 247n17

MacDonald, Henry, 208, 214, *216*, 218, 222
MacNab, Robert, 108
McKay, Chris, 94, 182, 249n63
McKay, David, *180*, 182–201
McLuhan, Marshall, 261n27
McSween, Harry, 182–183, 193
Machol, Robert, *137*
"macrobes," 87
Machtley, Ronald, 148–149
Mamikunian, Gregg, 53
Man in Space program, NASA, 30. *See also* Project Mercury
Marcy, Geoffrey, 168, 169, 171–172, 175
Margulis, Lynn, 3, 44, 47–49, 68, 72–73, 80, 84, 108, 114–*115*, 128, 195, 251n84; funding, 251n94
Mariner 2, *56*
Mariner 4, 58, 86–88, 102, 252n115
Mariner 6 and *7*, 90
Mariner 9, 90, 122
Mariner B, 32, 58, 82–84, 89. See also *Voyager*
Mariner Mack, Ruth, 37, 39
Mark, Hans, 78, 133, 138, 144
Mars: in American popular culture, 2; and astrobiology, 22, 177; gullies, *228*. *See also* Martian meteorite; *Viking*
Mars Global Surveyor, 179, 227
Mars Odyssey, 102, 227
Mars Pathfinder, 100, 179, 227, 264n60
Marshall Spaceflight Center, 213
mass extinctions, 118–125
Martian Chronicles, The, 2
Martian meteorite, 4, 179–201, 262n34; first collected, 183; main lines of evidence for biomarkers, 187
Martin, James, 93
Maynard Smith, John, 69–71, 116–117
Mayor, Michel, 171, 172
Mayr, Ernst, 108
MCSA (Multi-Channel Spectrum Analyzer), 140–141, 145–146
Meinschein, Warren, 74–75
Meselson, Matthew, 25
"metabolism first" approach to OOL, 63–64, 67–71

meteorite. *See* Martian meteorite; Murchison meteorite
Meteoritics and Planetary Science (journal), 195
Meyer, Michael, *38*, 48, 61, 94, 127, 190–191, 206–207, 208, 212, 213, 217–218, 222, 223
Microwave Observing Project (MOP), 144
Mikulski, Barbara, 149
Miller, Stanley, 2, 3, 15, 25–28, *27*, 48, 73, 110, 126–130, 246n13, 246–247n16; NAS nomination, 54–55; opposition to S. Fox, 41–43, 67, 71–72
Miller-Urey experiment, 2, 15–16, 25–*27*, 40, 56, 78, 88, 92, 109–110, 126–128, 199, 246n12, 282n87
Mitchell, Peter, 63, 254nn31, 36
Mittlefehldt, David, 183–184
Mojzsis, Steve, 200
molecular clouds, interstellar, 43, 49, 78
moon: analysis of rocks, 40, 73–76, 181–182, 184; formation of, 106
Moore, Carleton, 35, 51, 76–78, 249n63
Moore, Gordon, 153
Morgan, Thomas Hunt, 40
Morris, Dick, 190–191; and prostitute girlfriend, 190
Morrison, David, 164, 167, and astrobiology, 205, 207, 208, 209, 217, 218, 222, 323
Morrison, Philip, 16, 132, 144; SETI workshops, 136, *137*, 138, 156–159, 160, 219
Morowitz, Harold, 32, 61–65, 76, 117, 194; Onsager-Morowitz definition of life, 62
Mount St. Helens, 106
Muller, H. J., 30, 71
Muller, Richard, 121
Multivator, 32–35, 57–58
Munechika, Ken, 203
Murchison meteorite, 37–38, 66, 75–79
Murray, Bruce, *137*, 138
Mutch, Thomas A., 162
Mutch, Tim, 262n39
Myhrvold, Nathan, 154

Nanobacteria, 184, 186–188, 194–195, 199, 279n38
"NASA envy," 52, 54, 93
NSCORT, 126–128
National Academy of Sciences: COMPLEX, 160, 162–164; Space Science Board (SSB), 24–25, 30, 58–61, 162. *See also* EASTEX, WESTEX
National Aeronautics and Space Administration (NASA): astrobiology patron, 1; Solar System Exploration Committee (SSEC), 162, 165–166; Solar System Exploration Division (SSED), 167; and Space Science Board of National Academy of Sciences, 162. *See also* Ames Research Center; Goddard Spaceflight Cel er; Jet Propulsion Laboratory; Johnson Space Center
National Institutes of Health, 4, 30, 35, 48
National Research Council (NRC)/NASA Ames post-docs, 36–38, 73, 183, 249n63
National Science Foundation, 4, 30, 35, 47–48, 155, 172; "NSF culture" vs. that of NASA, 47–48, 94, 120, 264n68
Naugle, John, 93, 263n60
Naval Observatory, 157
Nealson, Kenneth, 270n47
nebular hypothesis, 10, 13, 14–15, 156
Newell, Homer, 136
NGST (New Generation Space Telescope), 174, 176, 177
Nicogossian, Arnauld, 208
Nixon, Richard, 76
Novick, Aaron, 60–61
Nuclear winter, 122–125, 260n4
"nucleic acid monopoly," 41–43, 69. *See also* "gene-first" approach, information first" approach

O'Keefe, Sean, 231
Oliver, Bernard, 16, 134–136, *137*, 143, *153*, 156, 157; and Cyclops project, 134–136; Deputy Chief of SETI Office at Ames, 147

one gene, one enzyme hypothesis, 58, 253n8
Oparin, Alexandr Ivanovich, 2, 15, 26–29, *28*, 40–41, 71–72
Oparin-Haldane theory, 15–16, 63, 72
Orbiting Stellar Interferometer (OSI), 169, *170*, 173. *See also* Space Interferometry Mission (SIM)
organics, exogenous delivery, 125–128
Orgel, Leslie, 42, 48, 72–73, 84, *91*, 99, 127–129; NASA funding, 124
origins of life, 15–16, 43–47, 53–54, 61–79, 105–114, 124–130, 195, 224–227. *See also* Oparin-Haldane theory
Origin of Life, The (Oparin), 2, 15, 246nn14–15, 257n75
Origins of Life (journal), 20, 53–54, 95
Origins of Life and Evolution of the Biosphere (journal), 20, 54
Origins program, 19, 172–178; and astrobiology, 206, 207, 214
Orion nebula, 172
Orion project, 158–159
Oró, Juan (John, Joan), 32, 71–72, 74, 84, 91, 98, 192; Martian peroxide theory, 88–89
Owen, Tobias, 125
Oyama, Vance, 36, 39, *81*, 83–84, 90, 95, 265n94
Ozma, Project, 31

Pace, Norman, 125, 194
Packard, David, 153
Paecht-Horowitz, Mella, 73
Paine, Thomas, 76
Papagiannis, Michael, 143
Pasteris, Jill, 199
Pattee, Howard, 64
Pearman, J.P.T., 58
Pepin, Robert, 156
Pering, Katherine, 37, 39, *75*, 259–260n105
Peterson, Allen, 145
Peterson, Etta, 39, *75*
Phillips, Charles R., 32, 253n11
Pierson, Thomas, 148
Pike, John, 152
Pillinger, Colin, 183

Pirie, Norman W., 40, 71, *74*, 256n57, 262n40
Pittendrigh, Colin, 35, 58–59, 249nn53, 60
Pizzarello, Sandra, 78–79
Planetary Biology Subcommittee, NASA, 64, 76
planetary protection, 2, 59–61
planetary science, 161–171
planetary systems, 4, 136; 155–178, 229–230; and cosmic evolution, 14–15; in Cyclops report, 135; detection techniques, 157–158, 229–230; and Hubble Space Telescope, 166, 169, 172, 174; and Origins program, 172–178; and planetary science, 161–171; and PSSWG, 167–168, 170; and SETI, 155–161; and SIRTF, 166; and TOPSSWG, 167–168, 170; turning point in acceptance of, 15; workshops, 156–161
plasmogeny, 71
Pollack, James, 122–126
Pollard, Ernest, 61–65, 108, 249n60
PAHs (polycyclic aromatic hydrocarbons), 185–188, 192–196; abiotic sources of, 185
Ponnamperuma, Cyril, *28*, 36–39, 43–48, 53–54, 56, 66, 74–*75*, 77–78, 98, 133, 257n77; leg injury, 66, 77. *See also* Kvenvolden, Keith, conflict with Ponnamperuma
porphyrins, 75
Precision Optical Interferometer in Space (POINTS), 169, *170*, 172
Proctor, Richard, 11–12
prokaryote-eukaryote distinction, 108. *See also* Van Niel, C.B.
"protein-first" approach (to OOL), 41–43 69
proteinoid microspheres, 40–43, 72, 124
proteinoids, 39–43, 50–52, 72. *See also* "thermal peptides"
protoplanetary disks, 172, 175
Proxmire, William, 142–144, 159
punctuated equilibrium, 106
pyrolytic release (PR) experiment, Viking, 88ff

Quastler, Henry, 64
Queloz, Didier, 171, 172
Quimby, Freeman, 37, 40, 43, 50–51, 61, 249n60

RNA World, 71, 128–130
Raup, David, 112, 115, 119–121
Reagan Administration, 122, 125
Reasenberg, Robert, 169, 173
Ranger 7, 31
Reeves, Hubert, 17
Reich, Wilhelm, 258n78, 261n27
Reynolds, Orr, 50, 64, 246n13, 249n60
Rich, Alex, 90, 93
Roberts, Richard, 208
Rohlfing, Duane, 249n63
Romanek, Chris, 183–201
Rosen, Robert, 65
Ross, Muriel, 204
Rothschild, Lynn, 222, 249n63
Roughgarten, Jonathan (now Joan), 65
Roussel UCLAF conference, 1973, Paris, 47, 73
Rubey, William, 109–110
Rummel, John, *38*, 48–49, 52, 94, 126–127, 182, 249n63; as Planetary Protection Officer, 61
Russell, Henry Norris, 15

Sagan, Carl, 3, 16, 17, 24–25, 31, 48, 53, 56–59, 71, 80–81, 83, 86–87, 100–102, 122–126, 133, 143, 144, 192, 211, 245n2, 246n13, 249n60, 252n1, 262n40, 265n86, 279n32
"Sagan standard of proof," 192, 198, 200
Sanchez, Robert, 71
Sapp, Jan, 47, 251n91
Scargle, Jeff, 166
Schmidt, Greg, 208, 211
Schneider, Stephen, 114, 261n16, 271n70
Schopf, J. William, 48, 110–114, 127–128, 185, 190–192, 199–200, 225
Schrödinger, Erwin, 69–70, 99
Schwartz, Alan, 40, 48, 54, 111–*112*, 249n63, 251n83
"science without a subject," 29–31, 55
Seeger, Charles, 136, 137, 138, *153*

Sepkoski, Joseph Jr., 119–121, 124, 271n68
SERENDIP, 147, 154
serial endosymbiosis theory (SET), 3, 47–48, 68, 256n57
Serkowski, Krzysztof, 158, 159
SETI (Search for Extraterrestrial Intelligence), 10, 18–19, 131–154; Allen Array, 154; and astrobiology, 222; cancellation of, 4; Congressional action on, 18, 141–142, 148–151; distinguished from CETI, 136; moved from Life Sciences to Space Science and NASA HQ, 147; Phoenix project, 153–154; and planetary systems, 155–161; sky survey, 144–145, 150, 153; societal implications, 149–150; "Square Kilometer Array," 154; targeted search, 144–145, 150, 153; and TOPS, 147; and Viking, 139. *See also* SERENDIP
SETI Institute, 20; origin of, 148; and project Phoenix, 153–154
Shao, Michael, 159, 169, 170–171, 173, 174
Shapley, Harlow, 17
Shergottite-Nakhlite-Chassignite (SNC) meteorites, 181–201, 278nn9, 16
Shklovskii, Joseph, 16–17, 133
Shock, Everett, 74, 126
Sillén, Lars Gunnar, 49, 84
SIM. *See* Space Interferometry Mission
Simpson, George Gaylord, 18, 29–31, 55, 57, 231
simulacra ("cell model experiments"), 71–72
Singleton, Rivers, 249n63
Sinton, William, 17
SIRTF, 173, 174
Slepecky, Ralph, 35
Smith, Adolph, 72, 249n63, 257–258nn77–78
Soffen, Gerald, 32, 52, 90—91, 93, 95, 97, 102; and astrobiology, 209–210, 213, 214, 219, 221
SOFIA, 173
Sogin, Mitchell, 125

Solar System Exploration Committee (SSEC), NASA, 162, 164, 165–166
Solar System Exploration Division (SSED), NASA, 167, 169
Space Infrared Telescope (SIRTF), 166, 167, 173
Space Interferometry Mission (SIM), 173, 174, 176, 177. *See also* Orbiting Stellar Interferometer (OSI)
Space Interferometry Science Working Group (SISWG), 173–175
space medicine, 132
Space Science Board (SSB). *See* National Academy of Sciences
Space Sciences, NASA Office of, Exobiology housed within, 36, 50–53
space station, 152, 166
space telescope, 159, 161. *See also* Hubble Space Telescope
spectroscopy, 12
Spencer, Herbert, 11
Spencer Jones, Sir Harold, 15
spontaneous generation, 11
Sputnik 1, 16, 23–24, 26
Sputnik 2, 23–24
Sridhar, K. R., 212
Stanier, Roger, 108
Stapledon, Olaf, 14, 83, 261n14
Steinman, Gary, 72, 258n78
Stent, Gunther, 25
Stillwell, William, 251n85
Stolper, Edward, 196
Stone, Ed, 152
Straat, Patricia, 95, 99–100
Strand, Kaj, 15, 157
stromatolites, 66–67, 110
Stull, Mark, 136, *137*, 138
Struve, Otto, 16, 17–18
Suess, Hans, 109
Surtsey, 66
Surveyor, 82
Swensen, George, *141*
synthetic ("constructionist") approach to OOL, 71–72, 257n74. *See also* "gemischers"
Szathmáry, Eörs, 69–71
Szent-György, Albert, 64

Tarter, Jill, 138, *141*, *151*, *153*, 249n63; NASA SETI Project scientist, 147–148
Tayor, Edwin, 65
Taylor, William, 64
Terrestrial Planet Finder (TPF), 175, 176, 177
theoretical biology, 63–65
"thermal peptides," 42
Thomas-Keprta, Kathy, *180*, 184–201
Tipler, Frank, 142, 143
Toffler, Alvin, 219
Toon, Owen, 122–125, 249n63
TOPS (Toward other Planetary Systems), and SETI, 147
Townes, Charles, 174
Townsend, Bill, 204
TOPS (Toward Other Planetary Systems), 147, 167–171
TPF. *See* Terrestrial Planet Finder
Troland, Leonard, 71
Turco, Richard, 122–125
Tyler, Stanley, 110

UFOs, 4, 131
Ulrich, Peter, 208
Urey, Harold, 2, 15, 25–27, 30–31, 33–35, 41, 109–110, 246n13
Uwins, Philippa, 194

Vali, Hojatollah, 186
van de Kamp, Peter, 15, 17, 156, 227
van Niel, C. B., 25, 63, 108
vents. *See* hydrothermal vents, undersea
Vernikos, Joan, 205
Viewing, David, 142
Viking spacecraft, 18, 73, 80–102, 179–180, 227; biology instrument, 80–*81*, 90–102
Vishniac, Wolf, 18, 30–35, 52, 57–59, 86–87, 93–94, 246n13
Vogt, Steve, 172
Von Neumann, John, 68–69
Voyager, 82–84, 89. See also *Mariner B*
Voyager 2, 227

Waddington, C. H., 258n83
Wald, George, 16, 30, 111
Walker, Gordon, 171
Walker, James C. G., 110
Wallace, Alfred Russel, 12–13
Walter, Malcolm, 66–67, 112–113
War of the Worlds, 2
Waterman, Alan T., 113, 254n16
Webb, James, 36, 52
Weber, Bruce, 254n29
Weiler, Edward, 160
Weiss, Armin, *74*
Welch, Jack, *141*,148
Welles, Orson, 2
Westall, Frances, 195
WESTEX, 25, 59–61
White, David, 73
Whitehouse, Martin, 200
Williams, Frederick, 64
Wilson, Edward O., 30
Wisniesk, Richard, 203
Woeller, Fritz, 37, 39, *81*, 265n94
Woese, Carl, 3, 47–49, 61–62, 67, 72, 106–109, *107*, 130, 224, 226; NASA funding amounts, 251n94
Wolf Trap, 30–35, 57, 93–94
Wolfe, Audra, 30, 245n1, 248n42
Wolfe, John, 136, *137*, 138, 141
"Worm," The, *180*, 192–194
Wright, Ian, 183

Yčas, Martynas, 64, 254–255n40
Yale University Biophysics Department, 61–62, 108
Yellowstone hot springs, 65–67
Young, Richard S., 35–*38*, 40, 43–51, 53–54, 61–62, 66, 71, 93–94, 112–114, 133, 182, 249n60
Yuen, George, *77*, 249n63

Zahnle, Kevin, 249n63, 270n55
Zare, Richard, 185
Zill, L. P., 37–39
Zuckerman, Ben, *141*

About the Authors

STEVEN J. DICK is the Chief Historian at NASA. Prior to that, he worked as an astronomer and historian of science at the U.S. Naval Observatory, ending as Chief of its Nautical Almanac Office. He obtained his B.S. degree in astrophysics (1971) and M.A. and Ph.D. degrees in history and philosophy of science (1977) from Indiana University and is well known as an expert in the field of astrobiology and its cultural implications. He is author of *Plurality of Worlds: The Origins of the Extraterrestrial Life Debate from Democritus to Kant* (1982), *The Biological Universe: The Twentieth Century Extraterrestrial Life Debate and the Limits of Science* (1996), and *Life on Other Worlds* (1998), the latter translated into four languages. He was also editor of *Many Worlds: The New Universe, Extraterrestrial Life, and the Theological Implications* (2000*)*. His most recent book is a history of the Naval Observatory, *Sky and Ocean Joined: The U.S. Naval Observatory, 1830–2000* (2003).

Dr. Dick served on Vice President Al Gore's panel to examine the societal implications of possible life in the Mars rock and is the recipient of the NASA Group Achievement Award "for initiating the new NASA multidisciplinary program in astrobiology, including the definition of the field of astrobiology, the formulation and initial establishment of the NASA Astrobiology Institute, and the development of a Roadmap to guide future NASA investments in astrobiology." He is on the editorial board of several journals, including the *Journal for the History of Astronomy* and the *International Journal of Astrobiology*. He has served as Chairman of the Historical Astronomy Division of the American Astronomical Society and as President of the History of Astronomy Commission of the International Astronomical Union. He is currently President of the Philosophical Society of Washington and a recent recipient of the Navy Meritorious Civilian Service Award.

JAMES E. STRICK is trained as a microbiologist and a historian of science. After completing his B.S. degree in biology (1981) then an M.S. degree (1983) at SUNY College of Environmental Science and Forestry, he taught high school and middle school biology and chemistry for ten years then returned to graduate study in the history of science at Princeton, completing an M.A. and Ph.D. degree (1997). His research interests include Darwin studies and the history of microbiology, especially ideas about the origin and nature of life. His first book, *Sparks of Life: Darwinism and the Victorian Debates over Spontaneous Generation* (2000), is a close-up look at heated debates about the origin of life among Darwin and his followers in the first twenty years after publication of *On the Origin of Species*.

Dr. Strick won the History of Science Society's 1994 Henry and Ida Schuman Prize. He has taught at Arizona State University, Johns Hopkins University, and Princeton and was a visiting senior fellow at the Center for History of Recent Science, George Washington University. He is currently Assistant Professor in the Program in Science, Technology, and Society at Franklin and Marshall College.